# The Safe Use of Cryogenic Technologies

A handbook for best practice and training

# The Safe Use of Cryogenic Technologies

A handbook for best practice and training

**Robert Done**
*Rutherford Appleton Laboratory, Harwell Science Campus, Chilton, Oxfordshire, UK*

**IOP** Publishing, Bristol, UK

ISBN 978-0-7503-3123-4 (ebook)
ISBN 978-0-7503-3121-0 (print)
ISBN 978-0-7503-3124-1 (myPrint)
ISBN 978-0-7503-3122-7 (mobi)

DOI 10.1088/978-0-7503-3123-4

Version: 20210501

IOP ebooks

British Library Cataloguing-in-Publication Data: A catalogue record for this book is available from the British Library.

Published by IOP Publishing, wholly owned by The Institute of Physics, London

IOP Publishing, Temple Circus, Temple Way, Bristol, BS1 6HG, UK

US Office: IOP Publishing, Inc., 190 North Independence Mall West, Suite 601, Philadelphia, PA 19106, USA

*Dedicated to my wife Sue Done*

# Contents

# Preface

Over more recent times, throughout the countries of the world, the use of cryogenic technologies has advanced exponentially, lending its purpose to an assortment of diverse applications, ranging from the prosaic freezing and long-term storing of samples and materials, all the way through to the cooling of advanced and inherently complex medical scanning equipment. Likewise, the hazards associated with this array of low-temperature practises are equally varied, and can in the extreme result in injuries and even fatalities if they are not correctly mitigated. As a common denominator, safe methods of work and the provision of personal protective equipment are often deployed as a standard means of reducing the risks of workplace exposure to cryogenic technologies. These are sensible policies in what can ostensibly be a high-risk working environment.

Yet, as demand for its applications has prevailed, the associated risks of cryogenic technologies has never been a significant deterrent for avoiding its use. In fact, its application within both research and industrial environments has continually grown as society's expedient developments have brought more and more advances to the general field with each successive year. Along with the variety of facilities and commercial enterprises eager to embrace the discipline, research institutions are pushing the boundaries of its use beyond what was ever originally conceived. As these developments establish in ever more locations throughout the continents, there has inevitably been a consequential growth in the likelihood of cryogenic accidents occurring as a result. Since the temperature range of cryogenic technologies has potential life threatening significance to the human body, a greater awareness of the associated risks and how to moderate hazards is crucial for their safe application. This is no feckless subject matter though, or one left to its own governance, but an industry fully supported by a wealth of mutual legislation in order to protect those using this type of technology.

The true challenge however for any practical user, is that current safety guidelines and literature appertaining to applications of cryogenic technologies can often be thought-provoking and perplexing to not only those unfamiliar with the specialist nature of the technology, but also those having already worked in the field for many years. The colossal list of potentially applicable rules and regulations covering the topic alone is wide-ranging in context and can often be challenging when attempting to determine specific relevance for any exact application. As a result, defining an overview of what they all mean and determining which of the rules and guidance notes require greater scrutiny, can be a daunting task. I know from my own experience that many critical sections can often be mistakenly overlooked without the reader's knowledge or intention. To further complicate matters, different laws and rules exist within different nations and transposing these into workable guidance is not always the easiest of tasks.

As such, it is envisaged that this book will go some way towards providing a simplified approach to assist with the decision-making task of ensuring cryogenic technologies are used in a safe and appropriate manner. It explains in some detail the relevance of each safety element and what statutory conventions are requisite for compliance within three specific economic regions; the United Kingdom, the European Union and the United States of America. A set of case-study examples at the end of the book also assist with the process, highlighting the method by which safety assessments of both existing and new facilities can be conducted. This will allow both the novice and the experienced cryogenic user to navigate and select those safety principles apposite to their relevant situation.

Deliberately throughout, I have avoided explaining in too much legal detail the elements of the regulations and directives, or putting them in full context of how they were originally generated and the manner in which they have evolved since their inception. The regulations and directives themselves already do this extremely well, offering a far better comprehensive insight to the various topics they cover. Instead, I have presented a brief explanation of the factors they encompass, highlighting those which have greater relevance when the use of cryogenic equipment is manifest. In turn, it is envisaged that these can be subsequently researched in greater scrutiny as and when required.

It must be noted, that in preparing this book, the timing coincided with significant political developments in Europe resulting in considerable changes in the United Kingdom's status within the European Union. As the United Kingdom ceased to be a functioning member of the European Union at the end of a defined transition period, many of the associated trade agreements and jurisdictive values also ceased to become obligatory. As much of the established workplace legislation in the United Kingdom had been created under the original member state obligations of the European Union, their eventual withdrawal from this membership may in the long term evoke changes to the established regulations and laws. As yet, this is an unknown, and I have therefore presented the current United Kingdom legislation in the context of the driving European Union directives which established them in the first place. Significantly, as most workplace standards have no direct relation to these regulations, these will not be affected and their use within the United Kingdom will continue as they are now. For standards used as part of conformity requirements for the European Union CE marking however, these will in the future be referenced in a completely different way as described in detail within the body of the text of this book.

In conclusion, the detailed contents of this book are derived from the many years of my personal experience in designing cryogenic equipment for scientific applications. The attendant need to understand not only the thermodynamic significance of the technology, but also the practical hazards it presents, has diligently informed my understanding of the potential risks apparent when routinely using extremely low-temperature equipment. However, confronted each time by the apparently interminable mass of regulations and industry guidance pertaining to the use of cryogenic equipment, I have often felt unsure as to which of these rulings are required for each specific application. Imbued by this sense of uncertainty, I have over my career collated as much information as possible relating to the latest germane guidance and legislation in order to inform my choices. Consequentially, the imperative of my designs has always been to ensure that the equipment would be thoroughly safe when utilised and that inherent risks are reduced as much as possible by following the established statutes and rulings. This is the premise of the following subject matter.

# Author biography

## Robert Done

Rob Done began his career as a mechanical engineering apprentice for UKAEA at Risley in Cheshire. On graduating, he joined the Science and Technology Facility Council, working first at the Daresbury site in Warrington and eventually moving to the Rutherford Appleton Laboratory in Oxfordshire. For the last 30 years he has been providing the mechanical engineering support for the neutron spallation source at the site, designing equipment which provides the experimental environments for user provided test samples. Much of this equipment operates at cryogenic temperatures. Rob routinely presents a series of cryogenic training courses for both internal and external clients. He is also a lead auditor for the Council's safety audit team.

# Chapter 1

# Introduction to general health and safety

## 1.1 The beginnings of cryogenics

Often perceived as a relatively modern technology, a very early use of a type of cryogenic cooling can actually be traced all the way back to the year 1845 when the English scientist Michael Faraday first managed to liquefy several gases, including chlorine and ammonia. Probably better known for his pioneering work in electromagnetic induction, he was recognised as a brilliant polymath at the time and became influential also in the field of chemistry as his various research aims developed. During that period, Faraday's studies in both physics and chemistry led him into investigating the possible change of state of gases into their liquid forms [1]. His procedure, somewhat rudimentary in form in comparison with modern cryogenic principles, was to cool the gases by immersion into a bath of ether and solid carbonic acid placed inside an air pump, so that both the surrounding air and gaseous carbonic acid could be rapidly removed. Carbonic acid is now better known as carbon dioxide and in its solid form is often described as dry ice. In Faraday's experiments, temperatures as low as −110 °C were recorded, which would be considered too high a temperature to actually be deemed cryogenic in today's physics world. However, his principle requirements of achieving lower than ambient temperatures in order to realise the liquefaction of the gases, could in essence, be considered the birth of cryogenics.

This was an era when adventure and scientific exploration led the way towards a journey of research, and safety would have been considered only out of the necessity of self-preservation rather than any legal requirements of a more holistic precept. The −110 °C achieved by Faraday is recognisably cold enough to cause permanent injury and would critically these days be considered a significant risk of causing harm. Yet, it is unlikely that the correct vestments of protective clothing would have been worn by Faraday and his assistants, or that his experiments would have been carried out inside the atmospherically controlled enclosure of something similar to a modern-day fume hood. Probably, he wore nothing more unassuming than an old morning suit to conduct his investigations in, or at best a blacksmiths apron and

doi:10.1088/978-0-7503-3123-4ch1

gloves to protect him from the cold materials being handled. If considered necessary, he may even have wrapped a scarf around his mouth and nose to prevent the accidental breathing in of any fumes being liberated from his experiments. It is easy therefore to imagine the near primitive nature of Faraday's laboratory workshop, with him possibly behaving capriciously, attempting a variety of tests, each carrying their own risk, all for the sake of achieving his ultimate aim of liquefying gases. It is also conceivable that his first attempts failed and through much dangerous trial and error he eventually reached his objective. How many accidents prevailed before he reached his destination is lost to the annals of history, but either through obsessive bravery, or ignorance of the consequences, Faraday and his team managed to eventually achieve their intended target.

Interestingly, during these nascent experiments, Faraday failed to liquefy the so-called permanent gases then known to exist [2]. Including hydrogen, oxygen, nitrogen and carbon monoxide, these gases were thought by the late 19th century scientists to be so 'permanent' that they could not be converted into any other state of matter. Modern knowledge of their liquefying temperatures explains that they were simply at a significantly lower temperature than Faraday was able to achieve with his early experiments. It was not until 1877 before the techniques were developed sufficiently to cool systems low enough which would bring about the liquefaction of these permanent gases (figure 1.1).

Faraday's seeming ambivalence towards not only his own safety, but for his co-workers as well in what he was undertaking, is not some egregious folly on his part. The laws of the day held little accountability for such practices, taking a lax view on the industrial processes at the time, not only for their short term consequences, but also for the long-term effects they could have on people's health. This was normally due to a combination of both technical ignorance and measured nonchalance, generally for the sake of profits needing to be made. The governing authorities had as yet to formulate a correlation between good health and safety and the mortality of their workers. A more cynical eye might perceive this naivety as a deliberately fiscally inspired objective, where the ease with which industry could replace staff outpaced the need to protect those workers during their employment.

To an enterprising scientist such as Faraday, who was seemingly intent on conducting his experiments without the burden of welfare protocols, the concept of a strict health and safety policy would be anathema. His carefree attitude towards risk

| Gas | Liquid | Liquid |
|---|---|---|
| Ammonia | 239.8 K | -33.3 °C |
| Chlorine | 239.1 K | -34.0 °C |
| Dry ice & ether | 163.1 K | -110.0 °C |
| | | |
| Oxygen | 90.1 K | -182.9 °C |
| Carbon monoxide | 81.6 K | -191.5 °C |
| Nitrogen | 77.3 K | -195.7 °C |
| Hydrogen | 20.2 K | -252.8 °C |

**Figure 1.1.** Showing the boiling point of the gases which Faraday was able to liquefy in comparison to the so-called permanent gases.

management during his investigations wasn't in the least assumed. Any concerns he might possess for his own wellbeing and that of his assistants would be fleeting in comparison to the drive he possessed to achieve his science. This was a time when discovery was dynamic and the boundaries of the real world slowly coming into focus as the knowledge of the physical sciences made greater comprehension with each new day. At what cost though? It is hard to contemplate the number of lives lost over these initial years, or indeed, the number of injuries incurred at the expense of this search for a greater understanding. Yet, without these pioneers, many of the important discoveries they made would never have materialised.

In a present sense however, it is extremely difficult to reconcile the fatalities and the debilitating injuries that would have been suffered by hundreds and thousands, as a means of justifying the end results. Thankfully, the modern world would not permit many of the transgressions undertaken by Faraday and his contemporary researchers, but that is not to say that risks are no longer taken in the name of exploration. Faraday's work, despite his obvious lack of regard towards safety when evaluating his actions against today's standard, was one of extraordinary achievement and bounty. By comparison, modern research has to abide by strict rules and be undertaken in such a way that the risks associated with its expression cannot result in injury. Yet accidents continue to happen, although thankfully not at the level they would have in the late 19th Century of Faraday's time.

## 1.2 Cultural perceptions of cryogenics technology

In many circumstances, the perception of cryogenic technology has been voraciously appropriated by the arts, and has been placed into modern-day literature and film to provide a futuristic insight of both the best and worst of scientific endeavour. Be it in the form of satirical cartoons, taking a humorous perspective of current know-how, or in films and novels where it is repeatedly employed as a means of freezing the human body for long-term space exploration, there is an inherent ignorance by those narrators of the definitive values in which the technology prevails. In some instances a blatant obfuscation of the truth behind the technology has been undertaken, distorting genuine fact from the commercially profitable fiction. Unwittingly, cryogenics as a technology is often confused in numerous circumstances with the low-temperature preservation techniques of cryonics. Whereas in fact, cryonics is merely a branch of cryogenics solely focusing on the low-temperature freezing and storage of legally dead humans in the hope that future developments in science and healthcare can restore them to life [3]. But cryogenics is much more than this, and yet it is often seen as an almost untouchable science that only the specialist elite fully comprehend, or something that an impending future world employs as an essential basis of civilisation's survival.

Yet the reality is divorced in so many ways from those creative thoughts produced for a populace in need of erudition. Admittedly, cryogenics can be a harsh and unforgiving environment, but it is not the singular property of the science fiction genre. It already exists, and in a multifaceted and ever changing way. It is an established, contemporary industry on a global business scale, fed by the need of pure research and practical enterprise, which both challenges and adopts its associated technology at all levels of conformity. From magnetically levitating trains and hospital scanning equipment, to the more prosaic technology of biological sample preservation, cryogenic technologies now touch more of our lives than ever before, playing a vital role in many circumstances, without our awareness. For those within the industry, they will already be aware of the difficulties and inherent risks associated with the use of cryogenic technology and that there exists real hazards and the potential to cause not only minor, but in certain situation, major incidents if things go wrong.

So it can be stated that, as with any development maturing into commercial usage, there are the inevitable application risks associated with its commerce. Safety is a paramount at all stages in the process, from the liquefaction of cryogens all the way to the practical use of superconducting magnets in a magnetic resonance imaging (MRI) scanner. As such, the related hazards for each separate procedure must be deliberated upon and mitigated as required. To fulfil this requirement, the designer of any cryogenic application has a duty of care to ensure the process is carried out in a benign manner, so as to prevent the likelihood of injury. This scrutiny will inevitably carry its own caveat. For as the usage of cryogenic equipment has evolved, so too does the overall risks. With more and more facilities throughout the world regularly employing cryogenic technologies for the cooling of an ever growing array of equipment, there is an inevitable increase in the likelihood of accidents occurring. The pure statistics prevail in that the more an item is used, the greater the chance an accident will take place as a result. No industry is immune to this fact, and incidents will unavoidably take place within any cryogenic related domain, from the smallest of cold-burns to the fatal consequences of asphyxiation. This probability is a reality that anybody working with cryogenic technologies needs to be aware of at all times, and to respect, whatever their level of involvement.

## 1.3 Health and safety principles

In all aspects of life, there is an expectation that what we undertake can be done in a safe manner and in such a way that it will cause no harm to those around us. This is the principle of much of the health and safety guidance throughout the world and often carries legal inference to support its implementation. Organisations are in turn encouraged to add credence to this doctrine by developing within their establishment a safety culture intolerant to potentially dangerous practices. By association, there are three main principles underpinning this ambition, with each carrying its own measured value to produce the overall outcome of improving health and safety standards. These are; the moral, the legal and the financial issues governing the day to day working practices within any organisation. To this end, it follows that the use of cryogenic equipment and materials would be naturally directed by the same three values [4].

### 1.3.1 Moral

It is an individual's moral obligation to prevent people getting hurt in every aspect of daily life. This can be seen when a parent takes charge of their children as they are about to cross a busy road, protecting them from possible harm by safely controlling the risky activity. Likewise, in the working environment, this responsibility governs the way people and businesses undertake safe working practices, with the aim of reducing the overall risk of anyone being injured or suffering ill health as a consequence of the tasks being undertaken.

Failure to meet this ethical obligation by businesses will generally indicate that poor health and safety standards are endemic within organisations and would inevitably result in people getting hurt during their routine work. As a result of this inadequate safety culture, if any such accidents are deemed serious enough, an injured employee will not only find their work life dramatically affected but, more probably as a consequence, their personal life as well.

An example where this could occur is when an individual undertaking a manual handling activity incorrectly performs the task in a way that they sustained a severe back injury. At an immediate level, they would no longer be able to execute the activity without suffering painful consequences. Typically, for this category of injury, they are also more likely to require some form of convalescence before they are able to return to their full duties. In reality, this may require a prolonged period of absence from their employment which could be weeks or even months depending on the severity of the original injury. Whatever the level of time away from their employment, the injury sustained at work is more than likely to also affect what they can do at home. Undertaking the most simplest of activities or attempting hobbies with a back injury could be restrictive and they may even find that undertaking relatively simple tasks, such as trying to sleep in a bed, becomes unbearable. As a consequence, their work life has clearly been changed, but equally so, their personal life has also been influenced by the injury.

The moral principle is prosaic, as we all have an ethical duty of preventing people from getting hurt. To successfully implement this, there is an expectation that we will all, without prejudice, actively reduce the risks associated with what we do, which will not only have an impact on our own wellbeing but also on those around us. We wouldn't for example purposely undertake a high risk activity near to a thoroughfare where the likelihood of the hazard coming into contact with pedestrians would be significantly high.

### 1.3.2 Legal

Depending solely on the moral standing of individuals for safety can be misguided. Too often influencing factors such as time constraints, costs and even negligence, can override the ethics of how something is undertaken. Safety very often gets pushed metaphorically to the back of the queue in these situations, leaving both workers and the general public vulnerable to exposure of unacceptable risks. As a consequence, laws have been established, making it illegal to override the safety protocols of a process, especially where there is an underlying inherent risk from that process. Many of these laws have been specifically put in place to prevent people from getting hurt and it would be unacceptable to break the law in such a way that it could put people at risk.

Health and safety law has been developed over time to formalise standards across all types of activities, creating a process that captures good practices but also strives to eliminate unsafe procedures. It has been written so that organisations are unable to undertake an activity in a way that it would be possible for their own staff or the public to get hurt. Per se, it places a legal responsibility on employers to suitably manage the risks associated with all their work activities and to provide both competent management and the necessary supervision of staff in order to prevent incidents occurring. This is the basis of the risk assessment process, where high risks are mitigated, and as a result, overall safety improved.

In the unfortunate event where injury has occurred, an organisation's legal duty is to report certain types of accidents to the respective enforcing authority. Examples in the UK would be Local Authorities or the Health and Safety Executive (HSE). In the USA, the Occupational Health and Safety Administration (OSHA) of each State would undertake this role, with the European Agency for Safety and Health at Work (EU-OSHA) being the central provider for health and safety guidance and information throughout Europe. These agencies have the responsibility for checking a company's compliance with the current Health and Safety law. Their primary aim is to reduce the number of fatalities, injuries and ill health occurring in the workplace. Part of their remit would be to visit the site where an incident has taken place and, if they find that the law has been broken, or that a significant health and safety risk remains, they have the power to close the facility and prosecute the organisation.

### 1.3.3 Financial

The final principle, and one which is often considered only as a minor irrelevance, or not even addressed in the first place, is the financial burden placed on an organisation if things go wrong. It is a fundamental consequence that an accident within any organisation will cost that business some amount of money and will therefore affect profitability. Some of these costs are immediate and obvious, such as the expenses due of an investigation, especially where a process may have to be halted in order for evidence to be gathered. Also, an injured person may as a consequence of incurring injury, have to take time off work in order for them to recover. As a result, their sick pay and any labour agency costs to cover for their absence will have to be found from a company's direct budget.

Less apparent however, and more generally longer term, financial costs would inevitably be incurred if a claim is submitted against the company's liability or if a prosecution is made by the enforcing authority and a resulting fine has to be paid. The less tangible however, and possibly the hardest to estimate in value, would be those monetary costs relating to the loss in a company's public relations due to the accident happening. Acquiring a poor health and safety reputation can result in damaging consequences to an organisation's image, which in the long term would have a direct, and possibly lasting correlation to their profits. The loss of public confidence in a company following any accident can be hugely significant. If a company's public image has been tarnished by such an event and the company brand accordingly considered unsafe, people are unlikely to buy from them in the future.

Taken collectively, it can therefore be seen from these three ideologies, that any organisation operating with significant hazards, which if realised as an accident or incident, could have serious physical or reputational risk implications. The various safety, health and environment laws are there to provide a means, through observance, for the workplace to become a safer place to be in. However, it remains incumbent on organisations to fuel this compliance through their safety culture. Married in with this, one of the cornerstones of any health and safety management system is risk management, including both assessment and control of risks, which organisations should adopt as a standard for every task they undertake.

## 1.4 The importance of safety when using cryogenic technology

As a factual indicator of its own technological success, more and more facilities globally now use cryogenic technology as a fundamental research tool or as part of their processing portfolio. In this ever growing field, the cryogenics principles are readily employed either as a primary means of cooling equipment down to incredibly low temperatures or as a secondary procedure which forms part of their overall manufacturing conventions. As a direct consequence, the more this hazardous technology is used, there will inevitably be a correlating increase in the likelihood of accidents happening.

It is a fundamental fact that the materials and systems used throughout the cryogenics industries function under conditions significantly below ambient temperatures. Incidents involving cryogenic technologies can therefore result in a variety of outcomes such as; cold contact burns, asphyxiation, explosions from pressurised gas build-up, and the embrittlement of materials, to name but a few. In addition, the use of liquid cryogens in the working environment, can readily expose people not directly involved in their use to serious risks.

As with any process, these incidents can range from the prosaic near-miss situations, all the way to life changing injuries or even fatalities. An unfortunate fact prevails that throughout the World, there are now deaths occurring annually which can be attributed directly to the use of cryogenic technology. In fact, by far the greatest cause of this loss of life are those brought about due to asphyxiation, directly associated with the use of liquid cryogens in confined spaces [5].

To put this in context, collective studies have been undertaken which demonstrate a direct relationship between the various types of incident outcomes which are commonly reported in all types of industries [6]. Figure 1.2 represents this as a triangle, with major

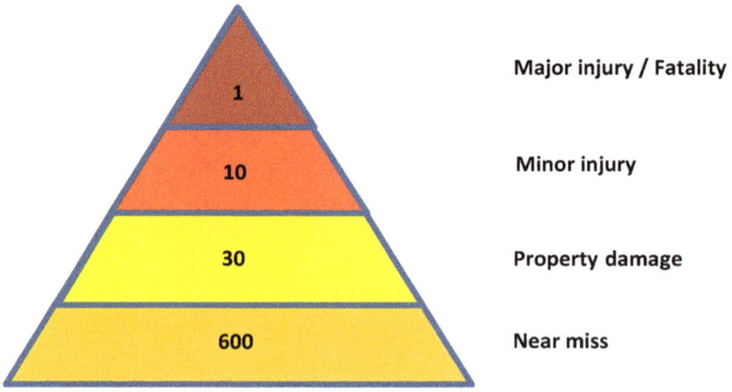


**Figure 1.2.** Showing the relationship between the types of incidents outcomes.

injuries/fatalities at the apex and near-miss incidents at the base. These are obviously averages, but a crucial association exists between the numbers of types of incidents reported. In this example, for every major injury/fatality reported, 640 other incidents would have occurred over the same period of time. By this token, it demonstrates that for every incident that occurs, there is a:

- 600/641 chance that it will result in a non-injury, near-miss incident
- 30/641 chance that it will result in a non-injury incident such as property damage
- 10/641 chance that it will result in a minor injury
- 1/641 chance the consequences will be a major injury or a fatality

The fact remains however, that users of cryogenic technology are not immune to these insightful statistics, and the first time a cryogenic related incident occurs, the individual involved may be lucky, and it might only result in a non-injury accident. There remains the reality though, that for every 641 incidents taking place, one of these will likely result in a life changing injury, or even a death. This is the sobering reality which anyone working in the field of cryogenic technology needs to be fully aware of and consider taking steps to provide remedial actions in order to reduce the likelihood of incidents occurring in the first place.

## 1.5 Safety management systems

As a means of mitigating the associated risks from using cryogenic technologies, there are a wealth of approaches available to the industry, many of which can be easily and effectively implemented, and generally with minimal costings involved. Collectively, these strategies form the basis of what is described as a company's safety management system. This is the administrative process intended to control all the individual elements relating to safety in the workplace. There are numerous elements to an effective safety management system and when used correctly they allow the management of the significant safety risks in many branches of industry. Users of cryogenic technology can readily adopt these systems in order to reduce risks down to acceptable and manageable levels. These include:

- Policy
- Organisation
- Planning and implementation
- Measuring performance
- Reviewing performance
- Auditing
- Legislation

### 1.5.1 Policy

A company's safety policy is generally a short statement, summarising their vision of how safety should be recognised for its importance, and affirming their overall approach to health and safety in the workplace. The policy is often endorsed by a senior manager within the organisation, effectually demonstrating the senior management team's commitment to the policy. It is dated in order to bring relevance to its contents in terms of a time period. Normally, the company policy is updated annually to reflect any changes in procedures over the last twelve months, which may have a significant impact on how a company undertakes its work.

### 1.5.2 Organisation

It is a sad reality that many health and safety incidents occur due to failures of management controls. The organisation section of a safety management system should therefore provide understanding of what particular individuals within an establishment are responsible for in order to maintain a good health and safety culture. It should also outline how every member of staff needs to work with their co-workers to achieve the expected health and safety standards defined within the policy.

Overall responsibility for health and safety remains with the senior employees of a company, but specific duties can be cascaded from the executive level and allocated to individuals within the organisation. By doing this, it increases the likelihood of the organisation maintaining control of its health and safety. It can also improve communication and co-operation of the various groups within a company, as individuals appointed to specific roles tend to take ownership of the task and expound the overall policy, sharing and learning from other groups who they are interacting with.

### 1.5.3 Planning and implementation

Since risk management is fundamental to a company's good health and safety performance, the planning and implementation of the safety policy should provide the more detailed information on how this is undertaken through a variety of systems and procedures. It should also define how the organisation intends to manage its risk, and more critically, how they will comply with any national or international legal requirements associated with health and safety.

### 1.5.4 Measuring performance

A company can only declare an excellent health and safety culture if they can demonstrate the effectiveness of their strategies. This is generally achieved by the measurement of their health and safety performance, through both active and reactive monitoring. Successful reactive systems involve a company responding to incidents in a way that any investigation can quickly identify weaknesses in the current safety systems. Any resulting actions, which should improve the robustness of a safety system, can therefore reduce the risk of similar incidents happening in the future. An active monitoring system in contrast, does not rely on an event such as an accident or an incident to take place for the provision of effective health and safety measures. This is the foundations of risk assessment, where likelihoods are assessed and control measures implemented as a result, even though an actual accident hasn't taken place. It could also be as simple as implementing regular safety tours, specifically looking for things that could go wrong and providing a remedy to stop them becoming manifest.

### 1.5.5 Reviewing performance

Regular reviews of an organisation's health and safety management system need to be undertaken in order to identify if the top-level policy and the accompanying arrangements require an update. Generally, any revisions tend to be relatively minor, capturing elements such as legislative updates or identifiable changes in a company's processes and outputs. As an organisation develops however, so too must its safety management system, reflecting its growth and changing dynamics.

### 1.5.6 Auditing

Interpreting the raw data of health and safety within an organisation can often fall foul of unconscious bias on the part of the executives. It is important therefore that an independent assessment is made of the safety management systems in the form of safety audits. By doing this, a test can be made to check if the defined procedures set out on the policy and safety documents are actually being followed in the correct way. It can also check the effectiveness of the policy and its continued relevance to the company processes which it covers.

### 1.5.7 Legislation

As an aid to implementing safety management systems, national and international legislation is often established in order to promote the health and safety at work. These laws place duties on specific people and groups within an organisation who have a direct influence on the health and safety standards in the workplace.

## 1.6 United Kingdom health and safety legislation

One of the core elements of health and safety legislation within the United Kingdom is the Health and Safety at Work Act (HASWA) [7]. As its title implies, it is made under a specific Act of the United Kingdom Parliament and is enforceable. It puts in place the statutory instruments relevant to any work situation, encouraging health and safety at work by defining the key duties people within an organisation have to comply with. This would include employers, managers, employees, contractors, manufacturers and suppliers. Under the HASWA, employers have a duty to ensure the health and safety and welfare of their employees while they are at work. The Act is implemented and overseen by the Health and Safety Executive (HSE) who are the United Kingdom's government agency responsible for both the regulation and the enforcement of workplace health, safety and welfare. They also undertake research into all levels of occupational risks in the United Kingdom.

As a working act, the HSWA is descriptive, and yet remains all-encompassing in order to fulfil its remit for all areas of work. To help employers fulfil their obligations under the act, direction is published in the form of the Managing Health and Safety Guidance (HSG65) [8]. There is no mandatory requirement to conform to the guidance, but it offers companies an essential overview of what they need to have in place in order to realise the legal requirements of the HSWA. It crucially stipulates that health and safety needs to be managed at the same important level as all other elements of an organisation's activities. It focuses on management controls, so that individuals are able to understand what it is they are responsible for. It also advocates that companies establish some means of measuring, reviewing and auditing their health and safety performance, in order to test that good health and safety procedures are being adequately followed.

## 1.7 United States of America health and safety legislation

In the United States of America, the Occupational Safety and Health Act (OSH Act) [9] is the labour law governing the federal law of occupational health and safety in both the private sector and the federal government. As an all-encompassing Act, its premise is to assure safe and healthy working conditions for all workers. It does this by the enforcement of the various standards developed under the Act, along with assisting the individual States in their efforts to implement the Act. Its main goal is to ensure that employers provide their employees with a working environment free from all known hazards. The Act was also the catalyst for the creation of the Occupational Safety and Health Administration (OSHA) who are primarily responsible for overseeing the application of the Act, along with the National Institute for Occupational Safety and Health (NIOSH), who are responsible for conducting research into work-related injuries and illnesses.

The policy structure within the United States of America does not regulate the Act nationally. The constitution and federal structure of the nation means that each of the 50 states that make up the United States of America have a relatively high degree of autonomy in how the OSHA standards are implemented. Effectively, each state can develop and enforce its own programme to suit local needs, although these would still have to be at least equal, or equivalent to the federal requirements in order to be deemed valid.

## 1.8 The European Union health and safety legislation

European Directives cover a raft of work-related health and safety requirements which are adopted by the various member states. In the European Union, the Health and Safety Framework Directive (Directive 89/391/EEC—OSH 'Framework Directive') [10] is the definitive piece of legislation which all member countries have to comply with. The objective of the Directive is to apply measures which encourages improvements in the health and safety of workers throughout Europe. It applies to all sectors of activities, both public and private. As it stands, the Directive is the basic safety and health legal Act which concerns the prevention and protection of workers against occupational accidents and diseases.

Linked meaningfully with the functioning of the Directive, the European Agency for Safety and Health at Work (EU-OSHA) are responsible for providing the related health and safety guidance and information throughout Europe. Their role is to work closely with the governments of the member nations, along with the employers and workers within the European Union. Where possible, they identify current and possible future risks associated with the working environment, sharing their findings and providing advice on good practice common to the European Union nations.

As the European Union is made up from a variety of nations, the laws for health and safety measures have been brought in line with existing nation laws and are therefore the same throughout Europe. The rules however, have to be altered in order to fit into the individual national laws of the member states. This can often mean that specific health and safety laws can differ between the various European Union nations and it is important for employees working across borders that they seek guidance and clarity of the current laws for the particular European Union country in which they are working.

## 1.9 Cryogenic temperatures placed in perspective

### 1.9.1 Temperature

In today's world, we often take it for granted that we have almost instantaneous knowledge of the outside temperature on any given day of the year, be it provided in the form of a weather forecast or as a direct reading from a nearby temperature measuring device. But it has not always been so, and this defining way in which hot and cold can be expressed was for many years utilised only by the science fraternities. In a physics sense, the temperature of an object is a discrete value of the amount of internal heat energy that object has. Under a set of thermodynamic principles, this energy can be transferred from one object to another, as those classed as hot will be able to impart some of this heat energy to another object which is colder.

The general way in which temperatures are measured is by the use of thermometers. Typically these are one of three defining types; liquid column, thermocouples or even infrared devices. Each of these versions has some means of sensing the temperature, the output from which will alter with a corresponding change in temperature. This could, for example, be an expansion or contraction of the column height in a mercury thermometer, or a change in the voltage produced by the junction of a thermocouple. The other important feature of thermometers is that they have some way of converting these changes into a recognisable numerical value. In a mercury column thermometer, this would be through a visible scale on the glass containment tube. In a thermocouple, the output voltage can be calibrated in such a way that it produces a direct temperature readout.

Whatever temperature scale these numerical values are based on inevitably depends on the application the thermometer is being used for. Historically, there have been many such scales adopted for specific practises, some used more commonly than others. Generally, the three most frequently used scales are: the Fahrenheit scale, denoted as ° F, the Celsius scale, denoted as °C and the Kelvin scale, denoted as K.

### 1.9.2 Temperature scales

The human body, looked upon purely as a thermodynamic mechanism, is a constantly dynamic system, relentlessly adjusting its temperature balance to suit the ambiance around it. Too small to see with the naked eye, our skin is riddled with fundamental sensors which continuously detect the relative temperature of our immediate surroundings. This sensitivity to temperature is important, as it allows the body to adapt and regulate its own internal thermal equilibrium based on the environmental conditions impacting upon it. We cannot however detect definitive temperature values, but instead, we rely on the body's sensors to tell us when our surroundings are discretely hot or cold, wherein signals are sent to the brain, which in turn instigates changes in the body to compensate for any temperature extremes incurred. This is the driving force behind the body's intuitive reaction to perspire when it is overheating and to shiver when it is cold. This is critical, as the physical stimulus triggering these reactions also provides us with a mechanism which automatically causes us to urgently withdraw from objects that are excessively hot or cold.

This awareness of the relative warmth or chill of our surrounds has served humans well, allowing our survival as a species in some of the harshest places in the world. Yet remarkably, even as late as the start of the 17th century, there was still no way to actually quantify the exact heat an object possessed. The first thermometer type devices at that period were known as thermoscopes, and rather than indicating a strict measurement of hot or cold, simply showed changes in temperature as they occurred. In their rudimentary form, these thermoscopes were built from a small vase filled with water. Attached to this vase was a thin vertical rising pipe which itself had a large, empty glass ball at the top. Changes of temperature of the air inside this upper ball would naturally exert either a positive or negative pressure on the water below, causing it to rise or fall in the thin column pipe (figure 1.3). Referencing these changes against the original water position would give the observer an immediate indication of whether the ambient temperature had indeed been increased or lowered over a given time [11].

In fact, it was not until 1612 before the Italian inventor and professor of science, Santorio Santorio adopted a more direct quantitative approach to the measurement of temperature. In that year, he produced a clinical thermoscope which was the first to include a numerical scale on the vertical column [12]. This allowed the user of the device to now define exact values of temperature change. This primitive form of thermometry still had its limitations however, as it could only define fixed points of temperature and not measure the specific quantity of heat in an object. Justifying its function, it was able to indicate if something was warmer or colder than another

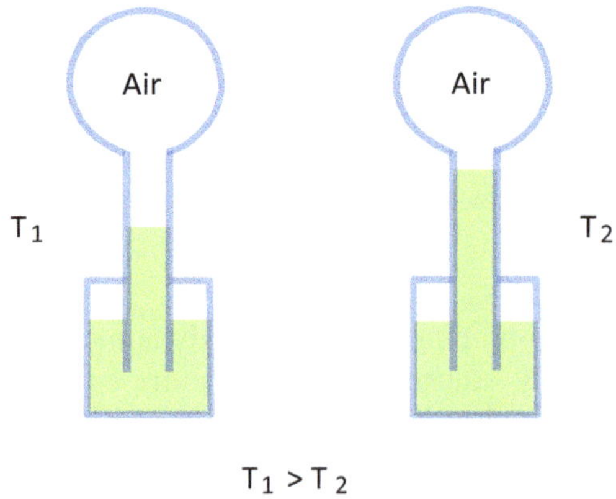


**Figure 1.3.** Showing how temperature change can be determined using a thermoscope.

thing, in essence matching scientifically what the human senses could already experience, but with a much greater accuracy.

In order to define what would be considered as reproducible temperature scales, there needs to be at least two fixed reference points of temperature which can always be duplicated over and over again to the exact same value. These two datum would in effect define a particular temperature range which can then be subsequently divided into a usable scale. Typically, these reference points are based on the change of state of a particular material, so that they generate a comparatively precise and reproducible temperature value. Examples would be the melting temperature of ice, or the solidification temperature of pure metals. However, whatever credentials are chosen, the defining conditions of these transitions need to be controlled and the material's composition made exact for each matching test. Any slight variance in either would inevitably result in a different temperature value being determined for each test undertaken.

### 1.9.3 Fahrenheit scale

The Fahrenheit scale, now infrequently used, was the foremost temperature standard employed throughout the world until the 1960s. It is based on a temperature scale proposed in 1724 by the German physicist, Daniel Gabriel Fahrenheit [13]. Earlier, he had invented both the alcohol and the mercury thermometers and he used these devices to determine his scale by referencing three fixed points of temperature. Utilising transition points available to him at the time, he defining a 0 degree point on his scale as the lowest temperature to which he could reproducibly cool brine, which is essentially salty water. The second reference point was selected as the temperature when ice was just starting to form on the surface of still water and was defined as 32 degrees. For the highest fixed point, he turned to the human body, defining 96 degrees on his scale as that of the average human core body temperature.

In more recent time, the Fahrenheit scale is generally defined by only two fixed temperature points which have a 180 degree separation. These datum are further defined as being measured specifically at sea level and subject to standard atmospheric pressure in order to remove any variance errors. On this version of the scale, the lower fixed point is the temperature at which water freezes into ice and is defined as 32 degrees. The higher fixed point is the boiling point of water and is defined as 212 degrees.

### 1.9.4 Celsius scale

The Celsius scale, was named after the Swedish astronomer, Anders Celsius who was born in 1701 and developed a similar temperature scale to that which is used routinely today [14]. Defined as the complete reverse of the modern scale, Celsius used both the boiling point of water as his 0 degrees, and the melting point of snow as his 100 degrees, for his two fixed points of temperature. It was not until after his death in 1744 that the scale was formally reversed and adopted in this revised form as a new alternative standard to the Fahrenheit scale. Throughout the 19th century, the scale was also referred to as the Centigrade scale due to it having 100 graduations between its fixed points. This was often confused with an angular measurement, also defined as centigrade. As a result, in 1948 the nomenclature of Celsius for the temperature scale was formally adopted in its place.

### 1.9.5 Kelvin scale

The Kelvin scale, was named after the Belfast-born engineer and physicist, William Thomson, who was born in 1824 and became the 1st Baron Kelvin. The title actually refers to the River Kelvin which flows near to the laboratory he used at the University of Glasgow and where he undertook important work into the formulation of the first and second laws of thermodynamics. In 1848 he proposed an absolute temperature scale which, unlike previous versions, would be independent of the physical properties of a specific substance and could therefore be reproduced without having to have exact conditions enacted each time [15]. His scale depends instead on the universal constant known as the Boltzmann constant, and as such, determines that the same amount of energy would be given out by an object whatever the unit change in temperature. It works on the principle that each incremental degree has an identical value wherever it is on the scale. He also postulated the existence of a temperature point at which no further heat can be transferred from an object. Today we know this as the absolute zero, which is defined as 0 K or −273.15 °C (figure 1.4).

## 1.10 Cryogenics put in temperature context

Our everyday lives revolve around the acceptance that certain temperature assumptions are met in order for us to infer reasoning. We know for example, that too little hot water inside a bathtub results in a miserably cold experience for the bather, and that too much hot water runs the risk of causing scalding. Likewise, only the foolhardy would venture outside without a coat on a frosty winter's day, or consider wearing a scarf and mittens on a gloriously sunny day. These are determinates which, as we grow older and gain more and more experience, form our knowledge of the physics around us.

The use of cryogenic technology however, may in many circumstances, be unfamiliar to the countless who are introduced to it through their employment. Our datum for comprehending how cold objects can become is probably based on the lowest temperatures our domestic freezer units can achieve, or images in our memories of particularly

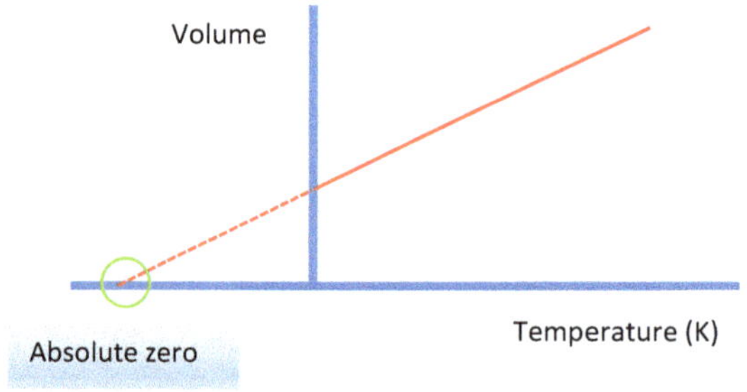


**Figure 1.4.** Showing the relationship between gas volume and temperature used to determine the absolute zero point.

cold winter days in the past. It is difficult at first for us therefore to automatically put in context where cryogenic technology sits compared with these modest references. This lack of understanding also makes it difficult for the newcomer to the discipline of cryogenics to adequately address any specific hazards associated with the use of cryogenic equipment under the extremely cold temperatures at which they operate. As a consequence, it is important that an understanding is gained as to where the technology sits on a generalised temperature scale in order to address these shortfalls in our comprehension.

Associated with this premise, when undertaking cryogenic calculations to determine fundamental properties such as heat transfer, it is more manageable if the working temperatures are defined as positive values throughout. An ideal measure for this application is the Kelvin absolute scale. Working with this Kelvin scale as a datum, key temperature events can be recognised and placed in context with each other using an imaginary thermometer as our typical measuring device (figure 1.5).

In an attempt to set these recognisable temperatures against the framework of cryogenic technology, the upper point on this imagined thermometer will be a relatively high value in comparison, progressing gradually downwards to the eventual absolute zero temperature as defined by the Kelvin scale.

### 1.10.1 The boiling point of water

The starting temperature on our imagined scale is a relatively familiar event sitting at 373 K which signifies the normal boiling point of water at 1 bar atmosphere and specifically located at sea level [16]. This is where the liquid state of water is changed into its gaseous form, in this case, steam. It has to be stated that this value however is

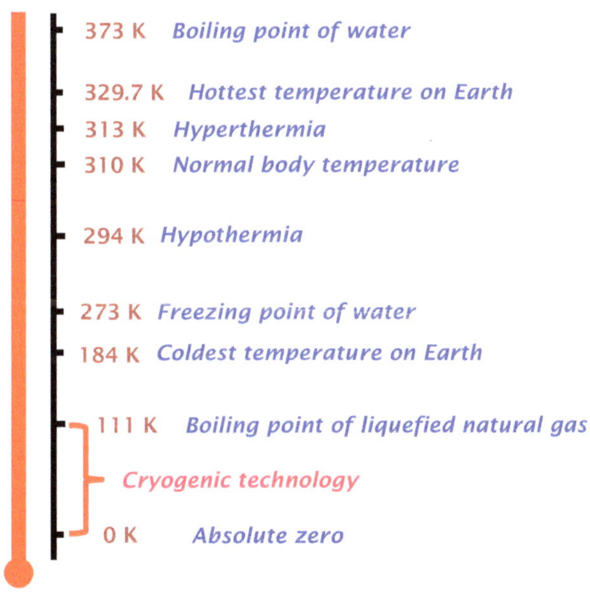


**Figure 1.5.** Showing the relative position of cryogenic technology against datum temperatures.

not a constant, as changing the elevation and hence the atmospheric pressure will change the temperature at which the water boils. If for example you are attempting to boil water on the top of a large mountain, the water will boil at a slightly lower temperature due to the gain in altitude. The boiling point will also depend on the purity of the water. If for example, an impurity such as salt is added, the water will now boil at a higher temperature.

### 1.10.2 Hottest temperature on Earth

Lowering the temperature slightly takes us down to a precise value of 329.7 K which is currently the hottest natural temperature recorded on Earth in the Death Valley desert in the USA in 1913 [17]. Although, according to the World Meteorological Organisation (WMO) [18] who monitor global temperatures, the validity of this record has now been questioned as conflicting evidence has since been submitted which challenges this historically recorded temperature. Traditionally, these highest temperature conditions were measured in the air, 1.5 metres above the ground, with the measuring device shielded from direct sunlight. There are also additional measurements taken directly at ground level. As technology has developed though, satellite measurements of the Earth's temperature are now becoming more available, each with a certain level of improved accuracy.

A more recent high temperature of 327.5 K was however recorded by an automated weather station at the Furnace Creek visitor centre at Death Valley National Park on the 16th August 2020 [19]. As the 1913 temperature is in question, the WMO have speculated that this may, in fact, be the hottest air temperature ever recorded on Earth.

### 1.10.3 Hyperthermia

Below this value, at 313 K sits the point where hyperthermia takes place [20] in humans. More commonly known as overheating, this is a particular medical condition in which the human body's temperature is elevated above its normal with damaging effect. At this point, the body absorbs more heat than it can dissipate and cannot perspire quickly enough in order to cool itself down. This is a dangerous condition for the body and can result in eventual death if not suitably treated in time.

### 1.10.4 Normal human body temperature

In contrast, 310 K is deemed the normal human body temperature, although it is in reality a range which will fluctuate on a daily basis depending on activity levels, age and general health [21]. This range is often quite wide and can be as much as 309.1 K at its lowest and 310.2 K at its highest. A recorded temperature beyond this normal range is often considered as a medical condition requiring some immediate treatment.

### 1.10.5 Hypothermia

Remaining with the human body, dropping down the temperature scale further gives 294 K, which is where the condition known as hypothermia will take place [22]. In this state, the human body dissipates more heat than it can absorb or generate. The observed symptoms for mild hypothermia would be involuntary shivering and a certain degree of mental confusion by the sufferer. Ironically, at a more moderate level of hypothermia, the shivering will cease and the confusion becomes more pronounced. In severe cases of hypothermia, there is an increased risk of the heart stopping and therefore fatalities occurring. Hypothermia is generally caused by the exposure to extreme cold conditions, such as those experienced during Arctic explorations. Users of cryogenic technology, which carries an association with cold temperatures, therefore need to be fully aware of the condition and be vigilant for any signs of the symptoms becoming manifest.

### 1.10.6 Freezing point of water

Beneath this temperature is 273 K, which is where liquid water solidifies, turning into ice [23]. This is known as its freezing point and is where the liquid to solid transition takes place. As with the boiling point of water, this freezing point can be modified if impurities are present in the water. Deliberately adding antifreeze mixture into the coolant system of a car, for example, will lower the freezing point allowing it to function at below normal freezing conditions. Adding a solvent such as salt to water will also lower its melting temperature, which is why common salt is often applied to icy roads during the winter months, thereby lowering the melting point of any ice present, allowing it to melt quickly from off the roads. Interestingly, very pure water can be supercooled to temperatures well below its normal freezing point. This occurs because ordinary water depends on small suspended particles on which to form the ice crystals as it freezes. In the absence of these impurities, the water will remain in its liquid state below the 273 K freezing temperature. Supercooling is an interesting phenomena, whereupon the water will freeze immediately on contact with any surface it comes into contact with.

### 1.10.7 Coldest temperature of earth

Below this temperature is 184 K, which is the lowest temperature recorded on Earth [24]. The temperature was logged at the Soviet Vostok Station in Antarctica in 1983 and was taken by ground measurements of the air above the surrounding ice surface. More recent remote sensing satellite observations have shown even lower temperatures on the ice surface in the higher regions of the Antarctic, but as yet these have not been verified, as their mode of measurement differs to that employed by ground and air measurements.

### 1.10.8 Boiling point of liquefied natural gas

Going even colder, 111 K is defined as the boiling point of liquefied natural gas (LNG) [25]. LNG is a naturally sourced gas and is predominantly methane, with a small amount of ethane in its mixture. In its liquid state, it takes up only 1/600 the volume it would in its gaseous state. It is therefore cooled down to form a liquid in order to allow its safe and more manageable storage and transport.

### 1.10.9 Absolute zero

At the lowest point on this imaginary thermometer is the theoretical absolute zero, 0 K [26]. At this point, the vibrational motions of the atoms in an ideal gas are assumed to be zero, effectively indicating that both the enthalpy and entropy of the gas will also be at their minimum values. Unlike elevated temperatures, which in theory have no upper limits, there can be no lower temperature than this 0 K value. The temperature is determined experimentally by extrapolating the ideal gas law which determines a change in volume of a gas as its temperature is reduced. As a reference temperature, it is often denoted as −273.15 °C.

### 1.10.10 Cryogenic temperature range

The temperature at which cryogenics is defined varies depending on the research organisation offering an explanation. A popular range is from 123.0 K, which is slightly above the boiling point of liquefied natural gas, down to 0 K. Others however, only consider temperatures below 93.0 K as being true cryogenics [27]. Their reasoning being that this is the temperature below which the boiling points of the original permanent gases lie, and therefore, gases such as helium, hydrogen, neon, nitrogen and oxygen can only be liquefied below this value. Adopting the same logical reasoning, the boiling point of the gases used commonly in refrigeration, such as Freon, are all above this 93.0 K temperature. Typically, the temperature range 273 K down to 123 K is defined as the refrigeration technology operating range.

To add further confusion to this general definition, the development of superconducting materials with critical temperatures above the boiling point of liquid nitrogen, have resulted in the creation of the new term, 'high temperature cryogenics' [28]. Principally, this describes a temperature range from just above the boiling point of liquid nitrogen of 77.36 K, up to 223 K, where these so-called high temperature superconducting materials operate.

Whichever definition and range is chosen in describing the technology, the field of cryogenics is fundamentally colder than most of us would ever experience naturally in our lives. Unless we are employed to work at the South Pole or in some vast refrigerated warehouse, it is unlikely, that we will ever encounter in our daily lives the extremely low temperatures associated with the cryogenics industry.

## 1.11 Humans and cryogenic technology

Throughout human evolution, as an emerging species, we have been able to learn how to cope with the extremes of the Earth's climatic conditions. As a result, humans have colonised even the most inhospitable regions of the world. Evidence prevails showing how even our Neolithic ancestors used insulating clothing in the form of furs to survive the harsh winters of their time, migrating to more favourable territories to escape the bitter cold period of the Earth's history during the last ice age glacial period [29]. That being said, in comparison to the low temperatures related to cryogenic technology, the temperature at which humans can comfortably exist on this planet, can typically range from 312 K, all the way down to 280 K [30]. With the obvious assistance of modern artificial heating and cooling, this has further been developed and many regions on the Earth thought previously uninhabitable are now comfortably populated.

But there remains an enormous gap between these relatively comfortable conditions for human existence to succeed and the temperature levels likely to be encountered when using cryogenic technology. Put into context against the cryogenic temperature definitions, it can be stated with a degree of qualification, that the temperature range of cryogenics technology carries enormous risks to each and every one of us, and has potential life threatening consequences to our bodies. This huge temperature difference between what we can endure in terms of extremes, and the base temperatures achieved with cryogenic equipment is beyond the normal tolerance of our physiques. Any direct physical contact for example, with cryogenic materials, can only be detrimental to the tissues making up the human form.

It is crucial also to recognise that as the technology has evolved, more and more facilities throughout the world now use cryogenic materials as part of the essential processes of deliberately cooling equipment. Also, the growth in its use over time increases annually, making it more likely that accidents will happen and that users will inevitably come into direct contact with the medium. This offers a direct correlation, in that if more accidents are expected, more injuries from the various strands and applications of cryogenics will result. This can be as superficial as a small cold burn obtained during a transfer process of liquid cryogens, to an individual suffering hypoxia due to breathing low oxygen content air in the working environment, due to the sudden release of evaporating gases.

Cryogenic technology is not the most user-friendly of environments to work in and yet, with the correct preparation and application of safety standards, it can be employed without accidents occurring. It is imperative therefore that this fact is recognised at the earliest stage of using the technology so that risks can be suitably assessed and all the associated hazards mitigated.

## Introduction to health and safety—summary

**Key learning points**

- One of the first recorded cryogenic experiments were undertaken by Michael Faraday in 1845 when he reached a temperature of −110 °C
- Cryogenic is often unwittingly confused with the low-temperature preservation technique known as cryonics
- There are three main principles which influence the safety culture within an organisation:
  - Moral
  - Legal
  - Financial

- There is a 600/641 chance that a cryogenic incident will result in a non-injury or a near-miss
- There is a 30/641 chance that a cryogenic incident will result in property damage
- There is a 10/641 chance that a cryogenic incident will result in a minor injury
- There is a 1/641 chance that a cryogenic incident will result in a major injury or a fatality
- Safe management systems are employed to protect those workers using cryogenic technology:
  - United Kingdom—Health and Safety at Work Act (HSWA)
  - United States—Occupational Safety and Health Act (OSH Act)
  - European Union—Health and Safety Framework Directive (Directive 89/391/EEC—OSH 'Framework Directive')
- Temperatures can be measured using a variety of temperature scales:
  - Fahrenheit (°F)
  - Celsius (°C)
  - Kelvin (K)
- The cryogenic technology range is generally stated as from 123 K down to 0 K
- Above 123 K is generally defined as refrigeration technology
- The temperature range of cryogenic technology has potential life threatening consequences to the human body

## References

[1] Faraday M 1859 *Experimental Researches in Chemistry and Physics* (London: Taylor and Francis)

[2] Wisniak J 2003 Louise Paul Cailletet—the liquefaction of the permanent gases *Indian J Chem. Technol.* **10** 223–36

[3] Thipse S S 2013 *Cryogenics: A Textbook* (Oxford: Alpha Science International Ltd)

[4] Institution for Occupational Safety and Health (IOSH) 2006 *Managing Safely Syllabus* (Clwyd Associates) Module 1

[5] Cable J 2006 Nitrogen: the silent killer *EHS Today Mag* September

[6] Anderson M and Denkl M 2010 The Heinrich Accident Triangle—too simplistic a model for HSE management in the 21st century? *Society of Petroleum Engineers Int. Conf. on Health and Safety and Environment in Oil and Gas Exploration and Production*

[7] Health and Safety at Work Act (HSWA) *1974 (Commencement No.1) Order 1974* (http:/Legislation.gov.uk) (Accessed 2020)

[8] Health and Safety Executive (HSE) *Managing for Health and Safety (HSG65) 2013* (www.hse.gov.uk/pubns/books/HSG65.htm) (Accessed 2020)

[9] United States Department of Labor—Occupational Safety and Health Administration *Occupational Safety and Health Act (OSHA Act) 1970* (www.osha.gov/laws-regs/oshact/completeoshact) (Accessed 2020)

[10] European Agency for Safety and Health at Work *Directive 89/391/EEC—OSH 'Framework Directive' 1989* (http:/osha.europa.eu/en/legislation/directives/the-osh-framework-directive/1) (Accessed 2020)

[11] Sherry D 2011 *Studies in History and Philosophy of Science—Thermoscopes, Thermometers and the Foundation of Measurement* (Amsterdam: Elsevier)

[12] Pearce J M S 2016 *A Brief History of the Clinical Thermometer* (Oxford: Oxford University Press)
[13] Encyclopaedia Britannica *Fahrenheit Temperature Scale* (www.britannica.com/science/Fahrenheit-temperature-scale) (Accessed 2020)
[14] Encyclopaedia Britannica *Celsius Temperature Scale* (www.britannica.com/technology/Celsius-temperature-scale) (Accessed 2020)
[15] Encyclopaedia Britannica *Kelvin Temperature Scale* (www.britannica.com/technology/Celsius-temperature-scale) (Accessed 2020)
[16] Engineering Toolbox *Boiling Point of Water* (www.engineeringtoolbox.com/boiling-points-water-altitude-d_1344.html) (Accessed 2020)
[17] World Meteorological Organization's World Weather and Climate Extremes Archive—Arizona State University *Highest Temperature Recorded on Earth* (https://wmo.asu.edu/content/world-highest-temperature) (Accessed 2020)
[18] World Meteorological Organization (WMO) (https://public.wmo.int/en) (Accessed 2020)
[19] The Guardian Newspaper *We May Have Just Seen the World's Highest Recorded Temperature Ever* (https://theguardian.com/environment/2020/aug/19/highest-recorded-temperature-ever-death-valley) (Accessed 2020)
[20] Healthline *Hyperthermia—Symptoms and Treatment* (www.healthline.com/health/hyperthermia#stages) (Accessed 2020)
[21] Martini F 2004 *Fundamentals of Anatomy and Physiology—Thermoregulation* International 6th edn (New York: Benjamin Cummings)
[22] Martini F 2004 *Fundamentals of Anatomy and Physiology—Hypothermia* International 6th edn (New York: Benjamin Cummings)
[23] thoughtco *The Freezing Point of Water* (www.thoughtco.com/the-freezing-point-of-water-609418) (Accessed 2020)
[24] Worldatlas *The Lowest Recorded Temperature on the Earth* (www.worldatlas.com/articles/the-coldest-temperatures-ever-recorded-on-earth.html) (Accessed 2020)
[25] British Cryogenics Council (BCC) 2017 *Cryogenics Safety Manual—Liquefied Natural Gas (LNG)* 5th edn (Oxford: Butterworth-Heinemann)
[26] Thipse S S 2013 *Cryogenics: A Textbook—Absolute Zero Temperature* (Oxford: Alpha Science International Ltd)
[27] Thipse S S 2013 *Cryogenics: A Textbook—Cryogenic Temperatures as Defined by the United States of America National Institute of Standards and Technology (NIST)* (Oxford: Alpha Science International Ltd)
[28] Thipse S S 2013 *Cryogenics: A Textbook—High Temperature Cryogenics* (Oxford: Alpha Science International Ltd)
[29] Miles D 2016 *The Tale of the Axe* (London: Thames & Hudson Ltd)
[30] Livescience *The Limits of Human Survival* (www.livescience.com/34131-infographic-limits-of-human-survival.html) (Accessed 2020)

# Chapter 2

# Relevant safety legislation

## 2.1 General health and safety law

Providing a safe working environment can, to some extents, be unpredictable. There are so many variables that could go wrong, either through the expression of immediate accidents or long-term faults that go undetected for years until they eventually develop into a significant event with overwhelming consequence. Likewise, so-called unknown hazards can lie dormant and unnoticed until other influencing factors create their maturity. These are the merits where injuries occur, not from neglect or complacency, but from the capriciousness of the workplace and those who populate them. The absolute currency of the Health and Safety Universe is that of an organisation completely devoid of risks, where accidents never happen, and commercial output is thriving. This is a utopian concept, whereas in truth, the ideals cannot prevail as it is common understanding that accidents are inevitable, especially where safety standards are found to be poor. There is a vast literature written on this prevailing subject and continuous research and studies undertaken to better understand how to prevent accidents happening in the first place and how to improve safety in the workplace setting. But accidents continue to occur at work, year in and year out, reminding everyone of the risks which we must endure as part of our daily lives.

In chapter 1, it was evidently seen how the use of cryogenic technology has significant ramifications to the health and welfare of individuals if an organisation's safety procedures are inadequately effective. It is imperative that the daily lives of workers using this specialist technology are not imperilled by what they undertake as part of their duties. For this purpose, all-encompassing legislation is prepared which enforces certain controlling values in the workplace and puts into effect penalties if noncompliance is found. By setting out these laws, the aim is to prevent accidents happening, or if they do, to reduce the likelihood of harm to those working with the cryogenic technologies. Equally so, it is a sobering fact that many of the very earliest safety laws were brought about as a direct result of some form of incident taking

doi:10.1088/978-0-7503-3123-4ch2  

place causing an injury, or in the worse cases, fatalities [1]. Fundamentally, for every safety device implemented in response to these original laws, someone previously has suffered harm because these devices were not originally in place to protect them. In the contemporary world however, the controlling powers are more prescient in their approach to safety legislation, anticipating risks rather than basing their judgements solely on data derived from accidents.

Putting this all in context, the following chapters look in detail at the present-day germane legislation for the technology in three primary economic areas of the world; the United Kingdom, the United States of America and the European Union. From the offset however, an important caveat has to be noted in that for the application of cryogenic materials and equipment there is no single definitive piece of safety legislation covering their use. Instead, the safety protocols are determined by the various regulations and standards of all of the associated equipment conventions. In order to work successfully with cryogenic applications, and for organisations to remain accident free, safe procedures have to be addressed through a collective of related general safety requirements, regulations, directives and codes of practice (figure 2.1). In fact, depending on the actual process being undertaken, there is often a long list of these interrelated influencing elements which all need to be considered as part of the overall legislative coverage.

*Examples in the United Kingdom would include the following:*

- The Health and Safety at Work Act—1974
- The Management of Health and Safety at Work Regulations—1999
- The Provision and Use of Work Equipment Regulations—1998
- The Control of Substances Hazardous to Health Regulations—2002

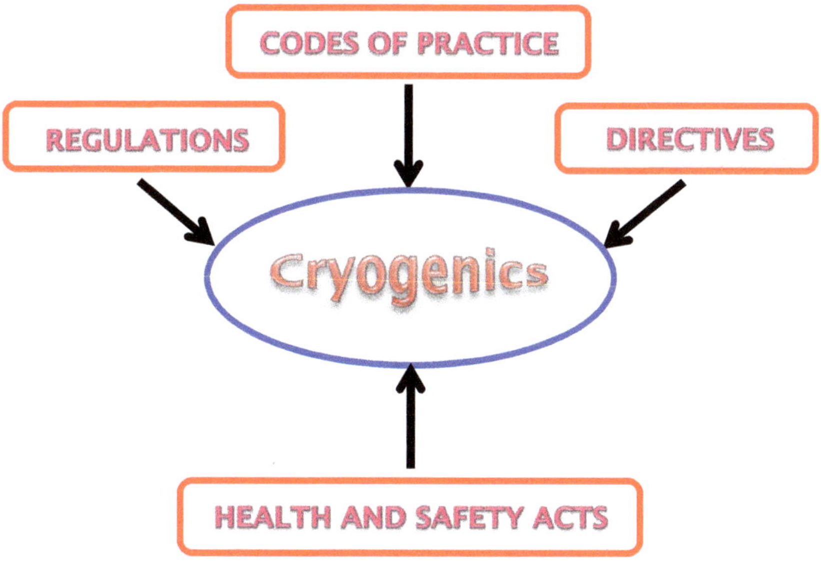


**Figure 2.1.** Showing the related general safety requirements for cryogenic applications.

- The Dangerous Substances and Explosive Atmospheres Regulations—2002
- Pressure Equipment Directive—2000
- Pressure Systems Safety Regulations—2000
- The Confined Spaces Regulations—1997
- Personal Protective Equipment at Work Regulations—1992

*Examples in the United States of America would include the following:*

- The Code of Federal Regulation—Title 16—Commercial practices
- The Code of Federal Regulations—Title 19—Labour
- The Code of Federal Regulations—Title 23—Highways
- The Code of Federal Regulations—Title 49—Transportation
- American Society of Mechanical Engineers (ASME)—Boiler and Pressure Vessel Code

*Examples in the European Union would include the following:*

- Directive 2009/104/EC—Use of Work Equipment
- Directive 2014/68/EU—Pressure Equipment
- Directive 2010/35/EU—Transportable Pressure Equipment
- Directive 2008/68/EC—Inland Transport of Dangerous Goods
- Directive 99/92/EC—Risks from Explosive Atmospheres
- Directive 92/58/EEC—Safety and/or Health Signs
- Directive 89/656/EEC—Use of Personal Protective Equipment
- Directive 2013/35/EU—Electromagnetic Fields
- Directive 90/269/EEC—Manual Handling of Loads

The process of determining which of the many rules, laws and regulations need to be adopted can be quite complex to initially recognise, especially for those unfamiliar with the process. Sometimes it can be a significant challenge to ensure that all the relevant standards and regulations applicable to a particular cryogenic use have been applied. Fortunately, many are instantiated through various cryogenic technology industry standards and codes of practice, which aptly provide guidance through typical examples representing the majority of customary applications of the technology. It remains a fact however, that creating a synthesis of these Codes and Regulations is not an informal task and organisations need to ensure that all relevant legal frameworks are implemented before their staff begin working with any form of cryogenic technologies (figure 2.1).

## 2.2 General workplace law in the United Kingdom

In the United Kingdom, laws associated with the workplace are implemented through a mechanism known as health and safety legislation [2]. There are both primary and secondary levels to this legislation, with the primary including the various Acts of Parliament, issued exclusively through the governmental processes. A typical example of this would be the Health and Safety at Work Act—1974. The secondary legislation comprises the statutory instruments which are more commonly referred to as regulations. Some regulations stipulate the requirements of specific duties being established within organisations and these are known as substantive regulations. Others are defined as administrative regulations, and typically describe the administrative requirements that all United Kingdom businesses must adopt in order to manage safety in the workplace. An example of this would be the enforcement powers given to the Health and Safety Executive (HSE) [3]. Regulations are also generally subject and category based, relating to a particular industry.

Many of the regulations empowered in law within the United Kingdom have European Union Directives as their genesis. It is a legal requirement for all Member States to implement these edicts by putting their own legal frameworks together in order to fulfil this obligation, but at the same time, allowing for national variance in the function of the Directives. An example would be where a land-locked European Union member state clearly would not have to implement any Directives relating to coastal or sea applications.

The United Kingdom's recent withdrawal from the European Union and the subsequent Departure Implementation Period will evidently negate any future impact on the United Kingdom of new Directives issued by the European Union, or even amendments to existing ones. It is possible therefore, that standards sitting within the United Kingdom Regulations framework may very well change as they no longer become a required compliance with the European Union feeder Directives. An example of this could be the standards defining exposure limits when organisation are working with dangerous substances. If the United Kingdom no longer have access to European Union studies which establishes these values, changes implemented by the European Union would not necessarily be mirrored in a revision of the United Kingdom standards.

It is difficult to be prophetic about what changes may take place, but looking at the situation sceptically, it is conceivable that existing regulations in the United Kingdom, which have a European Union Directive as its context, will at some point be adjusted, as they are no longer bound by any European Union constraints. That being said, trade from the United Kingdom into the European Union will continue, and in order to meet the safety demands of the European Market, there will have to be some form of compliance for products from the United Kingdom to be sold through that trading route. It is hard therefore to imagine that the Regulations controlling safety in the United Kingdom will be significantly changed. As they stand, they are functioning legal obligations which ensure safety in the workplace and it would only be sensible for the United Kingdom to recognise and reflect global changes as both processes and technology within all industries evolve.

### 2.2.1 The Health and Safety at Work Act—1974

The Health and Safety at Work Act is the primary legislation covering work-related health and safety across the United Kingdom [4]. Delegated authorities, such as the HSE, are responsible for providing enforcement of the Act. As a general health and safety provision, it encourages organisations to adopt an appropriate safety culture with the foremost aim of reducing occupational accidents. It is an all-encompassing statute covering the general principles of the workplace. As a law, compliance to it is obligatory and as such, it can be stringently enforced.

As with many laws, this particular Act can be challenging to understand, not least because of its legal content, but also because it covers so many individual topic areas. This makes the interpreting and eventual implementation difficult to achieve. To support this process, information in the form of guidance notes and Approved Codes of Practice (ACOPs) [5], is provided by the HSE. This guidance gives advice on how to comply with the law, explaining any specific requirements or specific technical information to which an organisation would legally need to observe as defined within the Act.

Taken individually, the ACOPs are not strictly laws and therefore do not formally need to be conformed to. Instead, they give recommendations of the methods to be used in order to comply with the regulations. As such, they are very often used as evidence that a legal standard has not been met if a breach of the health and safety regulation has ever occurred.

### 2.2.2 The Management of Health and Safety at Work Regulations—1999

The Management of Health and Safety at Work Regulations underpins the Health and Safety at Work Act by outlining a set of well-defined duties which employers have to comply with [6]. The core measure being to keep a workplace safe. One of its main responsibilities is the requirement of organisation to assess the risks to health and safety for not only their employees, but also anyone directly affected by the work they undertake. For companies employing five or more members of staff, this risk assessment process has to be formally recorded.

By defining this risk management process within the framework of the Regulation, it establishes the fundamentals of a good safety culture within any organisation. The findings of these assessments can also be subsequently used as the foundation for preventing risks from maturing. This is often referred to as the hierarchy of risk control and is a relatively simple method by which risks can be sensibly and economically mitigated [7]. There are many different versions of this grading system, ranging from as little as five to as many as nine tiers. What they all embody however, is the overall principle of sequentially considering the hazards to be found in the workplace and assessing their likelihood of causing harm. The general aim of this exercise is to reduce any high risks by considering each control measure in turn, with inclusive precautions at the top of the hierarchy and the more task specific towards the bottom of the list. A typical hierarchy would consider the following levels:

Eliminate the risk
Substitute the process
Combat at source
Safe systems of work and training
Personal protective equipment

Once any appropriate controls have been selected, the Regulation requires an organisation to establish methods of putting these into practice. At the same time, they will also need to create a system which monitors their effectiveness. One key area of this process is the setting up of health surveillance for employees, so that ill health can be identified at an early stage and any corrective actions immediately applied. Examples of where health surveillance may be justified would be if employees are routinely exposed to noise, vibrating equipment, solvents, fumes, dusts or even biological agents.

The Management of Health and Safety at Work Regulations also provides for staff to receive regular health and safety training applicable to their work activities, so that employees are never asked to undertake activities beyond their capabilities. The final part of the Regulation requires that emergency procedures, such as an evacuation protocol, be established, so that employees are never placed in serious danger should a situation, such as a factory fire, arise.

## 2.3 General workplace law in the United States of America

Much of the laws of the United States of America are essentially derived from what is known as the common law system which originated primarily from English law [8]. This legal arrangement was entrenched at the time of the American War of Independence, having been brought over from the United Kingdom with the early settlers. Since then however, American law has meaningfully developed, moving away from these ancestral roots. In essence, the law is now derived from five sources; constitutional law, statutory law, treaties, administrative regulations and common law. The most important of these is the United States Constitution which forms the basis of the federal government in the country. The constitution defines the limitations of federal law and includes; Acts of Congress, treaties which are endorsed by the Senate, regulations disseminated by the executive branch, and case law which are generated by the federal judiciary. Much of the United States law however falls under the remit of state law, which often differs greatly from one state to another.

### 2.3.1 Occupational Safety and Health Act—1970

Laws associated directly with the workplace in the United States of America are derived from the federal Employment and Labour laws. Formed to provide employment rights for workers, these federal laws are the driving force which cascades downwards, encouraging state law provision of at least a minimum, recognisable standard for the employee. Included as part of this structure are the various health and safety laws, the main one being the Occupational Safety and Health Act (OSH Act), signed into law in 1970 [9]. As an integral part of the federal law, the Act functions within both the private sector as well as the federal government organisations throughout the United States of America.

The core objective of the OSH Act is to ensure that employers adequately provide their employees with working conditions free from identifiable hazards. Its remit is all-encompassing, including such hazards as the potential exposure to toxic chemicals, excessive noise, extremes of temperature, or even unsanitary conditions. The Act also establishes employment protection to employees who complain about unsafe conditions, and also provides workers with the right to refuse to undertake work which they consider to be unsafe.

As part of its formation, the OSH Act created the Occupational Safety and Health Administration (OSHA) [10], an agency of the Department of Labour who are granted the authority to establish and enforce workplace health and safety standards throughout the United States of America. Reviews of the OSHA enforcement priorities, actions and specific cases are undertaken by an independent body known as the Occupational Safety and Health Review Commission. In addition, the Act also established the National Institute for Occupational Safety and Health (NIOSH) [11], who are a self-governing institute formed to ensure that safe working conditions are maintained through the undertaking of research into occupational safety and health and disseminating their findings accordingly.

## 2.4 General workplace law in the European Union

The starting point of all laws within the European Union are the General Principles, which are the fundamental legal principles governing the way in which the European Union functions [12]. They are part of the Primary European Union law, with which all member states have to comply with. Beneath this sits the European Labour Laws, which regulate transnational standards of employment for workers in all of the European Union member states. Within this framework, various safety and health at work measures have been established which form the basis of a series of European Directives.

These Directives define the minimum requirements and fundamentals values which are needed to provide safe working environments. They offer instruction on the prevention of incidents and how to undertake risk assessments, as well as defining the various responsibilities of both employers and their employees. These European directives are legally binding in the context of Article 153 of the Treaty on the Function of the European Union [13]. As such, they have to be adopted into national laws by the respective Member States. Offering support for this, a series of European guidelines are also available which explains how the Directives have to be implemented, as well as how European standards can be complied with. These standards are adopted by a series of European standardisation organisations.

### 2.4.1 The Safety and Health at Work Directive—1989 (Directive 89/391/EEC)

The Safety and Health at Work Directive [14] is known as a Framework Directive and establishes the basic requirements needed to prevent and insure against workplace risks within the European Union Member States. Its main objective is to introduce measures which encourage improvements in the health and safety of employees at work. It is supported by a vast range of specialised Directives which focus on specific high-risk activities and processes which may be found in the workplace.

As part of the management and compliance requirements of the Safety and Health at Work Directive, the decentralised European Agency for Safety and Health at Work (EU-OSHA) [15] have been established to provide all relevant information that can assist with the implementation of the various occupational health and safety related Directives across the European Union. Their work includes the collection, analysis and distribution of data which can eventually be used as guidance for future policy making in this field. Their main activities focus on the analysis and research of safety related topics, incident prevention, and raising awareness of the health and safety Directives particular to specific workplace environments.

## Relevant safety legislation—summary

**Key learning points**

*General Health and Safety Law*

- There is no single definitive piece of safety legislation covering the use of cryogenic technology
- Safe procedures have to be addressed through a collective of related general safety requirements, regulation, directives and various industry related standards and codes of practice

*General workplace law in the United Kingdom*

- Laws associated with the workplace are implemented through health and safety legislation either as Acts of Parliament or Regulations
- Many of the Regulations are derived from European Directives
- The Health and Safety at Work Act is the primary legislation covering work-related health and safety
- The Health and Safety Executive (HSE) are responsible for the enforcement of the Act
- The Management of Health and Safety at Work Regulations outlines the duties which employers have to comply with

*General workplace law in the United Stated of America*

- General laws are derived from; constitutional law, statutory law, treaties, administrative regulations and common law
- The United States constitution forms the basis of the Federal Government
- Many of the laws fall under the remit of State law which can differ greatly between States
- The Occupational Safety and Health Act (OSH Act) functions within both the private sector and the Federal Government Organisation
- The Occupational Safety and Health Administration (OSHA) are the agency responsible for the enforcement of the OSH Act

*General workplace law in the European Union*

- European labour laws regulate transnational standards of employment for workers in Member States through a series of Directives
- These Directives define the minimum requirements needed to provide safe working environments
- These Directives are legally binding to all Member States and have to be adopted into national laws
- The Safety and Health at Work Directive establishes the basic requirements to prevent and insure against workplace risks
- The European Agency for Safety and Health at Work (EU-OSHA) provide information to assist with the implementation of all the occupational health and safety Directives

## References

[1] Engage E H S 2020 *A Short History of Health and Safety* (www.effective-software.com/blog/a-short-history-of-health-safety)

[2] Health and Safety Executive (HSE) *The United Kingdom Health and Safety Legislation Law—Laws in the Workplace* (www.hse.gov.uk/legislation/) (Accessed 2020)

[3] GOV.UK *The Health and Safety Executive (HSE)* (www.gov.uk/government/organisations/health-and-safety-executive) (Accessed 2020)

[4] The Health and Safety Executive (HSE) Health and Safety at Work Act— 1974 (www.hse.gov.uk/legislation/hswa.htm) (Accessed 2020)

[5] Health and Safety Executive (HSE) *Legal Status of HSE Guidance and Approved Codes of Practice (ACOPs)* (www.hse.gov.uk/legislation/legal-status.htm) (Accessed 2020)

[6] Legislation.gov.uk *The Management of Health and Safety at Work Regulations*— 1999 (www.legislation.gov.uk/uksi/1999/3242/contents/made) (Accessed 2020)

[7] Institution for Occupational Safety and Health (IOSH) 2006 *Managing Safely Syllabus—Taking Action to Reduce Risks—Module* 3 (Coalville: Clwyd Associates)

[8] Wikipedia *Law of the United States* (https://en.wikipedia.org/wiki/Law_of_the_United_States) (Accessed 2020)

[9] Wikipedia *Occupational Safety and Health Act (United States)* (https://en.wikipedia.org/wiki/Occupational_Safety_and_Health_Act_(United_States)) (Accessed 2020)

[10] United State Department of Labor *Occupational Safety and Health Administration (OSHA)* (www.osha.gov/aboutosha) (Accessed 2020)

[11] The National Institute for Occupational Safety and Health (NIOSH) *Centers for Disease Control and Prevention* (www.cdc.gov/niosh/index.htm) (Accessed 2020)

[12] Wikipedia *General Principles of European Union Law* (https://en.wikipedia.org/wiki/General_principles_of_European_Union_law) (Accessed 2020)

[13] The Lisbon Treaty *Article 153 of the Treaty on the Function of the European Union* (www.lisbon-treaty.org/files/420-article-153) (Accessed 2020)

[14] European Agency for Safety and Health at Work *The Safety and Health at Work Directive*—1989 *(Directive* 89/39/*EEC)* (https://osha.europa.eu/en/legislation/directives/the-osh-framework-directive/1) (Accessed 2020)

[15] European Agency for Safety and Health at Work *The European Agency for Safety and Health at Work (EU-OSHA)* (https://osha.europa.eu/) (Accessed 2020)

# Chapter 3

# Relevant cryogenic safety legislation in the United Kingdom

## 3.1 The Provision and Use of Work Equipment Regulations (PUWER)—1998 [1]

The Provision and Use of Work Equipment Regulations imposes specific legal duties on organisations who use all types of work equipment. In order to comply, these organisations are expected to provide appropriate and safe equipment to their employees, which are properly designed for the intended use. They must also maintain and inspect the equipment to ensure that correct installation takes place at all times. Likewise, they have to pro-actively manage the risks associated with the use of the equipment in order to minimise the potential of failures occurring. In addition, the Regulations stipulate that work equipment must only be used by employees who have been given the required instruction in their use, and have undertaken any necessary training for that purpose.

The definition of work equipment within the Regulations is wide ranging, covering devices such as machinery and appliances, along with smaller items such as tooling and fixed installations. Their use has an equally extensive construal and includes as examples, the starting, stopping, programming and transporting of work equipment. For organisations involved in the supply of cryogenic equipment to industry, they must by association comply with these defining Regulations. Their equipment must be designed in such a way that it remains safe during use, mainly through the provision of safety measures which will prevent injuries occurring.

One of the main provisions laid out within the Regulations, is for organisations to install protective devices and safety controls to their work equipment. These measures are chiefly to prevent access to dangerous parts of machinery, such as rotating or reciprocating parts. Normally, this will be fulfilled by the provision of fixed or interlocked guards, the common denominator being that any moving parts need to be completely stopped before an individual can get near to any potential

doi:10.1088/978-0-7503-3123-4ch3  

hazard which the moving parts may cause. In the same token, inspection and testing of any such protective devices has to be done by a competent person who needs to comprehensively record all associated examination and test results.

Where guards are not possible, due to the nature of the work equipment, safe systems of work have to be provided. These ensure that robust procedures are employed as the primary risk reducing method, and that the likelihood of injury from any process undertaken with the work equipment is kept to an absolute minimum. This measure must also allow maintenance on the work equipment to be undertaken without anybody being exposed to an increased risk to their health and safety.

For companies marketing work equipment on the European Market, the Regulations oblige them to conform to European Community law and the Machinery Directive [2]. This requires equipment to be CE marked and be supplied with a declaration of conformity and all necessary instructions for its safe use.

Very often, work equipment such as lifting equipment and accessories, are already covered by other regulatory documents. Approved Codes of Practice (ACOPs) support this crossover, giving guidance on how the requirements of the inclusive regulations should be made in order to comply fully with the law.

## 3.2 The Pressure Equipment Regulations (PER)—1999 [3]

The starting point for the Pressure Equipment Regulations (PER) is the Pressure Equipment Directive (PED) 2014/68/EU, which is a European Single Market Directive for certain types of pressure equipment. This Directive establishes the standards for the design and fabrication of pressure equipment throughout the European Union. Its detailed definition covers the design, manufacture, conformity assessment and proof test requirements of all applicable pressure equipment with a maximum allowable pressure greater than 0.5 bar, prior to them being put into service. The designation of pressure equipment includes all pressure vessels, piping, safety accessories and pressure accessories within any given pressure circuit. As a result, the Directive is implemented as Law in the United Kingdom by the PER 1999 (SI 1999/2001).

The PER describes the essential safety requirements that all qualifying pressure equipment must satisfy and how the different parts within a circuit are classified for the purposes of the Regulations. It also describes the technical requirements that must be satisfied, such as the design scope and the European approvals for the material of construction, along with the procedure that must be followed to fulfil the conformity assessment.

An integral part of the Regulations are the application of a CE mark to all conforming pressure components. The use of this CE mark is linked directly to the process of conformity assessment. Part of this validation is the formal design requirements, where many of the approved design codes, such as PD 5500 [4], also include the principles of designing vacuum vessels. As vacuum systems are often an essential part of many cryogenic storage systems, where they often form insulating regions in Dewars, the important methodology of their safe design falls under this element of the Regulations. Unlike positive pressure systems, vacuum assemblies would primarily fail under collapse. As such, thorough calculations to determine the minimum wall thickness of specific assemblies are written into the design codes under the requirements of compliance to the PER.

## 3.3 The Pressure Systems Safety Regulations (PSSR)—2000 [5]

Following on from the PER, which is focused around the design, manufacture and initial testing of pressure equipment, the Pressure Systems Safety Regulations (PSSR) fulfils the safety requirements once the pressure equipment has been put into service. Its prime edict is to minimise the risks associated with working with systems or equipment which contain either a liquid or gas under pressure. Cryogenic systems generally have some form of liquid cryogen, or a pressurised charge of gas, such as helium forming their basic design. Therefore by default, they need to comply with the PSSR as soon as they are brought into use.

The Regulations describe the critical requirements relating to the installation of the equipment, either as a stand-alone component, or as part of a more substantial installation. They define the safe operating limits of the overall system, as well as the requirements of statutory inspections through examinations over a given time frame. The Regulations also define how there should be consideration made to the maintenance requirements of the pressure systems, making sure that provision is made to allow easy access to component parts, such as relief valves, which would need periodic checking, inspection and maintenance.

In conclusion, the Regulations also cover the situation where modifications or repairs are made to an existing system, describing in detail how a comprehensive review of the design and the safe operating limits would need to be made in these circumstances. It is not permissible for pressure systems to be changed without some reassessment having been made of their safety integrity before they are recommissioned back into service.

## 3.4 The Confined Space Regulations—1997 [6]

Working environments are by definition varied and can often contain multiple hazards which all individuals need to be aware of. Where work is undertaken in a confined space, these hazards often become exaggerated, and the risk of general injury or more specifically asphyxiation due to the lack of sufficient oxygen in the working environment, can be profound under these circumstances.

The Confined Space Regulations cover all work undertaken under confined space conditions. It begins by defining what a confined space is for the purposes of the Regulations, linking this with the need for a full risk assessment to be made of any activities being undertaken within these determined work areas. One of the overriding doctrines is the avoidance of entry into the confined space in the first place, eliminating the related risks by adopting other means of undertaking work in these areas. An example of this would be where meter readouts are positioned external to the confined space, so that individuals do not need to enter the area in order to record the information conveyed by the readouts.

In many circumstances, such as the need to undertake maintenance or repairs in a confined space, entry into these higher risk areas cannot be avoided. In these situations, the Regulations prescribes that emergency procedures are established, which would include the process of rescuing an injured person from within the confined space. This could simply be though the egress of easy-access doorways, but in more severe confined spaces, may be through narrow openings over a range of different levels. Whatever method is employed, the process should be familiar to the individuals undertaking the rescue work and the Regulations give guidance on how companies should routinely carrying out practice rescue drills, effectively rehearsing what would have to happen in a real life emergency situation.

Very often, there is a requirement for safety rescue equipment to be made available at all times during work in confined spaces. Typical examples of this equipment would be a tripod winch, used to lower and raise an individual into manhole service entrances. A rescue stretcher, used in emergencies to extract an injured person, along with breathing equipment, would also be required where the possibility of low oxygen levels being present in the confined space are highly likely.

Lone working is never recommended in confined spaces and the use of a watchman or a buddy system is a common practice. Here, one individual would be responsible for the actual undertaking of any maintenance or repair work in the confined space, with a second individual positioned in a safe location, monitoring what they are doing. In the event of an incident occurring, the second individual can either enact an immediate rescue, or at least, be in a position to raise the alarm so that the appropriate emergency services can be contacted.

It therefore follows that staff expected to work in confined spaces must be suitably trained and where possible, work alongside more experienced staff until they are sufficiently familiar with the surroundings and safety procedures. Likewise, any rescue teams should be competent to undertake a rescue from a confined space, this competency being gained from training and even routinely undertaking a practice rescue from a confined space. In fact, many training courses associated with the Confined Space Regulations will often require the successful completion of a practice rescue before training certificates can be awarded.

In terms of cryogenic technology, and in particular where liquid cryogens are used, the working environment can quickly change from a benign, safe setting, into one that falls under the remit of a confined space due to the release of gas from the cryogens. This release over time, could have the effect of reducing the overall oxygen concentration in the work area, making it an extremely hazardous environment to be working within. This potential for an area to change from being effectively safe to a higher risk confined space is often neglected when hazard identification is undertaken for a given work space. The Regulations provide direction on the ways in which these latent hazards should be accounted for when risk assessments in the workplace are conducted.

## 3.5 The Control of Substances Hazardous to Health Regulations (COSHH)—2002 [7]

The Control of Substances Hazardous to Health Regulations (COSHH) is the governing law aimed to prevent or reduce any potential exposure of individuals to so-called hazardous substances. It covers any substance that is deemed hazardous to health. These can be in a variety of forms, with some migrating from one state to another depending on the work process being undertake. The Regulations cover not only man-made materials, but also naturally occurring elements that can bring about harm to human health. These include:

- Chemicals
- Fumes
- Dusts
- Vapours
- Mists
- Nanotechnology
- Gases and asphyxiating gases
- Biological agents
- Germs that cause diseases

Many of these substances will often be applied directly as part of the work activities of a company. An example of this might be where adhesives, paints and cleaning agents are all used in the preparation of mechanical parts for assembly. Conversely, other substances may be generated during a work activity, such as the fumes created from a soldering or welding process. The so-called naturally occurring substances would include things like grain dust and even fine powders from materials like dry plaster, or other similar types of resources frequently used in the construction industry. Biological agents would typically be bacteria and other micro-organisms, which people working in the healthcare industry, or even testing laboratories, may be at risk of exposure to.

The main requirement within the Regulations is for employers to conduct specific risk assessments associated with their use of any of these substances. A fundamental understanding of how each of the substances can have a negative effect on the human body is therefore important in the preparation of these risk assessments. To do this, some knowledge of the health hazards associated with particular substances are required, generally gleaned from their related material safety data sheet (MSDS) [8]. This information will constructively guide the selection of appropriate control measures, the general aim being to reduce any potential harm to the health of individuals working directly with the substances. The Regulations also stipulate the need for emergency plans to be in place should there happen to be an incident involving any of these substances.

Put in perspective, of the overall number of reportable accidents in the United Kingdom, those related to the use of chemicals and substances are typically only around 13% of the annual total [9]. This does not however quantify the potential harm to health that an accident of this nature would cause and the relative importance of applying preventative measures. The effects of skin contact for example can cause superficial skin irritation at a minor level, or more significantly, give arise to inflammatory diseases such as dermatitis if exposure has been over a prolonged period of time. Allergies to certain substances may result in the development of asthma, a disease relating to the airways of the lungs, which leaves the sufferer short of breath. If on the other hand, toxic substances are being used, breathing in the associated fumes given off by them could result in an individual losing consciousness, which may in itself lead to hypoxia if medical treatment is not administered in sufficient time. By their very nature, the use of biological agents can result in personnel catching infections from bacteria and other micro-organisms. Taking a holistic view, and looking at the longer term effects of some of these substances, health issues may only become prevalent when cancers are detected many years after the initial exposure to the substance which caused the disease. As these examples highlight, the severity can in many cases become a significant factor to the health of a person involved in some form of accident where chemicals or substances are involved.

For any use of hazard substances, knowledge of the routes of entry into the human body is also critical, as these can inform the appropriate control measures needed for safe use of specific substances. The entry route for a particular chemical or substance will greatly depend on the form it is in. These include dusts, fumes, gases, mists, vapours and liquids. Generally, solids pose low health hazards and it is only when they undergo a process which produces more hazardous forms, such as machining dust, that they become a higher risk component. There are four different routes of entry into the human body for chemicals and substances [10].

### 3.5.1 Absorption

This is where substances have the potential to come into contact with any previously broken skin or the mucous membranes, such as those found in the mouth, nose and eyes. From this initial contact, the substances are then able to enter into the blood stream of the body or are suspended within the skin layers where they typically can cause skin rashes. In addition, particular types of chemicals are even able to directly penetrate unbroken skin.

### 3.5.2 Ingestion

This is the process by which substances are taken into the body openly through the mouth from where they will travel directly into the stomach. The majority of tissue damage due to ingestion is caused as these substances travel through the gastro-intestinal tract. Generally, ingestion is an inadvertent process rather than a deliberate action, brought about by the accidental transferring of chemical and substances from the hands into the mouth. This is often as a result of poor hygiene in the working environment, where the failure of workers handling substances to wash any contaminants from their hands before eating or drinking results in the unintentional consumption of the chemicals.

### 3.5.3 Injection

This is where substances enter the body through the skin due to the chemical or substance being applied under significant pressure. An example would be where compressed gases are being used and exposed skin comes into contact with a particular high-pressure release of the gas. It is also possible for injection to occur if an individual is accidentally stabbed by a sharp object contaminated with a particular substance. This can often occur in situations where used hypodermic needles are not correctly disposed of and an individual involuntarily comes into contact with the contaminated, sharp point of the needle.

### 3.5.4 Inhalation

This is the process by which substances enter the body through the mouth or nose during the normal breathing process. Contaminated air will transfer any suspended chemicals or substances directly to the lungs by this means. Larger, solid particles are generally prevented entry by the body's natural defences, such as nasal hairs. Smaller particles however, including vapours and fine dusts, will normally circumvent these protective mechanisms. By their very design, the lungs readily transfer these substances from the bronchioles directly into the blood stream. This in turn circulates the contaminants around the body to the major organs, such as the liver and kidneys. However, certain materials entering the body by inhalation are unable to be transferred into the blood stream. These remain in the alveoli, which are the series of air sacs at the ends of the bronchioles. Remaining there, these particles have the effect of either reducing lung capacity or damaging the surface of the air sacs which eventually results in symptoms of breathlessness for the sufferer.

In the main, these four routes of entry into the body for chemicals and substances, and their associated potential to cause harm, can be readily controlled. For about 500 registered substances, the Regulations defines a set of Workplace Exposure Limits (WEL) which are related directly to the risks associated with their use. By ensuring that work is only ever undertaken strictly below these limits guarantees a low risk to individuals coming into contact with the hazardous substances. This however necessitates the implementation of monitoring, where measurements have to be routinely made of the substance in the breathable air, local to where a task is taking place. In addition, there is a further control process which runs parallel with the COSHH Regulations. This is the Registration, Evaluation, Authorisation and restriction of Chemicals (REACH) [11]. Like COSHH, this requires the preparation of risk assessments, in this case, mainly undertaken by the manufacturer of the chemicals. It defines a so-called Derived No Effect Level (DNEL) [12] for the substances being made and supplied. It is important to recognise though that this level is meant to be used only as a benchmark standard and is not used as a definitive exposure limit. Instead, the DNEL is utilised in ascertaining the relevant risk management measures required for a particular task or procedure associated with that chemical or substance.

In terms of the application of the COSHH Regulations for cryogenic technology, many of the cryogens employed in order to produce low temperature conditions can create mists or clouds of cold water vapour. These can be harmful if they are involuntarily inhaled. More significantly however, expanding gas from cryogenic storage vessels can rapidly contaminate the surrounding working atmosphere. The immediate health effects of such in influx will depend on the actual gases being used, but can typically range from a mild headache or dizziness to the real danger of death by asphyxiation as the released gas eventually comes to dominate the work space volume.

Clearly, if part of the cryogenic process includes the use of chemicals, reagents or even solvents, compliance to the COSHH Regulations automatically becomes relevant and the stringent management of these substances needs to be implemented. In these situations, the risks to health needs to be assessed and a decision made on what precautions are required in order to minimise these risks, regardless of whether or not any particular substance being used has a defined WEL. In the United Kingdom, the Hazchem [13] warning plate system is adopted as standard on any containers used for the storage of hazardous substances. The plates are designed to provide important information on how the emergency services need to deal with any incident involving the substances within the containers. In addition, the *Dangerous Goods Emergency Action Code List* [14], provided by the National Chemical Emergency Centre [15], is an up to date list of dangerous goods typically used within the United Kingdom. The list is frequently employed as a compliance document by not only the emergency services, but also local governments involved in the controlling of emergencies involving complex chemical risks.

## 3.6 Personal Protective Equipment at Work Regulations—1992 [16]

As a fundamental mandate, controlling risks is essential for any safe practices. This can be done at higher levels, where the process itself is managed in such a way that risks are either completely removed or are suitably mitigated. Adopting processes such as safe systems of work will also highlight potential hazards, allowing suitable process control measures to be implemented where needed. One such control measure is the wearing of personal protective equipment (PPE) so that individuals are kept safe from any direct contact with whatever hazards are prevalent in their workplace situation. We all instinctively adopt this method of protection by wearing gloves and a hat in the winter months in order to stop our hands and our heads from suffering the effects of the cold weather at that time of the year. Likewise, any keen gardener will also employ a similar technique, by wearing heavy-duty gloves if the task of cutting back brambles or rose bushes are ever required. The PPE in this case preventing the inevitable injury which would occur if the hands came into contact with any of the sharp thorns on the plants in question.

In fact, the Personal Protective Equipment at Work Regulations covers all personal protective equipment which by definition is intended to be worn or held by a person at work, and which serves to protect them against one or more risks to their health and safety. Common examples seen throughout industry would include hard hats on building construction sites, eye protection where the risk of injury to the eyes is prevalent, such as during the turning of metal on a lathe, and the use of appropriate footwear to protect against injury to the feet in situations where heavy loads are being moved manually or by the use of lifting aides. In the United Kingdom, employers are not allowed to charge for any PPE that is used by their staff for undertaking their work.

As with many of the primary Regulations, they very often overlap with other specialist Regulations and where possible, these requirements within the Personal Protective Equipment at Work Regulations are deliberately omitted so that they are not duplicated. An example of this is where hearing protection would be covered by the Control of Noise at Work Regulation, even though in essence the hearing protection is an actual piece of PPE. Another area where this overlap is more profound is where ionising radiation has to be considered. Here, the necessary controls applied in order to prevent exposure to any ionising radiation very often override other Regulations, especially where exposure levels would be increased if full implementation of the other Regulations were made. An example of this would be where the installation of scaffolding in order to work at height in a radiation environment would take some considerable length of time and therefore would increase exposure levels to anyone undertaking the construction. Conversely, simply using a set of ladders to fulfil the same task would significantly reduce the exposure levels, even though under normal circumstances it would be deemed a higher risk to use the ladders rather than the scaffolding.

Two very obvious types of PPE are adopted for use with cryogenic technology; those which offer protection to the hands and those which provide eye protection. As it is the hands which are most likely to be closest to any cryogenic hazard, it is always important to use the correct hand protection, especially when handling liquid cryogens due to the extremely cold temperatures involved. Fundamentally, any gloves or gauntlets used must protect the user from any accidental cryogenic liquid spills. In emergencies, they must also be easy to remove without causing further harm. The European standard, EN 511—*Protective Gloves Against Cold* [17], is the safety standard which defines the minimum requirements for any gloves designed and marketed for cryogenic use. The standard states the level of testing undertaken on the material combinations making up these gloves, especially their insulating properties, which are primary for the safety of the eventual user.

Likewise, when handling the same cryogenic liquids, it is equally important to protect the eyes against any accidental splashes, especially during filling or decanting operations. Normally this is enacted by providing safety goggles or even a full-face visor to anybody working with these very cold materials. It is important to note that any PPE provided to protect the eyes mustn't impair the vision of the wearer, otherwise this in itself would increase the risks associated with their use. For this reason, eye protection must always have anti-mist properties and should be designed to allow for prescription glasses to be worn underneath them. The European standard EN 166—*Personal Eye Protection* [18], is the technical safety standard which defines the minimum requirements for any eyewear intended to protect against hazards likely to damage the eyes. This would not however include eye protection needed for working with x-rays or laser emissions, where more specific requirements associated with these particular hazards would be implemented.

## 3.7 The Dangerous Substances and Explosive Atmospheres Regulations (DSEAR)—2002 [19]

The Dangerous Substances and Explosive Atmospheres Regulations (DSEAR) is the law concerned with protecting against the risks to safety from fire, explosions and any substances which are corrosive to metals. The Regulations apply to both employers and the self-employed, placing duties on them to protect people from these risks. This explicitly includes members of the public who may be put at risk by any associated work activity undertaken by self-employed individuals or larger organisations. An example being where it would be unacceptable to undertake an activity in a densely populated area when there is the likelihood of an explosion occurring due to the process involved.

The main requirements of the Regulations is for a risk assessment to be carried out for any work activities involving dangerous substances or where there are explosive atmospheres. Controls should then be implemented to eliminate, or at least reduce the associated risks. Following this, areas where explosive atmospheres may occur need to be classified into zones and protected from any potential ignition sources. Equipment used in these zones would then need to comply with the ATEX Directives [20] which are two detailed European Directives; one specifically for the manufacturer of this type of equipment and the other for the users of the equipment.

The Regulations also stipulates that appropriate rescue equipment, as well as rescue plans and procedures should be in place to deal with accidents and emergencies in any work area where the use of dangerous substances is being undertaken. All employees should also be provided with information about the dangerous substances they are likely to be using and must be suitably trained to deal with the associated risks.

At a first consideration, many organisations may determine that they have no need to comply with DSEAR as they not operating with any dangerous substances or explosive atmospheres. However, the actual dangerous substances defined within the Regulations are more than likely to be present in most workplace environments. They would typically include things like solvents, paints, flammable gases, dust from machining and sanding processes, pressurised gases, and any substances which are corrosive in nature to metals. It is highly unlikely that dusts of any type play a major role within any branch of cryogenic technology, but both flammable and pressurised gases are a common feature, especially when they are in their cold, liquid state. Cryogens such as liquid oxygen and liquid hydrogen in particular are used as spacecraft rocket fuels because of their high combustibility, deliberately ignited to generate the required propulsive thrusts in space where the absence of oxygen is apparent.

## 3.8 The Manual Handling Operations Regulations (MHOR)—1992 [21]

Many work activities often require the movement of equipment or products from one location to another in order for particular processes to be commenced. For relatively low weight components this is more commonly undertaken as a manual handling activity where the items would be physically lifted and carried, or a trolley employed which would allow them to be pulled or pushed to their desired location. These lifting, carrying, pulling and pushing activities are covered by the Manual Handling Operations Regulations, which defines the load being handled as any object, person or animal which has a defined mass.

The foremost guidance within the Regulations advocates that individuals should avoid hazardous manual handling as far as is reasonably practicable, especially if there is a possibility of injury resulting from them doing so. If a particular manual handling operation cannot be avoided, then there is a requirement that a risk assessment is made of the activity before any load is moved. This assessment should look explicitly at the load being handled, taking into account its weight, shape and centre of gravity, the proposed method of handling the load, and the start and finish locations of the handling operation. This will highlight the higher risk elements of the process so that additional control measures can then be implemented with the aim being to reduce the risk of injury throughout the movement of the load.

One such risk control measure is the use of manual handling equipment to assist with the routine movement and manipulation of heavy or awkward loads. Use of these aids has become more common in the workplace in recent times, as these days there are a wealth of multi-purpose trolleys and lifting apparatus which can be employed for both general and explicit manual handling applications. There are also specialist equipment, such as tipping frames for small cryogenic Dewars, which are targeted at individual process uses.

Unfortunately, manual handling injuries are a major cause of occupational ill health in the United Kingdom, with a large proportion of both reported injuries and back injuries being caused by individuals having undertaken some form of manual handling activity [22]. However, it is not necessarily the single manual handling activity alone which is the reason for these injuries. A frequent cause for work-related upper limb disorders is brought about by repetitive movements and overuse of the muscles and tendons in the body. Even when working with relatively light weights, repeated body movements can put strain on the upper limbs. In these situations, an assessment of the duties being performed is critical, with the aim of making the task and the working environment more ergonomic so that extreme repetitive movements can be eliminated. If the task cannot however be altered, a change of duties to reduce workers discomfort may be necessary. A typical example of where this type of injury may occur would be where the manual handling task is of a particular repetitive nature, such as loading and unloading of small, lightweight items on and off a conveyor belt for prolonged periods of time without the individual taking a break.

It is important to remember that employees in any work situation have a duty to take reasonable care of their own health and safety, and that in most workplaces, manual handling of some nature will happen every day. Adopting the correct lifting and handling postures is therefore of vital importance for any manual handling activity, since moving even the lightest of loads with an inappropriate method can be detrimental to the body and could cause injury. One of the core solutions to the reduction of likely injuries and to ensure safety in the workplace is that manual handling training is provided within any organisation to endorse the Regulations and to teach the recommended best practices. This training should effectively provide guidance on the correct methods of undertaking manual handling activities and needs to apply to everyone in a company. Under the MHOR there is a legal obligation for organisations to ensure all employees are suitably trained and therefore have a level of competency to allow them to undertake all forms of manual handling.

The MHOR also provides useful advice on the recommended weight limits for both male and females undertaking specific manual handling operations. Significantly, the information is presented as guidance only as each individual may have strengths or weaknesses which would add further limitations to what they could comfortably lift. It defines a range of lifting positions from lower leg height to shoulder height, relative to the starting and finishing position of the load being manipulated.

The obvious manual handling of equipment encountered as part of cryogenic technology tend to be dominated by the movement of liquid cryogen storage vessels and any associated cryostats. Generally bulky in form, Dewars in particular are supplied in a variety of capacity sizes. Their actual weight will also depend on how empty or full they are at any given point in time. It is also important to remember that as they contain a large mass of liquid, any movement of their contents as they are being manually handled may have an effect on the centre of gravity, causing an imbalance of their load. The larger Dewar types tend to have handling rails and wheels installed as standard, thereby reducing the risks associated when they need to be moved. Smaller storage vessels and equipment such as vacuum pumps and compressors associated with cryogenic technology can be physically smaller in volume, but sometimes heavier in weight by virtue of their integral component parts.

## 3.9 The Lifting Operations and Lifting Equipment Regulations (LOLER)—1998 [23]

The movement of any load with a total mass in excess of the manual handling guidance weights will require the use of some form of lifting equipment. An example where this would be applicable for the use of cryogenic technology would be where a near full Dewar of liquid nitrogen, which could weigh as much as 100 kg, needs to be moved onto an elevated location. Clearly the mass is above what would be safe to attempt through manual handling alone, even if several people were to assist in the lift. To transfer the Dewar safely to the planned elevated position would therefore require the use of some form of lifting equipment, such as a crane, and by association any lifting accessories needed to attach the Dewar to that crane. All such lifting equipment is controlled by the Lifting Operations and Lifting Equipment Regulations (LOLER) which describes the legal duties placed on not only the individuals using the lifting equipment, but also the companies who own that equipment. It also defines how all lifting operations undertaken must be properly planned by a competent person and carried out in a safe manner.

The Regulations also stipulates that all lifting equipment must be fit for purpose, in that it is sufficiently rated to be able to lift or lower any given load and is in good working condition at the time of use. Generally, lifting equipment is supplied with the Safe Working Load (SWL) marked on it so that it is clearly evident what its specific rating is. This SWL is based on lifting equipment standards which would always include the provision of a proof test of the equipment prior to initial use. In addition, the Regulations also requires that each individual item of lifting equipment owned by a company has a unique identification number marked on it, and that it is subject to a periodic statutory inspection. These inspections are a means of capturing any defects to the equipment which may over time render it unusable.

Implementing the guidance within the Regulations is part of the overall requirement for managing and controlling the risks when using lifting equipment in order to avoid any injury to staff or damage to equipment. Supporting the Regulations is the *Safe Use of Lifting Equipment: Approved Code of Practice* [24]. This document is not law, but has special status as part of the Health and Safety at Work Act which supports; LOLER, the Management of Health and Safety at Work Regulations and PUWER, in relation to the use of lifting equipment and any associated lifting operations.

By definition, lifting operations can involve both the lifting and lowering of loads. The load itself is defined as the item or items being lifted or lowered, which would generally be equipment of some sort, but in the case of passenger lifts, would also include a person or people. It follows therefore, that lifting equipment is defined as the work equipment employed for the lifting and lowering of these loads. This would include lifting accessories and attachments used for anchoring, fixing or supporting larger items of equipment. Selecting the correct lifting equipment for use in any lifting operation is critical and they must be of adequate strength and remain stable throughout their application.

Throughout the use of any lifting equipment, the aim is to reduce the risk of the load, which is being moved, from striking a person, or falling free or even being unintentionally released during any lifting operation. Where people are being lifted, the Regulations makes provision for additional requirements to prevent people from being injured in or by the carrying device. Personnel lifts generally have more frequent thorough examinations, specifically looking for the potential failure of essential parts used in the lifting process.

Representative examples of lifting equipment would include; overhead cranes, motor vehicle lifts, goods and passenger lifts and telehandlers and fork lifts. The usual types of lifting accessories associated with lifting operations would include; fibre or rope slings, chains, hooks, eyebolts, spreader beams and magnetic and vacuum holding devices.

As with the manual handling of cryogenic technology, the use of lifting equipment must also consider the dynamics of the load being moved. Cryogen storage Dewars and many types of cryostats contain liquids which will readily move about internally when they are disturbed. It is important therefore that any lifting operation considers the likely change in the centre of gravity of things like storage vessels and cryostats when they are being relocated from one position to another. Generally, certain elements of laboratory cryogenic technology are also likely to be delicate in nature and wouldn't be able to withstand too much shock-loading when they are being lifted. Manual operated cranes, and those with a remotely operated coarse movement setting should be avoided in order to prevent unplanned damage to any internal parts of this type of equipment when they are being moved.

## 3.10 The regulations for transporting dangerous goods by air, sea, inland waterways, road and rail transport

For many industries, it is unlikely that subsequent processes of a product are carried out directly at the point of origin of their raw materials. Some form of movement of materials and parts from their source to their destination of use is all too common. As such, the United Kingdom's public highways, rail network and inland waterways are often utilised as the obvious means of transporting throughout the country a whole variety of materials and component parts, from the benign to the more hazardous. In addition, if products are being brought into or being sent out of the country, the use of air and sea transport would have to be undertaken.

Transporting goods by any of these methods involves the risk of accidents at some point during the process and a series of International Regulations and European Directives are applied in the United Kingdom through national laws and regulations for each mode of transport in order to reduce and mitigate these risks. Fortunately, the majority of transported goods are not deemed sufficiently dangerous to merit extra special precautions during their carriage. However, certain goods have properties which makes them potentially dangerous to transport and here, the full implementation of the applicable regulations applies. The purpose of these transport regulations is to provide protection to everyone directly involved in the carriage of these specific dangerous goods, such as the consignors or the carriers. They also offer the same level of protection to anyone who might become involved if there were an incident, such as the emergency services or even the general public.

The definition of dangerous goods includes liquids or solid substances, along with any device containing them, which are either prohibited from being transported by the relevant regulations, or authorised under strict conditions stated within the regulations. They also usually include hazardous substances which would include the following properties; flammable, harmful, toxic, an irritant, corrosive, an oxidiser, explosive, radioactive, carcinogenic, or hazardous to the environment. Dangerous goods are classified depending on their predominant hazard and are identified by a United Nations (UN) four digit number [25]. Examples for cryogenic applications would be UN 1977—Refrigerated liquid nitrogen and UN 1963—Refrigerated liquid helium.

Running parallel with this, any packaging for the carriage of dangerous goods has to be designed and manufactured to UN specifications and standards [26]. In order to comply, the packaging must pass a set of practical tests which reproduce the likely handling characteristics during their use, such as being dropped or being stacked beneath heavier packaging. The packaging must also be suitable for the substance contained within it, complying with the different UN package specifications relating to various listed substances. All packaging must then be correctly labelled and marked with the appropriate information relating to both its contents and the packaging itself. Any UN approved packaging must be certified by a national competent authority. In the United Kingdom, the Vehicle Certification Agency (VCA) Dangerous Goods Office has this responsibility [27].

The enforcement process for the safe carriage of dangerous goods is undertaken by several competent authorities. The relevant enforcing authority will depend on the mode of transport being employed. The exception to this being the civil carriage of UN Class 7, radioactive materials, which are exclusively regulated by the Office for Nuclear Regulation (ONR) [28]. There may also be additional security regulations applied for certain categories of nuclear materials.

### 3.10.1 The Carriage of Dangerous Goods by Road Regulations—1996 [29]

The Carriage of Dangerous Goods and Use of Transportable Pressure Equipment Regulations (CDG), along with the European agreement 'Accord européen relatif au transport international des marchandises dangereuses par route' (ADR) [30], collectively regulates the carriage of dangerous goods by road in the United Kingdom. The ADR defines the classification process, packaging, labelling and certification of any goods which are defined as dangerous. It also specifies vehicle and transport tank requirements for use on the public highways, along with other operational necessities. As it is derived from International Conventions, it allows both overseas goods vehicles to transport dangerous goods using the United Kingdom roads and likewise, United Kingdom vehicles to transport dangerous goods using the roads in other countries.

It is a known fact that transporting any goods by road carries the risk of traffic accidents. To put this in perspective, in 2018 alone there was over 160 000 casualties in reported road traffic accidents on the United Kingdom's roads [31]. Accidents which resulted in damage only and did not cause personal injury are not included in this total. As such, the total number of traffic accidents would therefore be considerably higher. However, this is not to say that every time a goods vehicle uses the road network it is likely to be involved in a traffic accident. Generally, the roads are a safe place to be, but consideration of the likely consequences should a collision occur with a goods vehicle has to be recognised. If any goods being transported are considered dangerous, the likely effects would be more profound and there are also risks of further incidents resulting specifically from;

- Spillage
- Fire
- Explosions
- Chemical burns
- Environmental damage

The Carriage of Dangerous Goods by Road Regulations sets out what needs to be done in the event of any of these incidents occurring, and more critically, what procedures need to be followed. It also defines what secondary hazards may exist if sufficient control measures are not adequately in place at the initial outset of transportation. In addition, it details the relevant documentation which needs to be established before any dangerous goods can be transported by road to the intended destination.

The enforcement for the safe carriage of dangerous goods for road transport in the United Kingdom is undertaken collectively by; the Health and Safety Executive (HSE) [32], the Driver and Vehicle Standards Agency (DVSA) [33] and the highways police [34]. For international road movements, each national authority enforces its own requirements under the ADR European Agreement.

### 3.10.2 The Carriage of Dangerous Goods by Rail Regulations—1994 [35]

The movement of dangerous goods along the United Kingdom's rail network is controlled by the Carriage of Dangerous Goods by Rail Regulations. As with the Carriage of Dangerous Goods by Road, this is derived from international agreements and European Directives. Appendix C of the Convention Covering International Carriage by Rail (COTIF)—International Carriage of Dangerous Goods by Rail [36], are the uniform rules set down to control the international carriage of dangerous goods by rail in the territories of the contracting States. In addition, the Carriage of Dangerous Goods and Use of Transportable Pressure Equipment Regulations 2009 [37], apply for movements of dangerous goods using the United Kingdom's rail network.

Once again, as the United Kingdom Regulations are derived from International Conventions, this allows both overseas dangerous goods to be transported along the United Kingdom rail network and for United Kingdom dangerous goods to be transported using the rail networks in other countries. All packaging of dangerous goods transported by rail must therefore comprehensively comply with the UN packaging and identification specifications.

Possibly considered safer than road travel, the use of the mainline railways in the United Kingdom isn't without its infrequent mishaps. In the period 2018–19 there were 517 reported mainline rail accidents, of which 26 were classed as Potentially Higher Risk Train Accidents [38]. These included incidents where trains collided with vehicles at level crossings or with other trains, along with one derailment. It is evident therefore, that accidents on the rail network do happen, and as such the movement of any dangerous goods need to be adequately prepared for this real possibility.

The Carriage of Dangerous Goods by Rail Regulations defines the construction requirements of freight and tank containers used on the United Kingdom railways for the carriage of dangerous goods. They ensure that any such freight and tank containers are designed correctly and are suitable for the purpose for which they are to be used. It also details the minimum testing, examination and maintenance requirements for any tank containers and their wagons intended to be use on the rail network. Information relating to the specific dangerous goods being moved needs to be documented and displayed on their packaging, along with the arrangements and safe systems which will adequately deal with any emergency involving the dangerous goods. The Regulations also stipulates the segregation requirements of any dangerous goods from each other, which if accidentally mixed could create a greater hazard then each of the individual dangerous goods is liable to create on their own.

The enforcement for the safe carriage of dangerous goods by rail in the United Kingdom is undertaken collectively by; the Health and Safety Executive (HSE) [39], the Office of Rail and Road (ORR) [40], the Office for Nuclear Regulation [41] and the Department for Transport [42]. Many individual nations around the world have also structured their own carriage of dangerous goods by rail regulations to harmonise with the UN Convention concerning international carriage by rail. Therefore, the specific requirements for international rail movement of dangerous goods originating from the United Kingdom should already automatically conform to a well-defined set of regulations. Each country will have their own specific enforcement authority, the organisation of which will generally follow an internationally agreed model.

### 3.10.3 The Carriage of Dangerous Goods by Inland Waterway Regulations [43]

In the United Kingdom there are over 4000 miles of inland waterways [44], used mostly these days by private pleasure vessels, although a small amount of commercial movement of goods are still undertaken utilising the canal and river networks. It follows therefore, that some of these goods will fall into the category of dangerous and that crucial safety management has to be undertaken for their safe carriage on these inland waterways. It is highly unlikely that dangerous substances relating to cryogenic technology would be frequently moved via this method, but the option exists, and the very specific hazards relating to this mode of transport have to be seriously considered at the same level as any other method of transference. The construction requirements of any container and what safety equipment has to be available on each vessel carrying dangerous goods depends on the nature of the waters in which they operate. For this purpose, the United Kingdom inland waters are categorised into; canals, non-tidal rivers, tidal rivers, large deep lakes and lochs, and estuaries [45].

The International Carriage of Dangerous Goods by Inland Navigation (ADN) [46] document regulates this form of transport. For vessels only undertaking the carriage of dangerous goods on the United Kingdom inland waterways, the ADN does not directly apply, as there is no physical connection between these waterways and those in Europe and the rest of the world. Under these circumstances the Carriage of Dangerous Goods and Use of Transportable Pressure Equipment Regulations 2009 [47] comes into effect in place of the ADN. However, if vessels are to also operate on the community waterways of the European Union Member States, they will need the relevant country's water authority to issue a Community Inland Navigation certificate as part of the ADN requirements. Apart from this specific operation, ADN is only applied in the United Kingdom to the training and examination system for safety advisors relating to the carriage of dangerous goods by inland navigation.

As part of the ADN, the specific *Annex A, Provisions Concerning Dangerous Substances and Articles* [48], outlines the detailed requirements for the construction and operation of inland navigation vessels involved in the movement of dangerous goods. It also defines the inspection requirements and the certification of approval process for these vessels. As these are waterborne vessels, many of the requirements focus on ensuring that features such as the bulkheads and cargo tanks are watertight. In addition, the European Commission Directive 2006/87/EC *on Technical Requirements for Inland Waterway Vessels, annex III* [49], describes further design necessities for the constructions of inland waterway vessels.

Water, from whichever waterway source, has to also be considered in relation to the dangerous goods themselves, as certain substances which on coming into contact with the water, could react in such a way that they would create a greater hazard on-board any inland vessel. Flammable reactions are one typical example which can clearly have devastating effects if brought about within the confines of any water-borne vessel. The Regulations therefore define what special equipment has to be provided in the form of PPE for the crew, flammable gas detectors, breathing apparatus and even suitable escape devices for each person on the vessel. The commercial operation of any vessel on the United Kingdom inland waters must comply with secondary construction regulations and a set of standards for the safety equipment which needs to be carried on-board at all times [50].

The navigation authorities overseeing inland rivers, canals, estuaries and harbours in the United Kingdom are The Canal and River Trust for England and Wales [51], The British Waterways, trading as Scottish Canals in Scotland [52] and the Environmental Agency [53]. Any craft using the United Kingdom's inland waterway network must be registered with these authorities. In addition, the Maritime and Coastguard Agency (MCA) [54] are responsible for implementing the United Kingdom government's maritime safety policy [55]. The Navigation Safety Branch of the MCA [56] are charged with managing the policies and processes specifically for the use of the inland waterways. There is also a joint venture between British Waterways and the Environmental Agency known as the Boat Safety Scheme [57] which has been created in order to minimise the specific risks of fire and explosions on boats. This is done through the provision of a set of safety requirements which have to be met before any vessel can obtain a navigational licence [58].

### 3.10.4 The Carriage of Dangerous Goods by Sea Regulations [59]

Guidance relating to the transporting of dangerous goods by sea is provided globally by the International Maritime Organisation [60] and their International Maritime Dangerous Goods code (IMDG) [61]. This code is effectively adopted by all operators in the world who make shipments of dangerous goods and hazardous materials on journeys involving a sea crossing. By definition, this also includes ferry services transporting goods through seaways. In the United Kingdom, the Merchant shipping (Dangerous Goods and Marine Pollutant) Regulations 1997 [62] and the Dangerous Substances in Harbour Areas Regulations 1987 [63] are also applicable for these same voyages.

Maritime accidents are not rare. In 2019 there was an average of 80 reported accidents per month at sea, with over 60 ships being lost over the full twelve month period with over 270 fatal maritime casualties during that same time frame [64]. These noteworthy figures include accidents on merchant and fishing vessels as well as recreational craft. As a large percentage of the global trade of goods is now transported by sea, with a corresponding increase in the number of vessels at sea at any given time, it is inevitable that accidents will occur. Yet it needs to be recognised that the magnitude of any particular incident at sea can be enormous due to the immense capacity of some of the vessels used to transport goods from one nation to another. The environmental impact alone which any maritime accident can have to the local ecology can be significant, and the Regulations are as much aimed at making seagoing vessels not only safe for the crews but also innocuous for the environment should incidents occur. Oil and chemical spills in particular have serious impact on marine life, generally caused by ships sinking or vessel collisions resulting in the unintended release of these substances. Cargo vessels alone can carry thousands of tonnes of heavy oil as their freight, along with their own diesel fuel, which released into the sea can progressively spread over an enormous area. Any breech of cargo holds can therefore result in an environmental disaster for local ecology on a huge scale.

The shipping of dangerous goods by sea clearly has its own unique and specific hazards, none more so than the effects the weather has on the movement of vessels and its stability, and as a consequence, on the goods themselves being held on-board. It is important therefore that the IMDG code is followed in order to ensure that any dangerous goods transported by this method are correctly packaged so that they can be transported safely, and in the correct manner. To assist with this process, the code classifies the dangerous goods into nine categories, each with their own level of danger and the consequential packaging, stowage, segregation, handing and emergency response requirements. These classifications include dangerous goods which are deemed explosive, potentially toxic and those which present the threat of corrosion or erosion. There are also included dangerous goods which on reacting with water can emit toxic gases.

The main aim of the IMDG code is to protect crew members who are manning vessels transporting the dangerous goods, but it also provides the methods which ultimately prevent any marine pollution during these journeys. In addition, it ostensibly involves all industries and mariner services connected with shipping, such as harbour controls. It follows therefore, that in preparing for the movement of dangerous goods by sea, both the loading and unloading, as well as the stowage plans have to consider the safety and seaworthiness of the vessel and what emergency procedures have been established which would come into effect in order to control any potential accidents.

In the United Kingdom, the Maritime and Coastguard Agency [65] are the body responsible for enforcing compliance with both the Regulations and the IMDG code for dangerous goods being moved by sea. Their role is fundamental in ensuring that nothing is transported by ship through United Kingdom waters which is considered hazardous in nature, without having the proper controls in place. Enforcement of the Dangerous Substances in Harbour Areas Regulation [66] is undertaken by the HSE [67] along with the United Kingdom's Statutory Harbour Authorities [68].

### 3.10.5 The Carriage of Dangerous Goods by Air Regulations [69]

In this modern world of economics and trade, many millions of items are transported by air each year, of which a large proportion are considered as dangerous goods. As air travel is now such a global article, with flights routinely crossing national borders, the guidance for the carriage of dangerous goods by this method is provided at an international level. As such, the International Civil Aviation Organisation (ICAO) [70] produce a set of instructions known as the *Technical Instructions for the Safe Transport of Dangerous Goods by Air* [71] which cover the principal requirements for the safe movement of this type of goods. In turn, the International Air Transport Association (IATA) [72] publish the *Dangerous Goods Regulations* [73] in accordance with these technical instructions. The United Kingdom is a contracting state of the ICAO and are therefore required to comply with both the technical instructions and the Regulations. Individual airline operators must also comply with the IATA's Regulations, although some airlines and countries are permitted the partial suppression of the Regulations and implement what is known as State and Operator Variations [74] in order to comply with higher level laws within their own country.

As with all other means of transporting dangerous goods, the main crux of the *Technical Instructions for the Safe Transport of Dangerous Goods by Air* includes the classification of the substances and the specific packaging requirements for each of these categories. In addition, air operators have to have a set of acceptance procedures established which includes the inspection of packages for leakages or damage throughout their movement by air. As the loading and unloading of dangerous goods on aircraft can be the most hazardous phase of the transport process, the technical instructions also define the training requirements of staff involved in this procedure. In the United Kingdom, this training has to be approved by the Civil Aviation Authority [75].

In order to ensure safety of any movement of dangerous goods by air, the rules and regulations need to be applied globally. International law has therefore been created to support the IATA in this quest. They in turn work closely with the various governments throughout the world as well as the ICAO, so that the effectiveness of the Regulations remain of value and risks associated with this method of transport can be effectively reduced. In fact, as a reference for the safe shipping of dangerous goods by air, the Regulations have become the only recognised standard adopted by the various airlines throughout the world. Even so, some items may be of such a high risk for them to be transported by air, that they will either be forbidden or at least restricted in such a way to further mitigate these risks.

In the United Kingdom, the Civil Aviation Authority are the body responsible for enforcing compliance with these air transport regulations. Their role is to not only ensure that the United Kingdom aviation industry meets the highest safety standards based on these international regulations, but that security risks are managed effectively and are not compromised by the transport of any dangerous goods by air.

### 3.10.6 Radiation screening at ports and airports

In the United Kingdom, part of the government's counter-terrorism strategy is evoked in the form of radiation screening of incoming freight, vehicles, passengers and pedestrians, with the aim of deterring the illicit trafficking of radioactive and nuclear materials by terrorist groups or criminals. The programme, known as Cyclamen [76], is managed by both the Home Office [77] and the United Kingdom Borders Agency (UKBA) [78]. It is the responsibility of the Home Office to implement the programme at the various United Kingdom ports and airports, with the UKBA being responsible for operating the actual detection equipment, looking for radiological or nuclear materials. Other countries throughout the world have adopted similar procedures.

The process works by having fixed radiation detection equipment installed in all United Kingdom ports and airports. In addition, there is detection equipment screening traffic running through the Channel Tunnel [79]. The detectors are of a passive construction, which are able to identify any radiation emitted from a vehicle moving through the ports and airports, without themselves emitting radiation. Some radioactive materials may however be carried legally by approved and regulated operators, and some foodstuffs and ceramics emit natural levels of radiation. These legitimate imports are addressed and processed accordingly.

## 3.11 The Working at Height Regulations—2005 [80]

The majority of cryogenic installations are relatively small and compact in size. It is not unusual however for some systems to have a grander infrastructure and can be considered in the category of plant equipment, covering a sizable footprint, as well as occupying several floors of equipment locations. The original construction of such a structure, but more importantly, the maintenance and inspection of elements of this plant may require some form of working at height either from purpose-built scaffolding or working from mobile platforms such as mobile elevating work platforms (MEWPS) [81] and cherry-pickers [82].

One of the key points of the Working at Height Regulations is the avoidance of risks from working at height. This could be as simple as repositioning valve locations or inspection hatches so that working at height is not necessary in the first place. Where working at height cannot be avoided, the Regulations gives guidance on how to minimise the distance and consequence should a fall occur. This is normally provided by the installation of fall-arrest type equipment which would limit the distance an individual would drop, if for example they were climbing external parts of a large plant assembly.

The selection and inspection of work equipment for work at height is important. This would cover things like harnesses and safety handrails on structures. Inevitably, during any work at height activity, parts of equipment or tooling being used may fall from the plant. It is important that provision of protection from any such falling objects is made. This could be implemented through the mandatory wearing of hard hats in the working zone, or even the construction of protective roofing in areas where walkways go through these areas.

There is also the requirement within the Regulations of having some means of rescuing an injured person when they are working above ground level. It is important that this rescue is routinely practiced, so that in the event of a real situation where it is necessary to rescue somebody from a high level, the practicalities and difficulties of the process are well understood beforehand.

## 3.12 The Control of Electromagnetic Fields at Work Regulations—2016 (CEMFAW) [83]

An electromagnetic field (EMF) is a physical field produced by any electrically charged object. They are created every time a piece of electrical equipment is used, irrespective of the voltage or current being applied. Because modern day workplaces are often likely to contain a variety of electrical equipment, such as computers and electrical hand tools, EMFs are considered to be present in most working environments and can in certain circumstances be considered a workplace hazard. The majority of these EMFs are however sufficiently small enough to be of minimal concern regarding the health of workers in their vicinity. If they are of a higher intensity though, employers are obliged to take action to ensure their employees are protected from the adverse effects they pose. For cryogenic technology applications in particular, many are directly associated with the production of high electromagnetic fields, such as those generated in magnetic resonance imaging (MRI) scanners [84] which can be as high as 7 Tesla, as well as laboratory superconducting magnets used routinely up to 14 Tesla.

It is a fact that the modern world is also surrounded by natural magnetic fields generated by the Earth itself, along with those created by solar storms where the activity of the Sun interferes damagingly with the Earth's magnetic field. Studies have been undertaken to determine the effect on the human body due to exposure to both these natural and also man-made electromagnetic fields [85]. Research has also been carried out to study the possibility that electric or magnetic fields can cause a greater risk of certain cancers or other serious illnesses to mature, such as Alzheimer's disease [86]. That being said, as it stands, there is currently no definitive scientific evidence as yet of the long-term effects due to the exposure of EMFs.

In the workplace, the effects of any EMF depend on their frequency range and intensity and the duration an employee is exposed to them. To a certain level, the frequency at which an EMF is pulsed determines if it is likely to be harmful or not. Normal frequencies of the human brain range from 8 to 22 Hz, and there is evidence that the potentially higher frequencies used in high-field magnets can cause unwarranted electrical currents inside the human body, disturbing the natural resonant frequency, which in time could lead to cellular damage [87]. The intense fields used by medical scanners for example, have been found in certain circumstances to trigger nausea and dizziness, but these effects are generally temporary. At the other end of the spectrum, it is believed that the lower frequency electric and magnetic fields typically found in the home are unlikely to cause any harm. In reality, employees suffering the damaging effects from EMFs are extremely rare, and in most workplace situations this would not occur anyway, as the levels of EMF would already be at safe levels or would have been suitably assessed and managed. The employment sectors in the United Kingdom where significant exposure to EMFs is possible are mainly the healthcare sector, energy distribution industries, broadcasting, telecommunication, and research Universities. Here potential exposures are higher and yet the risks are generally well understood and managed in a way that incidents involving contact with EMFs are uncommon.

The European Physics Agents EMF Directive—2013/35/EU [88] lays out the minimum health and safety requirements regarding the exposure of workers to the risks arising when using this type of equipment. This is implemented in the United Kingdom by the Control of Electromagnetic Fields at Work Regulations (CEMFAW). The Regulations and the supporting guidance document, *Guide to CEMFAW Regulations—HSG* 281 [89], provide information on how organisations can protect their employees from any risks to their health and safety arising from any likely exposure to EMFs. They explain how employers can conduct a risk assessment of the electromagnetic equipment they are using, in particular on how to calculate the EMF strengths. Where possible, exposure to these EMFs should be eliminated, or at least reduced to below defined exposure limits. In the unfortunate event where an employee has been exposed to EMFs above these limits, employers must provide medical examinations and health surveillance and implement an action plan to ensure future compliance with the exposure limits. Providing information and training to employees, specific to the EMF risks present in the workplace is also a key element of the Regulations. They also contain a schedule which provides details of the safety conditions which must be met for full compliance. Employees considered within the Regulations at a greater risk of working with EMFs are; expectant mothers, those fitted with artificial cardiac pacemakers or anyone with surgical metallic implants.

## 3.13 Control of Major Accident Hazards Regulations (COMAH)—2015 [90]

The EU Directive 2012/18/EU [91], which is sometimes given the name Seveso III Directive, defines the hazard controls needed in the event of a major chemical accident taking place at a site or facility where dangerous substances are being held. It is implemented in national legislation of the member states and enforced by national chemical safety authorities. It is executed in the United Kingdom under the Control of Major Accident Hazards Regulation, which aims to prevent and mitigate the effects of any major accident at a facility involving dangerous substances. This includes any likely damage or harm to people, as well as that caused to the environment due to the accidental release of any dangerous substances. Mainly these are applicable to the chemical industry, where they are more likely to routinely process dangerous substances. However, the use of cryogenic technology, and in particularly the storage of liquid cryogens of sufficient quantity, would have to conform to these Regulations.

The Regulations focus on the prevention of accidents happening in the first place, done through the implementation of suitable control measures for any site facility storing dangerous substances. Every operator of such a facility must prepare a major accident prevention policy, which references the safety management systems already in place and employee training records. The policy must also address what emergency plans have been adopted in the event of an accident occurring. Any prevention strategy should be based on reducing human risks to as low as is reasonably possible and using the best available technology to reduce environmental risks. The Regulations are divided into two operating tiers, a lower and an upper level, which are defined by the quantity of dangerous substances being held at any particular site. Intrinsically, operators at all establishments must notify certain basic details to the enforcing competent authority, such as the details of the dangerous substances on an organisation's site and the name and company position of the person in charge and what activities take place at the site.

Additional specific duties for upper tier operators include them having to prepare a more detailed safety report, which demonstrates they have established arrangements to control major accident hazards. This safety report should also include possible major accident scenarios and what the likelihood is of them occurring. It would also detail the measures, such as fire-fighting and safety relief systems already adopted, which would limit the consequences of any such major accident if it ever did occur.

The competent authority in the United Kingdom for enforcing the Regulations is the HSE [92] or the Office for Nuclear Regulation (ONR) [93] for nuclear establishments. In addition, the separate devolved national agencies also have an enforcement role. In England, this will be the Environment Agency (EA) [94], in Scotland, the Scottish Environmental Protection Agency (SEPA) [95] and in Wales, the Natural Resources Body for Wales (NRW) [96].

## 3.14 The Dangerous Substances (Notification and Marking of Sites) Regulations (NAMOS)—2013 [97]

There is a stringent legal requirement in the United Kingdom for all sites storing 25 tonnes or more of dangerous substances to provide the local fire and rescue services, and the enforcing authority, with information about these products. The Dangerous Substances (Notification and Marking of Sites) Regulations defines the information that must be displayed in order to inform of the presence of the dangerous substances, such as access to the storage areas and specific location on the site. It also details what notification needs to be provided in advance to the emergency services about the substances being stored in any particular facility. The way in which organisations need to implement the Regulations is defined in the supporting guidance information, HSR29 [98]. The NAMOS Regulations use the Carriage of Dangerous Goods and Use of Transportable Pressure Equipment Regulations [99] as their legal basis, which adopts the European Agreement known as ADR [100] for the listing of dangerous substances. The enforcing authority for the regulations would be the HSE [101] and any local authority departments [102].

In terms of cryogenic technology, the NAMOS Regulations would apply where any site or facility stored over 25 tonnes of liquid cryogens or even liquefied flammable gases. Supplying the relevant information about these substances in advance to the emergency services will be vital if they ever had to respond to a cryogenic related incident on a particular site. Providing them with sufficient warning of the hazards will allow the correct emergency equipment to be brought to the site in the event of an accident occurring on the site.

## 3.15 Planning permission [103]

Planning permission in the United Kingdom is a requirement which allows for the building on land, or the change of use of land or buildings. This stems from the Town and Country Planning Act of 1947 [104] which granted planning title to all pre-existing land use and buildings at that time. Since then, any subsequent developments require formal planning permission before any work can commence. Planning permission applications are made through Local Planning Authorities, which is generally the local borough or district council [105], or the National Park Authority [106], for any planning proposals covering land in the United Kingdom's National Parks [107].

Obtaining Planning Permission also requires compliance with Building Regulations [108], which are a set of rules describing the construction standards put in place to protect public health, safety and general welfare. This would include all necessary fire management requirements. They relate not only to the construction process itself but also the eventual occupancy of buildings and structures. The Building Regulations are supported by Approved Documents [109] which provide practical guidance on how compliance can be achieved. Generally, any new building must be inspected during construction and at the end of the build in order to ensure compliance with these Buildings Regulations has been achieved.

In terms of cryogenic technology, it follows that any initial construction or planned changes to a building housing cryogenic equipment, would require planning permission before the work can begin. This would cover the location and layout of any proposed construction, taking into account things like access requirements for cryogenic equipment. However, obtaining planning permission in compliance with Building Regulations alone would not ensure the intended cryogenic workplace is intrinsically safe, as these approvals are only related to specific issues such as construction and fire-safety standards and not the specialised hazards associated with cryogenic technology. A more detailed assessment of the proposed building would have to be undertaken, considering additional features, such as the requirement of forced ventilation in areas where quantities of cryogens would be routinely handled.

## 3.16 Approved Codes of Practice (ACOPS) [110]

The huge range of regulations put in place in the United Kingdom are often complemented by an HSE [111] published Approved Code of Practice (ACOP). These are guidance documents which explain how the implementation of the regulations are to be achieved and can be used by organisations seeking compliance with the conditions associated with any specific regulations. They describe the preferred methods to be used, including any standards which need to be met, and detail the obligatory duties imposed by the Health and Safety at Work Act [112]. The ACOPs are approved by the HSE, with consent of the United Kingdom's Secretary of State [113]. As such, they have specific legal standing with reference to the regulations, covering not only hazards, but also workplace practices. Failure to follow an ACOP is not an offence, but it would be necessary to show that equally effective methods had been adopted as an alternative in order to demonstrate compliance with the regulations.

Examples of ACOPs specifically associated with cryogenic technology would include:

L5 The control of Substances Hazardous to Health Regulations

L22 Provision and Use of Work Equipment Regulations

L23 Manual handling—Manual Handling Operations Regulations

L24 Workplace health, safety and welfare—Workplace (Health, Safety and Welfare) Regulations

L25 Personal protective equipment at work

L74 The Health and Safety (First-Aid) Regulations

L101 Confined Spaces Regulations

L111 A guide to the Control of Major Accident Hazards Regulations

L113 lifting Operations and Lifting Equipment Regulations

L122 Pressure Systems Safety Regulations

L131 Approved classification and labelling guide—Chemicals (Hazard Information and Packaging for Supply) Regulations

L135 Storage of dangerous substances—Dangerous Substances and Explosive Atmospheres Regulations

L136 Control and mitigation measures—Dangerous Substances and Explosive Atmospheres Regulations

L138 Dangerous Substances and Explosive Atmospheres Regulations

## 3.17 Related hazards

Taken individually, all these regulations and guidance documents cover very specific hazard controls which are likely to be associated with the use of cryogenic technology. They do not however address the many related hazards which may result from the application of the range of materials and techniques connected with the technology. An example of this would be where an individual is expected to work alone in a laboratory and the presence of cryogenic equipment has been identified as a primary hazard. The individual undertaking tasks without co-workers being present becomes a secondary hazard in what is potentially a hazardous environment. In these circumstances, a lone working policy would be required, along with some guidance on the risk controls needed for that individual to safely undertake cryogenic related tasks on their own in the laboratory.

Typical examples of these related hazards are:

- Lone working
- Extreme cold

- Expansion of liquefied cryogens into gas
- Vacuum systems
- Asphyxiation due to oxygen depleted atmospheres
- Oxygen enriched air

In addition, there are also so-called psychological hazards associated with all types of workplace and processes which would need to be considered in relation to the use of cryogenic equipment. These would be aspects such as those hazards indirectly caused by stress, long working hours, travel, social conduct and other human factors. Many reported workplace absences are no longer deemed to be associated to physical health issues and are instead related to some form of stress or other psychological concerns [114]. As a collective therefore, these particular invisible hazards need to be considered in terms of how they can be addressed and included as part of risk assessments and subsequent control.

As developments over time are made into the general applications of cryogenic equipment, further regulations will naturally become pertinent. In the past, the technology was traditionally limited to the direct cooling of technical apparatus. In more recent times, there has been a steady growth in the development of other types of equipment such as superconducting magnets which have to be maintained at extremely low temperatures in order to function successfully. As a consequence, the associated portfolio of hazards has also grown, bringing with them the accompanying regulatory direction. By using this long list of statutory information, along with any industry guidance, all recognisable hazards can effectively be identified for any application of cryogenic technology. Once this has been achieved, the implementation of safe methods of work, along with the correct provision of things like personal protective equipment, can be employed to reduce the risk of exposure to any cryogenic related hazard.

## 3.18 Associated safety organisations in the United Kingdom

### 3.18.1 General

Any breach of the United Kingdom's safety regulations is defined as a crime and is punishable by conviction or a fine. Individuals or entire organisations can be punished for any such contraventions, although a company itself cannot be imprisoned. Individual managers within a corporations could however be convicted if found guilty of a severe enough offence. All violations of the regulations are subject to trial in the Magistrates Courts [115], with hearings for more serious offences being carried out in the Crown Courts [116].

Supporting the work of the Law in helping to prevent infringements happening in the first place are specific associated organisations. These include; professional bodies, trade associations and regulatory bodies, established with the aim of managing the interests of any individuals in that profession along with protecting the public in general from any associated hazards resulting from those specific industries. They do not necessarily exist as approved bodies, but in general they represent the expertise within their own industry themes. Trade associations and industry trade groups for example, are normally created and funded by the businesses that operate in that field.

In addition, national bodies [117] have been set up which provide accreditation of standards which meet the stringent requirement of the various safety regulations. These associations are often established through Government departments and are therefore independent of the industries they are providing advice and qualification to.

### 3.18.2 British Standards Institution [118]

The British Standards Institution (BSI) is the national standards body of the United Kingdom, who produce technical standards on a whole range of products and services. They also provide audits and certification for businesses and those trade associations who implement management system standards in order for them to achieve compliance with their own codes of practices. BSI publish all European Standards as an identical British Standard which have the prefix BS EN [119]. They can also adopt and publish international standards which would have the prefix BS ISO or BS IEC. Examples related to cryogenic technology would include:

**ISO 9001** Quality
**ISO 14001** Environment
**ISO 45001** Occupational Health and Safety

The BSI also undertake the testing of products or services where safety is overriding, assessing if they meet the requirements of the related specifications and standards and issuing any compliant products and services with a Kitemark™ certification mark [120]. Typical examples of products covered by the Kitemark™ service quality trade mark are; crash helmets, smoke alarms, fire extinguishers, ladders, and certain services, such as electrical installations. The Kitemark™ on its own is not a legal requirement, but is a recognised and trusted identification of the safety of the product bearing the mark.

The BSI can also conduct independent tests for certifications such as the CE marking requirements [121] for products intended for sale within the European Union. They also have Notified Body status for a variety of European Union Directives, such as pressurised equipment and personal protective equipment, which allows them to provide third-party certification to United Kingdom companies making any related products.

### 3.18.3 The Royal Society for the Prevention of Accidents (RoSPA) awards [122]

The Royal Society for the Prevention of Accidents (RoSPA) is an independent United Kingdom charity formed with the unassuming aim of saving lives and preventing life-changing injuries occurring as a result of accidents. Established at the time of the First World War in response to an unprecedented increase in road accidents, they presently campaign not only on road safety issues, but also on those subjects related to occupational health and safety. They have a distinctive perspective in this field, providing accident prevention services and support, as well as assisting organisations to become safer and healthier places for employees to work in.

Businesses recognise that any accidents at work and work-related ill health can have a negative effect on their practices and productivity. They are also aware of the impact these events may have on the quality of life of their staff. Good safety performance is therefore a valued philosophy endorsed by RoSPA through their running of a recognised Health and Safety Awards scheme [123]. Through the scheme, organisations from every conceivable sector of industry can benchmark their safety performance each year. Presented in the form of merit based awards of Gold, Silver and Bronze, they are a method of recognising how an organisation is presently performing but also that they have an ongoing commitment to raising Health and Safety standards in their facilities.

### 3.18.4 The Health and Safety Executive (HSE) [124]

The Health and Safety Executive in the United Kingdom is a government agency responsible for the regulation and enforcement of workplace Health, Safety and Welfare. They also undertake a specific amount of research into occupation risks within the various United Kingdom industries. The HSE was established as part of the Health and Safety at Work Act—1974 [125], taking on the mantle of many pre-existing regulatory bodies who were already established prior to the Act's inception. A key part of their role is the investigation of industrial accidents of any size, including those defined as major incidents. They also provide assistance and advice in matters relating to the operation and implementation of the fundamental objectives of the Health and Safety at Work Act with the aim of preventing work-related death, injury and ill health.

The HSE are an enforcement agency and as such, have the powers to prosecute where they find the law has been broken, or on investigation, that there remains a significant health and safety risk within an organisation. Enforcement officers belonging to the HSE have the right to enter any premises following a significant accident occurring there. Any salient information relating to such an accident must by law be provided to the HSE Inspectors by the respective organisation.

### 3.18.5 Public Health Agencies in the United Kingdom [126]

The Public Health Agencies in England, Wales, Scotland and Northern Ireland function at an individual national level and form part of the Department of Health and Social Care provision in the United Kingdom [127]. Working closely with the National Health Service [128], their aim is to protect and improve health in the United Kingdom. To achieve this, they provide advice on all health related issues along with undertaking research into specific health topics, such as health protection. They provide both national and local governments with expertise and professional health information. Their remit includes areas such as chemical and environmental hazards, working with hazardous materials, major incident responses, health improvement and health screening.

The Public Health Agencies establish arrangements for health protection concerns and emergencies by bringing together local partners, such as the Clinical

Commissioning Groups [129], to agree regional priorities. In this role, they are able to provide fact-based evidence pertaining to local health needs alongside the more general advice which they offer on how overall health can be improved. They are there to protect the United Kingdom from existing public health hazards, but they have to also respond to new public health challenges as they emerge. In doing this, they work closely with similar international health agencies.

In more recent times, Public Health England has been replaced by the National Institute for Health Protection (NIHP) [130]. This new institute brings together the work previously undertaken by Public Health England, along with that of the NHS Test and Trace [131], established as a response to the global Covid-19 pandemic [132]. In addition, the analytical capability of the United Kingdom Joint Biosecurity Centre [133] becomes part of the same leadership team, with the aim of strengthening public health and protecting people from external health threats, including pandemics, biological weapons and infectious diseases.

### 3.18.6 The United Kingdom Accreditation Service (UKAS) [134]

The United Kingdom Accreditation Service is the national accreditation body set up to assess the competence of those organisations providing certification, testing, inspection, and calibration services throughout industry. They are formally recognised by the United Kingdom government and their role is to evaluate the various conformity assessment bodies and provide them with accreditation where they are found to meet the relevant, national, European Union and international standards. They deliver this accreditation based on the various elements of the quality and safety standards within United Kingdom Regulations, European Union Directives and Internationally harmonised standards.

### 3.18.7 The United Kingdom UKCA Mark [135]

The United Kingdom Conformity Assessed (UKCA) Mark is a new United Kingdom product marking used for certain products being placed on the United Kingdom Market as the country has now left the European Union. Significantly, the marking is not recognised on the European Market and such products requiring a CE marking [136] will still need this proviso in order for them to be sold in the European Union. However, if any of these products require the services of a notified body for third-party assessment of conformity, the United Kingdom notified bodies are no longer recognised by the European Union and a notified body in one of the remaining 27 member states has to be used instead.

Most products previously covered under United Kingdom Regulations and European Union Directives for CE marking will fall under the scope of the UKCA marking. As such, the BSI [137] currently offer UKCA marking for; construction products, personal protective equipment, gas appliances, pressure equipment, lifts, marine equipment, transportable pressure equipment and measuring equipment.

# Relevant cryogenic safety legislation in the United Kingdom—summary

**Key learning points**

*The Provision and Use of Work Equipment Regulations (PUWER)*—Legal duties for organisations who use work equipment

*The Pressure Equipment Regulations (PER)*—The design, manufacture, conformity assessment and proof test of pressure equipment operating above 0.5 bar pressure

*The Pressure Systems Safety Regulations (PSSR)*—The safety requirements of pressure equipment once put in service

*The Confined Space Regulations*—Covers all work undertaken under confined space conditions

*The Control of Substances Hazardous to Health Regulations (COSHH)*—The law aimed to prevent or reduce exposure of individuals to hazardous substances

*The Personal Protective Equipment at Work Regulations*—Covers all personal protective equipment which serves to protect individuals against risks to their health and safety

*The Dangerous Substances and Explosive Atmospheres Regulations (DSEAR)*—The law concerned with protecting against the risks from fire, explosions and substances corrosive to metals

*The Manual Handling Operations Regulations (MHOR)*—The law covering all manual handling activities

*The Lifting Operations and Lifting Equipment Regulations (LOLER)*—The legal duties on individuals and companies who use lifting equipment

*The Carriage of Dangerous Goods by Road Regulations*—The legal duties for organisations who transport dangerous goods using the United Kingdom roads

*The Carriage of Dangerous Goods by Rail Regulations*—The legal duties for organisations who transport dangerous goods using the United Kingdom rail network

*The Carriage of Dangerous Goods by Inland Waterway Regulations*—The legal duties for organisations who transport dangerous goods using the United Kingdom inland waterways

*The Carriage of Dangerous Goods by Sea Regulations*—The legal duties for organisations who transport dangerous goods by sea to and from the United Kingdom harbours

*The Carriage of Dangerous Goods by Air Regulations*—The legal duties for organisations who transport dangerous goods by air to and from the United Kingdom airports

*The Working at Height Regulations*—The law covering all work at height activities
*The Control of Electromagnetic Fields at Work Regulations*—The minimum health and safety requirements regarding exposure of workers to EMFs

*Control of Major Accident Hazard Regulations (COMAH15)*—The law regarding the necessary hazard controls in the event of a major chemical accident

*The Dangerous Substances (Notification and Marking of Sites) Regulations (NAMOS)*—The legal requirements for the provision of dangerous substance information for sites storing 25 tonnes or more

*Planning Permission*—The legal requirement which allows for the building on land or the change of use of land or buildings

*Approved Codes of Practice (ACOPs)*—Guidance documents explaining how the implementation of the Regulations can be achieved

**Associated safety organisations in the United Kingdom**

*British Standards Institute (BSI)*—The national standards body who produce technical standards on a range of products and services

*The Royal Society for the Prevention of Accidents Awards (RoSPA)*—An independent charity with the aim of saving lives and preventing injuries occurring

*The Health and Safety Executive (HSE)*—A government agency responsible for the regulation and enforcement of workplace health, safety and welfare

*The Public Health Agencies*—National level agencies forming part of the Department of Health and Social Care provision in the United Kingdom

*The United Kingdom Accreditation Service (UKAS)*—The national accreditation body who assess the competence of organisations providing certification, testing, inspection and calibration services

## References


[1] Health and Safety Executive (HSE) *The Provision and Use of Work Equipment Regulations —1998* (www.hse.gov.uk/work-equipment-machinery/puwer.htm) (Accessed 2020)

[2] European Agency for Safety and Health at Work *The Machinery Directive* (https://osha.europa.eu/en/legislation/directives/directive-2006-42-ec-of-the-european-parliament-and-of-the-council) (Accessed 2020)

[3] Legislation.gov.uk *The Pressure Equipment Regulations*—1999 (www.legislation.gov.uk/uksi/1999/2001/contents/made) (Accessed 2020)

[4] Wikipedia *PD 5500* (https://en.wikipedia.org/wiki/PD_5500) (Accessed 2020)

[5] Legislation.gov.uk *The Pressure Systems Safety Regulations*—2000 (www.legislation.gov.uk/uksi/2000/128/contents/made) (Accessed 2020)

[6] Legislation.gov.uk *The Confined Space Regulations*— 1997 (www.legislation.gov.uk/uksi/1997/1713/contents/made) (Accessed 2020)

[7] Legislation.gov.uk *The Control of Substances Hazardous to Health Regulations*—2002 (www.legislation.gov.uk/uksi/2002/2677/made) (Accessed 2020)

[8] Health and Safety Executive (HSE) *Material Safety Data Sheet (MSDS)* (www.hse.gov.uk/chemical-classification/labelling-packaging/safety-data-sheets.htm) (Accessed 2020)

[9] Institution for Occupational Safety and Health (IOSH) 2006 *Managing Safely Syllabus—Chemicals and Substances—Module* 5 (Coalville: Clwyd Associates)

[10] BCF Group *Routes of Entry for Substances into the Human Body* (www.thebcfgroup.co.uk/coshh/coshh-hazardous-substances-routes-of-entry.php) (Accessed 2020)
[11] Health and Safety Executive (HSE) *Registration, Evaluation, Authorisation and Restriction of Chemicals (REACH)* (www.hse.gov.uk/reach/) (Accessed 2020)
[12] Wikipedia *Derived No Effect Level (DNEL)* (https://en.wikipedia.org/wiki/Derived_no-effect_level) (Accessed 2020)
[13] Safeopedia *HAZCHEM Warning Plate System* (www.safeopedia.com/definition/356/haz-chem) (Accessed 2020)
[14] National Chemical Emergency Centre (NCEC) *Dangerous Goods Emergency Action Code List* (https://the-ncec.com/en/news-en/the-dangerous-goods-emergency-action-code-list-2019) (Accessed 2020)
[15] Wikipedia *National Chemical Emergency Centre* (https://en.wikipedia.org/wiki/National_Chemical_Emergency_Centre) (Accessed 2020)
[16] Legislation.gov.uk *Personal Protective Equipment at Work Regulations*—1992 (www.legislation.gov.uk/uksi/1992/2966/contents/made) (Accessed 2020)
[17] EN European Standards *The European Standard, EN* 511—*Protective Gloves against Cold* (www.en-standard.eu/din-en-511-protective-gloves-against-cold/) (Accessed 2020)
[18] EN European Standards *The European Standard EN* 166—*Personal Eye Protection* (www.en-standard.eu/bs-en-166–2002-personal-eye-protection.-specifications/) (Accessed 2020)
[19] Legislation.gov.uk *The Dangerous Substances and Explosive Atmospheres Regulations (DSEAR)*—2002 (www.legislation.gov.uk/uksi/2002/2776/contents/made) (Accessed 2020)
[20] Health and Safety Executive (HSE) *ATEX Directives* (www.hse.gov.uk/fireandexplosion/atex.htm) (Accessed 2020)
[21] Health and Safety Executive (HSE) *The Manual Handling Operations Regulations (MHOR)—1992* (www.hse.gov.uk/msd/backpain/employers/mhor.htm) (Accessed 2020)
[22] Health and Safety Executive (HSE) *Risk at Work—Manual Handling* (www.hse.gov.uk/toolbox/manual.htm) (Accessed 2020)
[23] Health and Safety Executive (HSE) *The Lifting Operations and Lifting Equipment Regulations (LOLER)—1998* (www.hse.gov.uk/work-equipment-machinery/loler.htm) (Accessed 2020)
[24] Health and Safety Executive (HSE) *Safe Use of Lifting Equipment: Approved Code of Practice* (www.hse.gov.uk/pubns/books/l113.htm) (Accessed 2020)
[25] ChemEurope.Com *United Nations (UN)—Four Digit Number* (www.chemeurope.com/en/encyclopedia/UN_number.html) (Accessed 2020)
[26] Health and Safety Executive (HSE) *Carriage of Dangerous Goods—Packaging* (www.hse.gov.uk/cdg/manual/packaging.htm) (Accessed 2020)
[27] Vehicle Certification Agency (VCA) *Dangerous Goods Office* (www.vehicle-certification-agency.gov.uk/dangerous-goods/) (Accessed 2020)
[28] *Office for Nuclear Regulation (ONR)* (http://onr.org.uk/) (Accessed 2020)
[29] Legislation.gov.uk *The Carriage of Dangerous Goods by Road Regulations*—1996 (www.legislation.gov.uk/uksi/1996/2095/contents/made) (Accessed 2020)
[30] The United Nations Economic Commission for Europe (UNECE) *European agreement 'Accord européen relatif au transport international des marchandises dangereuses par route' (ADR)* (www.unece.org/trans/danger/publi/adr/adr_e.html) (Accessed 2020)
[31] Department for Transport 2019 *Reported Road Casualties in Great Britain—2018 Annual Report*

[32] GOV.UK *The Health and Safety Executive (HSE)* (www.gov.uk/government/organisations/health-and-safety-executive) (Accessed 2020)
[33] GOV.UK *Driver and Vehicle Standards Agency (DVSA)* (www.gov.uk/government/organisations/driver-and-vehicle-standards-agency) (Accessed 2020)
[34] GOV.UK *Highways England Traffic Officers* (www.gov.uk/government/organisations/highways-england/about-our-services) (Accessed 2020)
[35] Legislation.gov.uk *The Carriage of Dangerous Goods by Rail Regulations*—1994 (www.legislation.gov.uk/uksi/1994/669/contents/made) (Accessed 2020)
[36] The International Rail Transport Committee (CIT) *Convention Covering International Carriage by Rail (COTIF)—International Carriage of Dangerous Goods by Rail—Rail Transport Law* (https://cit-rail.org/en/rail-transport-law/cotif/) (Accessed 2020)
[37] Legislation.gov.uk *The Carriage of Dangerous Goods and Use of Transportable Pressure Equipment Regulations* 2009 (www.legislation.gov.uk/uksi/2009/1348/contents/made) (Accessed 2020)
[38] Office of Rail and Road (ORR) *2019 Rail Safety Statistics 2018–19 Annual Statistical Release*
[39] GOV.UK *The Health and Safety Executive (HSE)* (www.gov.uk/government/organisations/health-and-safety-executive) (Accessed 2020)
[40] GOV.UK *The Office of Rail Regulation (ORR)* (www.orr.gov.uk/) (Accessed 2020)
[41] ONR *The Office for Nuclear Regulation (ONR)* (www.onr.org.uk/) (Accessed 2020)
[42] GOV.UK *The Department for Transport* (www.gov.uk/government/organisations/department-for-transport) (Accessed 2020)
[43] Legislation.gov.uk *The Carriage of Dangerous Goods by Inland Waterway Regulations* (www.legislation.gov.uk/uksi/2009/1348/2019-04-21) (Accessed 2020)
[44] GOV.UK *The United Kingdom Inland Waterways and Categorisation of Waters* (www.gov.uk/guidance/inland-waterways-and-categorisation-of-waters) (Accessed 2020)
[45] GOV.UK *The United Kingdom Inland Waters and How They Are Categorised* (www.gov.uk/guidance/inland-waterways-and-categorisation-of-waters) (Accessed 2020)
[46] The United Nations Economic Commission for Europe (UNECE) *The European Agreement Concerning the International Carriage of Dangerous Goods by Inland Waterways (ADN)* (www.unece.org/trans/danger/publi/adn/adn_e.html) (Accessed 2020)
[47] Legislation.gov.uk *The Carriage of Dangerous Goods and Use of Transportable Pressure Equipment Regulations 2009* (www.legislation.gov.uk/uksi/2009/1348/contents/made) (Accessed 2020)
[48] The United Nations Economic Commission for Europe (UNECE) *ADN, Annex A, Provisions Concerning Dangerous Substances and Articles* (www.unece.org/fileadmin/DAM/trans/danger/publi/adn/adn2000/English/ADN2000) (Accessed 2020)
[49] Legislation.gov.uk *The European Commission Directive 2006/87/EC on Technical Requirements for Inland Waterway Vessels, Annex III* (www.legislation.gov.uk/eudr/2006/87/body/adopted) (Accessed 2020)
[50] GOV.UK *Inland Waterway Vessels: Construction and Maintenance Standards* (www.gov.uk/guidance/standards-for-the-construction-and-maintenance-of-inland-waterway-vessels) (Accessed 2020)
[51] *The Canal and River Trust* (https://canalrivertrust.org.uk/) (Accessed 2020)
[52] *Scottish Canals* (www.scottishcanals.co.uk/) (Accessed 2020)

[53] GOV.UK *The Environmental Agency* (www.gov.uk/government/organisations/environment-agency) (Accessed 2020)
[54] GOV.UK *The Maritime and Coastguard Agency (MCA)* (www.gov.uk/government/organisations/maritime-and-coastguard-agency) (Accessed 2020)
[55] GOV.UK *The United Kingdom's Maritime Safety Action Plan* (www.gov.uk/government/publications/maritime-safety-action-plan) (Accessed 2020)
[56] GOV.UK *The Navigation Safety Branch of the MCA* (www.gov.uk/guidance/navigation-safety-for-uk-seafarers) (Accessed 2020)
[57] *The Boat Safety Scheme* (www.boatsafetyscheme.org/) (Accessed 2020)
[58] GOV.UK *Marine Navigational Licence* (www.gov.uk/guidance/do-i-need-a-marine-licence) (Accessed 2020)
[59] GOV.UK *The Carriage of Dangerous Goods by Sea Regulations* (www.gov.uk/guidance/moving-dangerous-goods) (Accessed 2020)
[60] The International Maritime Organisation (IMO) (www.imo.org/) (Accessed 2020)
[61] The International Maritime Organisation (IMO) *The International Maritime Dangerous Goods code (IMDG)* (www.imo.org/en/OurWork/Safety/Pages/DangerousGoods-default.aspx) (Accessed 2020)
[62] Legislation.gov.uk *The Merchant shipping (Dangerous Goods and Marine Pollutant) Regulations* 1997 (www.legislation.gov.uk/uksi/1997/2367/made) (Accessed 2020)
[63] Legislation.gov.uk *The Dangerous Substances in Harbour Areas Regulations* 1987 (www.legislation.gov.uk/uksi/1987/37/contents/made) (Accessed 2020)
[64] GOV.UK *Marine Accident Investigation Branch (MAIB) Accident Reports* (www.gov.uk/government/organisations/marine-accident-investigation-branch) (Accessed 2020)
[65] GOV.UK *The Maritime and Coastguard Agency (MCA)* (www.gov.uk/government/organisations/maritime-and-coastguard-agency) (Accessed 2020)
[66] Legislation.gov.uk *The Dangerous Substances in Harbour Areas Regulations* 1987 (www.legislation.gov.uk/uksi/1987/37/contents/made) (Accessed 2020)
[67] GOV.UK *The Health and Safety Executive (HSE)* (www.gov.uk/government/organisations/health-and-safety-executive) (Accessed 2020)
[68] GOV.UK *Guidance on Harbour Closure Orders and Pilotage Function Removal Orders* (https://assets.publishing.service.gov.uk/government/uploads/system/uploads/attachment_data/file/453865/annex-e-draft-guidance.pdf) (Accessed 2020)
[69] Legislation.gov.uk *The Carriage of Dangerous Goods by Air Regulations* (www.legislation.gov.uk/uksi/2002/2786/made) (Accessed 2020)
[70] *The International Civil Aviation Organisation (ICAO)* (www.icao.int/Pages/default.aspx) (Accessed 2020)
[71] The International Civil Aviation Organisation (ICAO) *The Technical Instructions for the Safe Transport of Dangerous Goods by Air* www.icao.int/safety/DangerousGoods/Pages/technical-instructions.aspx) (Accessed 2020)
[72] *The International Air Transport Association (IATA)* (www.iata.org/) (Accessed 2020)
[73] The International Air Transport Association (IATA) *Dangerous Goods Regulations* (www.iata.org/en/publications/dgr/) (Accessed 2020)
[74] The International Civil Aviation Organisation (ICAO) *State and Operator Variations* (www.icao.int/safety/DangerousGoods/Pages/StateVariationPage.aspx) (Accessed 2020)
[75] *The Civil Aviation Authority* (www.caa.co.uk/home/) (Accessed 2020)

[76] parliament.uk *Programme Cyclamen* (https://publications.parliament.uk/pa/cm200809/cmselect/cmpubacc/336/9030910.htm) (Accessed 2020)

[77] GOV.UK *The Home Office* (www.gov.uk/government/organisations/home-office) (Accessed 2020)

[78] GOV.UK *The United Kingdom Borders Agency (UKBA)* (www.gov.uk/government/organisations/uk-border-agency) (Accessed 2020)

[79] Wikipedia *The Channel Tunnel* (https://en.wikipedia.org/wiki/Channel_Tunnel) (Accessed 2020)

[80] Legislation.gov.uk *The Working at Height Regulations*—2005 (www.legislation.gov.uk/uksi/2005/735/contents/made) (Accessed 2020)

[81] Health and Safety Executive (HSE) *Mobile Elevating Work Platforms (MEWPS)* (www.hse.gov.uk/construction/safetytopics/mewp.htm)

[82] Health and Safety Executive (HSE) *Preventing Falls From Mobile Elevating Work Platforms (MEWPS) and Selection and Use of Fall Protection Equipment* (www.hse.gov.uk/foi/internalops/ocs/300-399/314_20.htm)

[83] Health and Safety Executive (HSE) *The Control of Electromagnetic Fields at Work Regulations*—2016 (www.hse.gov.uk/radiation/nonionising/emf-regulations.htm) (Accessed 2020)

[84] National Health Service (NHS) *Magnetic Resonance Imaging (MRI) Scanners* (www.nhs.uk/conditions/mri-scan/) (Accessed 2020)

[85] Public Health England *The Effect on the Human Body due to Exposure to Both Natural and Man-Made Electromagnetic Fields—Public Health England Guidance: Electric and Magnetic Fields—Health Effects of Exposure*—published 1 July 2013 (www.gov.uk/government/publications/electric-and-magnetic-fields-health-effects-of-exposure/electric-and-magnetic-fields-assessment-of-health-risks) (Accessed 2020)

[86] GOV.UK *The Effects of Electric and Magnetic Field Exposure* (www.gov.uk/government/publications/electric-and-magnetic-fields-health-effects-of-exposure/electric-and-magnetic-fields-assessment-of-health-risks) (Accessed 2020)

[87] GOV.UK *Cellular Damage due to Electric and Magnetic Field Exposure* (www.gov.uk/government/publications/electric-and-magnetic-fields-health-effects-of-exposure/electric-and-magnetic-fields-assessment-of-health-risks) (Accessed 2020)

[88] European Agency for Safety and Health at Work *The European Physics Agents EMF Directive—2013/35/EU* (https://osha.europa.eu/en/legislation/directives/directive-2013-35-eu-electromagnetic-fields) (Accessed 2020)

[89] The Health and Safety Executive (HSE) *Guide to CEMFAW Regulations—HSG* 281 (www.hse.gov.uk/pubns/books/hsg281.htm) (Accessed 2020)

[90] The Health and Safety Executive (HSE) *Control of Major Accident Hazards Regulations* (COMAH)—2015 (www.hse.gov.uk/comah/background/comah15.htm) (Accessed 2020)

[91] European Chemicals Agency (ECHA) *Directive 2012/18/EU* (https://echa.europa.eu/regulations/clp/understanding-seveso) (Accessed 2020)

[92] GOV.UK *The Health and Safety Executive (HSE)* (www.gov.uk/government/organisations/health-and-safety-executive)

[93] *Office for Nuclear Regulation (ONR)* (www.onr.org.uk/) (Accessed 2020)

[94] GOV.UK *The Environmental Agency* (www.gov.uk/government/organisations/environment-agency) (Accessed 2020)

[95] *The Scottish Environmental Protection Agency (SEPA)* (https://sepa.org.uk/) (Accessed 2020)
[96] Legislation.gov.uk *The Natural Resources Body for Wales (NRW)* (www.legislation.gov.uk/wsi/2012/1903/made) (Accessed 2020)
[97] The Health and Safety Executive (HSE) *The Dangerous Substances (Notification and Marking of Sites) Regulations (NAMOS)*—2013 (www.hse.gov.uk/pubns/indg467.htm) (Accessed 2020)
[98] The Health and Safety Executive (HSE) *NAMOS Regulations Supporting Guidance Information, HSR29* (www.hse.gov.uk/pubns/books/hsr29.htm) (Accessed 2020)
[99] Legislation.gov.uk *The Carriage of Dangerous Goods and Use of Transportable Pressure Equipment Regulations* 2009 (www.legislation.gov.uk/uksi/2009/1348/contents/made) (Accessed 2020)
[100] The United Nations Economic Commission for Europe (UNECE) *European Agreement 'Accord européen relatif au transport international des marchandises dangereuses par route' (ADR)* (www.unece.org/trans/danger/publi/adr/adr_e.html) (Accessed 2020)
[101] GOV.UK *The Health and Safety Executive (HSE)* (www.gov.uk/government/organisations/health-and-safety-executive) (Accessed 2020)
[102] GOV.UK *Local Authority Departments* (www.gov.uk/government/organisations) (Accessed 2020)
[103] GOV.UK *Planning Permission* (www.gov.uk/planning-permission-england-wales) (Accessed 2020)
[104] Legislation.gov.uk *The Town and Country Planning Act of* 1947 (www.legislation.gov.uk/ukpga/1947/51/enacted) (Accessed 2020)
[105] GOV.UK *Understanding How Your Council Works* (www.gov.uk/understand-how-your-council-works) (Accessed 2020)
[106] GOV.UK *The National Park Authority, Broads Authority and AONB Conservation Boards Appointments* (www.gov.uk/government/news/national-park-authorities-broads-authority-and-aonb-conservation-boards-appointments) (Accessed 2020)
[107] *The United Kingdom's National Parks* (www.nationalparks.uk/) (Accessed 2020)
[108] GOV.UK *Building Regulations* (www.gov.uk/topic/planning-development/building-regulations) (Accessed 2020)
[109] GOV.UK *The Building Regulations Approved Documents* (www.gov.uk/government/collections/approved-documents) (Accessed 2020)
[110] Health and Safety Executive (HSE) *Legal Status of HSE Guidance and Approved Codes of Practice (ACOPs)* (www.hse.gov.uk/legislation/legal-status.htm) (Accessed 2020)
[111] GOV.UK *The Health and Safety Executive (HSE)* (www.gov.uk/government/organisations/health-and-safety-executive) (Accessed 2020)
[112] The Health and Safety Executive (HSE) *Health and Safety at Work Act—1974* (www.hse.gov.uk/legislation/hswa.htm) (Accessed 2020)
[113] GOV.UK *The United Kingdom's Secretary of State for the Home Department* (www.gov.uk/government/ministers/secretary-of-state-for-the-home-department) (Accessed 2020)
[114] GOV.UK *Workplace Health: Applying All Our Health—Guidance* (www.gov.uk/government/publications/workplace-health-applying-all-our-health/workplace-health-applying-all-our-health) (Accessed 2020)
[115] The Magistrates Courts—Courts and Tribunals Judiciary (www.judiciary.uk/you-and-the-judiciary/going-to-court/magistrates-court/) (Accessed 2020)

[116] GOV.UK *The Crown Courts* (www.gov.uk/courts/crown-court)
[117] GOV.UK *Standards and Accreditation* (www.gov.uk/guidance/standards-and-accreditation) (Accessed 2020)
[118] British Standards Institution (BSI) (www.bsigroup.com/) (Accessed 2020)
[119] BSIgroup *BSI and International Standards* (www.bsigroup.com/en-GB/About-BSI/Our-partners/industry-associations/) (Accessed 2020)
[120] BSIgroup *BSI Kitemark™ Certification Mark* (www.bsigroup.com/en-GB/kitemark/) (Accessed 2020)
[121] GOV.UK *The CE Marking Requirements* (www.gov.uk/guidance/ce-marking) (Accessed 2020)
[122] The Royal Society for the Prevention of Accidents (RoSPA) (www.rospa.com/) (Accessed 2020)
[123] The Royal Society for the Prevention of Accidents (RoSPA) *Health and Safety Awards Scheme* (www.rospa.com/Awards) (Accessed 2020)
[124] GOV.UK *The Health and Safety Executive (HSE)* (www.gov.uk/government/organisations/health-and-safety-executive) (Accessed 2020)
[125] The Health and Safety Executive (HSE) *Health and Safety at Work Act—1974* (www.hse.gov.uk/legislation/hswa.htm) (Accessed 2020)
[126] GOV.UK *Public Health Agencies in the United Kingdom* (www.gov.uk/government/publications/the-health-and-care-system-explained/the-health-and-care-system-explained) (Accessed 2020)
[127] GOV.UK *The Department of Health and Social Care Provision in the United Kingdom* (www.gov.uk/government/organisations/department-of-health-and-social-care) (Accessed 2020)
[128] The National Health Service (NHS) (www.nhs.uk/) (Accessed 2020)
[129] The National Health Service (NHS) *The Clinical Commissioning Groups* (www.england.nhs.uk/ccgs/) (Accessed 2020)
[130] GOV.UK *The Future of Public Health: The National Institute for Health Protection (NIHP) and Other Public Health Functions* (www.gov.uk/government/publications/the-future-of-public-health-the-nihp-and-other-public-health-functions/the-future-of-public-health-the-national-institute-for-health-protection-and-other-public-health-functions) (Accessed 2020)
[131] GOV.UK *The NHS Test and Trace: How It Works* (www.gov.uk/guidance/nhs-test-and-trace-how-it-works) (Accessed 2020)
[132] BBC *Covid-19 Pandemic: Tracking the Global Coronavirus Outbreak* (www.bbc.co.uk/news/world-51235105) (Accessed 2020)
[133] GOV.UK *United Kingdom Joint Biosecurity Centre—Our Plan to Rebuild: The UK Government's COVID-19 recovery strategy* (www.gov.uk/government/publications/our-plan-to-rebuild-the-uk-governments-covid-19-recovery-strategy) (Accessed 2020)
[134] The United Kingdom Accreditation Service (UKAS) (www.ukas.com/) (Accessed 2020)
[135] GOV.UK *Using the UKCA Mark from 1st January 2021* (www.gov.uk/guidance/using-the-ukca-mark-from-1-january-2021) (Accessed 2020)
[136] GOV.UK *The CE Marking Requirements* (www.gov.uk/guidance/ce-marking) (Accessed 2020)
[137] British Standards Institution (BSI) (www.bsigroup.com/) (Accessed 2020)

# Chapter 4

# Relevant cryogenic safety legislation in the United States of America

## 4.1 The United States of America safety legislation

Safety in the United States of America is fundamentally considered under the combined ethical and legal remit of minimising or even eliminating hazardous conditions which could cause bodily injury. The associated safety precautions are covered by two discrete demographic sections; occupational safety and public safety. The occupational safety section is particular to those risks which are likely to be present in the workplace, and includes a variety of locations such as offices, manufacturing plants, farms, commercial and retail facilities, and construction sites. The public safety section is specifically focused on the hazards encountered in the home, or during travel and recreational activities.

The Occupational Safety and Health Act—1970 (OSH Act) [1] is the primary federal law in the United States of America controlling the Health and Safety within both the private sector and federal government organisations. Its fundamental aim is to reduce workplace hazards and to establish safety and health programs suitable for both employers and their employees. It places legal obligations on employers to provide their workers with adequate training and equipment in order to ensure they are as safe as possible in whatever they are being asked to undertake. This is the general duty clause of the OSH Act, whereby employees should be provided with a place of employment which is free from the types of hazards that are likely to cause death or serious physical harm. This is defined as a common-law and it follows therefore that employers have a duty under this law to take reasonable care of the safety of their employees. In addition, statute law imposes other general and specific duties on these employers relating to their particular industry. The statute law also allows for the creation of the various government bodies who regulate workplace safety and provide them with the necessary powers to enact this responsible role.

doi:10.1088/978-0-7503-3123-4ch4  

The OSH Act is overseen by the Occupational Safety and Health Administration (OSHA) [2] and is primarily managed by the application of occupational safety and health requirements through both federal level legislation [3] and State specific plans [4]. Employers can also adopt additional voluntary systems in order to establish and maintain safe and healthful work for their staff and their associated places of employment. This is endorsed by the OSHA through their recommendation that employers conduct risk assessments of the work processes undertaken in their organisations. The OSH Act is subdivided into sections of which the following are central to its function and thereby encompass specific cryogenic technology risk management.

- **Section 6** of the Act specifies the occupational safety and health standards which must be achieved for compliance to be enacted. These are fundamental requirements with the standards being specific for particular workplace environments.
- **Section 8** covers the requirements for the reporting of work-related incidents where fatalities or multiple hospitalisations occur. As a federal level law, these requirements are mandatory to all workplace situation. Section 8 also permits OSHA inspectors entry to facilities where reported incidents have occurred in order to inspect and investigate the causes and the effects of those incidents.
- **Section 18** authorises the provision of individual States across the United States of America to adopt their own occupation safety and health structures, so long as they are at least as effective as those prescribed in the Act. This allows for the incorporation of regional and technological workplace variations particular to parts of the United States of America. States which adopt this process are defined as OSHA States [5].

## 4.2 The Occupational Safety and Health Administration (OSHA) [6]

The OSHA is the agency set up, as part of the Occupational Safety and Health Act, with the authority for issuing workplace health and safety regulations in the form of Standards. These Standards are generally focused more on the practicalities of mechanical and chemical hazards in the workplace rather than the procedures, systems and management roles associated indirectly with the hazards. In practical terms, they typically provide; defined limits on hazardous chemical exposure, the stipulated requirements for the use of personal protective equipment, the requirements for preventing falls when employees are working at height and the safety procedures needed to reduce hazards from the operating of dangerous equipment.

However, before any Standard can be issued, they must go through an extensive process of public engagement [7]. As part of this consultation, the OSHA must demonstrate that a significant risk to workers does in fact exist and that there are suitable control measures which employers could enact in order to protect their workers from these risks. This covers a variety of employment areas such as construction, general industries, maritime and agriculture. Cryogenic technology typically would be covered by the specific product industry, including product areas like pressurised gases and liquid cryogens. The Standards are written in such a way that they safeguard any workers from a wide range of serious hazards which they are likely to encounter in their daily working environments. The OSHA also provide compliance assistance for the implementation of the Standards along with training courses covering all areas of health and safety.

The OSHA's role further encompasses the inspecting and assessing for compliance against these Standards within organisations. The OSHA inspectors implement this through strict enforcement of the standards and the application of legal sanctions [8] where infringements are found. They also have the mandate, where appropriate, of imposing fines for any identified violation and it is the OSHA, rather than the courts, who ultimately determine the level and amount of these non-compliance penalties. This value of these fines will depend on the severity of the violation and the likelihood that its outcome would result in serious harm occurring. Infringements are therefore often classed either as a Serious Violation, or a Wilful Violation.

*A Serious Violation* [9] by definition, is where there is a high probability that death or serious physical harm would result from the identified hazard which brought about any incident, and that the employer was aware of its existence. This carries a mandatory penalty, often in the thousands of dollars.

*A Wilful Violation* on the other hand, is where the employer knows that what they are doing is considered to be a violation and that the specific hazardous condition exists which brought about an incident, but made no reasonable effort to eliminate it. This carries a more substantial fine, with a minimum penalty in the thousands of dollars and a maximum in the tens of thousands. Where a Wilful Violation resulted in the death of an employee, the transgression is considered a criminal offence and punishable by a court-imposed fine which can be in the hundreds of thousands of dollars, or even by imprisonment.

The specific hazards associated with cryogenic technology covered by the OSHA Standards would include:

- **Fall hazards**, such as those likely to be encountered for work undertaken at elevated heights on things like cryogenic plant installations. The Standards require that employers provide the correct fall protection equipment, such as harnesses and fall-arrestor lines, along with guardrails on all platforms at raised heights.

- **Confined space hazards**, which include all enclosed and poorly ventilated work environments where there is a possibility of oxygen depletion to take place. Implementation of the relevant Standards ensures the safety of any workers, who through the course of their employment, have to enter specifically defined confined spaces to undertake tasks.

- **Dangerous atmospheres hazards**, where there is the potential for the release of dangerous gases and fumes into the working environment, which can in turn cause explosive conditions if the concentration percentages are sufficiently high. The Standards describe the need for the provision of safety equipment for personnel, such as respirators, in workplace areas where dangerous atmospheres or oxygen depleted atmospheres exist or can come about during processes being undertaken in those areas.

- **Chemical hazards**, which includes harmful chemicals and substances in their solid and liquid forms as single elements or in combinations whereby the hazard may be more significant. It also includes chemicals and substances being used directly as an agent or indirectly as part of an overall production process. The Standards outline the methods of preventing employee exposure to any of these harmful chemicals and substances. Where exposure cannot be avoided, Permissible Exposure Limits (PELs) [10] are defined within the Standards for specific chemicals and substances, regulated at a level which protects workers from the health effects of exposure to all currently known hazardous substances. There are also a set of non-binding occupational exposure limits, known as recommended exposure limits (RELs) within the Standards which are generally below the PELs and are considered to provided better protection to employees working continually with various types of hazardous substances.
- **Machine equipment hazards**, which encapsulates all types of machine equipment from small manual devices to large, power-driven assemblies. The standards indicate the requirements of employers to install suitable guards to all so-called dangerous machines and equipment used in the workplace where employee interactions are possible.

## 4.3 The Code of Federal Regulations (CFR)—2019 [11]

Sitting beneath the Occupational Safety and Health Act, the Code of Federal Regulations are the general and permanent rules and regulations published by the executive departments and agencies of the federal government [12] of the United States of America. They are broken down into 50 specific titles representing areas where the federal regulations have to be applied. For cryogenic technology these would include associated titles for the direct use of the technology, along with things like the transportation requirements linked with movement of cryogenic equipment and products within the United States of America and to other destinations throughout the world;

Title 14—Aeronautics and Space
Title 16—Commercial Practices
Title 23—Highways
Title 29—Labour
Title 33—Navigation and navigable waters
Title 49—Transportation

### 4.3.1 Code of Federal Regulations—Title 14—Aeronautics and Space [13]

CFR Title 14—Aeronautics and Space, are the rules covering the regulations of all aviation activities taking place within the United States of America. They are established by the Federal Aviation Administration (FAA) [14] and cover a variety of activities such as; aircraft design, pilot training and the actual airline flights. Their principal aim is to provide safe aviation throughout the country by the elimination of all unnecessary risks, encompassing all aspects of aviation from design all the way through to specific flight undertakings. The regulations apply not only to pilots of the aircraft and their flight attendants, but also to any carrying passengers and the general public affected in anyway by the aviation undertakings being considered. The CRF Title 14 is divided into various parts, with each one dealing with a specific type of aviation activity. The separate parts do not distinguish the types of aircraft, but rather the type of activity undertaken with whatever aircraft is being flown. As a result, the regulations for commercial aviation are intensive and very detailed pilot training is required for this application. The various parts relevant to the transport of cryogenic technology by aircraft would include:

#### *4.3.1.1 Part 21*

Covers the general certification procedures required for all component parts of any aircraft which ensures their airworthiness. This includes not only certification at the manufacture and initial installation phase, but also the continuous inspection and test of these parts over the lifetime of the aircraft they are part of.

#### *4.3.1.2 Part 25*

Covers the airworthiness standards specifically for transport aircraft, which includes cargo jets as well as the more conventional propeller-driven airplanes. It sets out the necessary criteria required at the initial design and construction stages, all the way through to the operational flight period and any subsequent inspections and maintenance requirements. Specific topics covered by this section in addition to the general requirements included in Part 21 would be the type of emergency provisions required on the cargo aircraft and what fire protection needs to be in place for the transport of any dangerous substances.

#### *4.3.1.3 Part 43*

Covers the general maintenance requirements, along with any preventative maintenance provisions needed in order to retain the airworthiness of all types of aircraft. It details the schedule of general maintenance for particular types of aircraft and the frequency at which these need to be undertaken. Part 43 also defines what rebuild and alterations can be undertaken within the framework of the CFR Title 14 Regulations.

#### *4.3.1.4 Part 61*

Covers the certification requirements for all pilots, irrespective of the aircraft they are operating. It defines the initial training they will need to obtain their pilot's licence and any refresher training requirements over the period of their flying careers.

#### *4.3.1.5 Part 63*

Covers the certification requirements for flight engineers and flight navigators. It defines the initial training they will need in order to qualify for their roles and any obligatory refresher training requirements needed to retain these credentials.

*4.3.1.6 Part 91*

Covers the general operating rules for all aircraft. Many of these will be standard procedures adopted by all airplanes, with supplementary rules applying to the more specific categories, such as those distinguishing between passenger and cargo planes.

*4.3.1.7 Part 117*

Defines the flight and duty-time limitations and rest requirements for all flight-crew. Many of these will be standard depending on the flight durations, but others may be more detailed where long-haul flights are involved, or where multiple landings and take offs are required during a single journey.

### 4.3.2 Code of Federal Regulations—Title 16—Commercial Practices [15]

CFR Title 16—Commercial Practices, are the rules covering the Regulations regarding all commercial practices in the United States of America. It is structured into 2 subdivisions; Volume 1, for industry commercial practices which is regulated by the Federal Trade Commission [16], and Volume 2, for consumer goods which is regulated by the Consumer Product Safety Commission [17]. Each volume is further divided into parts, with each separate one focusing on specific commercial practices. The various parts relevant to cryogenic technology would include:

*4.3.2.1 Part 1120*

Covers products deemed to be substantially hazardous and the necessary standards of safety required to mitigate their inherent risks. This would include any product which has a defect that would pose a substantial risk of injury to the public if they were used. An example being an extension cord where the cable wires have become exposed.

*4.3.2.2 Part 1500*

Covers the enforcement regulations for hazardous substances and mixtures. It includes labelling requirements of particular substances and those determined to be 'special hazards' within the CFR Title 16 Regulations. It also lists those hazardous substances specifically banned from commercial use.

### 4.3.3 Code of Federal Regulations—Title 23—Highways [18]

CFR Title 23—Highways, sets out the rules and regulations defined by federal agencies regarding the use of the highways in the United States of America. The Federal Highway Administration (FHWA) [19], is the organisation who oversees the federal funds provided for the construction and maintenance of the National Highway System [20]. They are a division of the United States Department of Transportation [21], and ensure that construction standards are met in order to guarantee that safe highways and road-bridge designs are being produced. They also sponsor particular research focused on roadway safety, such as traffic signs and road surface markings.

For any transportation of cryogenic technology and equipment utilising the United States of America highways, not only does the highways have to be safe to use, but also the transport vehicles themselves need to be of a sufficient standard to ensure safe movements of the equipment at all times. To this end, federal motor vehicle safety standards form part of the CFR Title 23 and are enforced in the United States of America by the National Highway Traffic Safety Administration [22]. The standards specify the design, construction of automobile safety related components and are the federal implementation of the UN Regulations decreed by the World Forum for Harmonisation of Vehicle Regulations [23].

### 4.3.4 Code of Federal Regulations—Title 29—Labour [24]

As one of the laws passed by the United States House of Representatives and Senate [25], CFR Title 29 covers the obligations an employer has in providing a place of employment that is free from recognisable hazards which could cause death or serious harm to their employees. The Code also describes the occupational safety and health standards they need to adopt in order to obligate this process. It defines in detail the respective State's powers to inspect establishments in order to audit the implementation of these standards, and what record keeping needs to be legally maintained by employers. There are enforcement procedures within the Regulation where clear violations of the CFR Title 29 would result in a citation being issued and penalties applied accordingly.

CFR Title 29 also includes the all-encompassing requirements of training and employee education, which are key factors to maintaining safety in the workplace. This instruction would include compliance assistance programs for the application of any detailed occupational safety and health requirements related to a company's specific responsibilities and their processes. The Regulations themselves are subdivided into standards with Part 1910 relating to the specific application of the general occupational safety and health standards.

For the implementation of the particular occupational safety and health standards associated with the use of cryogenic technology, and more specifically cryogenic liquids, CFR Title 29 outlines the directives for the handling of hazardous materials in the workplace. The Federal Regulation also explains the response requirements to hazardous materials related incidents. These are implemented through a separate set of regulations known as the Hazardous Waste Operations and Emergency Response (HAZWOPER) regulations [26].

#### *4.3.4.1 Part 1910—Occupational Safety and Health Standards [27]*

##### ***4.3.4.1.1 Subpart A—general***

Covers the general requirements for occupational safety and health standards such as the compliance duties organisations must have in place for their employees. This would include general housekeeping procedures for the majority of places of work, along with the more stringent safety and health requirements for more hazardous work environments.

#### *4.3.4.1.2 Subpart D—walking-working surfaces*

Covers the general requirements for workplace walking and working surfaces needed to ensure safe movement by all staff around all types of working environments. Subpart D also includes the requirements dealing with the higher risks associated with the use of ladders and scaffolding as would be needed for any work at heights activities. It outlines not only the specifics for the equipment being used to ensure they are safe, but also the maintenance and periodic testing of that equipment and the initial installation requirements of any scaffolding structures. For anyone using ladders and scaffolding there are also training requirements which provide knowledge and understanding of how they can be used safely. The subpart also details the provision of fall protection systems and falling object protection when working from things like elevated platforms, along with the general training of employees expected to work at elevated heights as part of their normal duties.

#### *4.3.4.1.3 Subpart E—exist routes and emergency planning*

Sets out the design and construction requirements for exit routes within buildings to ensure that evacuations are made in a safe and secure manner. It also details what emergency action plans and fire prevention plans organisations must have established for the workplace in order to reduce the risks associated with the eventuality of an emergency evacuation situation.

#### *4.3.4.1.4 Subpart F—powered platforms, man-lifts and vehicle-mounted work platforms*

Covers the principal requirements for working at height activities, but this time by the exclusive use of powered platform type equipment. As these tend to be specialist pieces of equipment, it includes the overall training requirements of any operator of the powered platforms, which covers both the actual operating of the equipment itself, but also the general work at heights requirements when using this type of platform. Subpart F also covers the mandatory maintenance and periodic test requirements of the work equipment being used to allow their continued safe employment.

#### *4.3.4.1.5 Subpart G—occupational health and environmental control*

Covers a range of occupational health and environment controls which need to be adopted in the workplace in order to ensure staff are suitably protected at all times from any prevailing hazards. A typical example would be the noise protection requirements, where occupational noise exposure is recognised as being above standard defined operational values. Subpart G also includes the workplace ventilation requirements, especially where there is a risk of the working environment being compromised by the release of gases or fumes. For cryogenic technology applications, this is a very relevant occupational health control, especially where liquid cryogens are being processed or stored and there is a dominant risk of the release of evaporated gases from the cryogenic liquids reducing the oxygen concentrations in the working environment.

#### *4.3.4.1.6 Subpart H—hazardous materials*

Where the general requirements for the use of any so-called hazardous materials are explained. For the particular area of cryogenic technology, it is also where the use of compressed gases are outlined, detailing specifics such as hydrogen and oxygen in terms of their hazardous nature. This section also includes any flammable liquids used which may be employed in the workplace. In addition, it sets out the process safety management systems needed for the application of any highly hazardous chemicals, as well as the hazardous waste operations and emergency response needed if handling these types of materials. Part of this obligation is the use of Material Safety Data Sheets (MSDSs) [28] as a means of avoiding chemical risks in the workplace by providing this information to all employees who may be working with hazardous chemicals.

#### *4.3.4.1.7 Subpart I—personal protective equipment (PPE)*

States the general requirements for all types of PPE which include; eye and face protection, respiratory protection, foot and hand protection, and head protection. All of these are typical examples of the type of PPE employed for the handling of cryogenic technology. Subpart I provides direction on workplace hazard assessment which will subsequently guide the selection of personal protective equipment to suit the respective application. This subpart also includes personal fall protection systems typically used for work at height and lays out the test methods and procedures for these systems to ensure they are kept in safe working order. Much of the guidelines are non-mandatory and are therefore provided as guidance only in the selection of the correct PPE for the application being employed.

#### *4.3.4.1.8 Subpart J—general environmental controls*

Where the main safety control requirements pertaining to the working environment are detailed. It includes the specifications for things like accident prevention signs and tags, which give advanced warning of the presence of localised hazards in the workplace. For cryogenic technology applications, subpart J is where work in confined space safety systems are to be found, in particular those requiring entry permits [29] as a means of controlling access to these hazardous environments.

#### *4.3.4.1.9 Subpart K—medical and first aid*

The section involved with the provision of suitable workplace medical services and first aid within organisations. It defines what medical and first aid provisions need to be delivered which are applicable to the particular hazards being employed. For cryogenic technology for example, these requirements would include not only the medical support and first-aid equipment provisions for the extremely cold technology involved, but also the suitable training of personnel to treat cold burns, or even the effects of hypothermia.

#### *4.3.4.1.10 Subpart L—fire protection*

Provides the scope and application of fire prevention systems which must be established by organisations in any workplace under their control in order to ensure personnel are protected from fire at all times. This section also includes the standards which need to be met for any portable fire suppression equipment and also any fixed fire suppression equipment, such as permanently installed sprinkler systems. It details the criteria for the fitting of fire detection systems in any workplace establishment, along with the testing and maintenance requirements of such detection systems to ensure they are operating correctly. It also includes the test methods for any fire-protective clothing used at work, defining how manufacturers of the clothing undertake the tests and what the pass/failure measures are.

##### *4.3.4.1.11 Subpart N—materials handling and storage*

The section covering general handling of materials and goods and includes any lifting, lowering, pushing or pulling activities. It has a general scope for how certain products must be stored to ensure safety. In addition, it has more specific topic areas covering the use of overhead and gantry cranes, slings, and the use of powered industrial trucks for the general movement of items around the workplace.

### 4.3.5 Code of Federal Regulations—Title 33—Navigation and Navigable Waters [30]

CFR Title 33—Navigation and Navigable Waters, sets out the rules and regulations defining the use of navigable waters within and around the United States of America. The Code in its entirety is derived from consultation with a broad spectrum of related industry specialists which includes; classification societies, engineering companies, deck officers on oceangoing vessels and qualified marine engineers. It is formally split into three chapters covering specific establishment responsibilities, which are then further divided into numerical parts which outline the detailed requirements applicable to the Code:

Chapter 1—United States Coast Guard
Chapter 2—Army Corps of Engineers
Chapter 3—Saint Lawrence Seaway Development Corporation

#### *4.3.5.1 Chapter 1—United States Coast Guard*

##### *4.3.5.1.1 Part 1*

Covers the general provisions of CFR Title 33, such as the definitions and the associated roles for the use of navigable waters around the United States of America. It outlines the duties and responsibilities placed on the United States Coast Guard in relation to vessels using these waters and the infrastructure supporting vessels, harbours, seaways and inland waterways.

##### *4.3.5.1.2 Part 2*

Details the jurisdiction of the Code and who has the incumbent authority for its implementation. As a Federal regulation, the implementation of CFR Title 33 is mandatory for all States in order to ensure safe use of the navigable waters within and around the United States of America.

##### *4.3.5.1.3 Part 6*

Specifies the fundamentals required for the protection and security of all waterborne vessels, including all classifications of pleasure vessels and cargo vessels. It endorses the safety and health requirements for the many specific types of vessels which are permitted to use the navigable waters of the United States of America. It also includes details of the safety and health requirements of the harbours and waterfront facilities supporting the marine industry which will employ their own specific measures to ensure safety and security of their facilities for any vessels docking there.

#### *4.3.5.1.4 Part 62*

Defines the system requirements for navigation aids covering all waterways in and around the United States of America. It outlines their design standards and what level of maintenance and inspection is required in order to ensure continual safe use of the respective waterways.

#### *4.3.5.1.5 Part 64*

Explains how the marking of sea-level structures, sunken vessels and other obstructions have to be applied wherever marine vessels are using waterways containing such obstacles. By this method, the risks of collision between marine vessels and these obstructions are suitably controlled.

#### *4.3.5.1.6 Part 80*

Covers the international navigation rules applicable for the navigable seas immediately around the United States of America. It also defines the so-called collision regulation boundaries of the United States of America [31]. These are legitimate demarcation lines defining the waters upon which all mariners have to comply with the International Regulations for Preventing Collision at Sea—1972 (COLREGS) [32]. Mariners using the waters inside these demarcation lines must comply with the Inland Navigation Rules covered by Part 89. Part 80 details the locations for all the coastal region demarcation lines for each of the various individual United States territories.

#### *4.3.5.1.7 Part 89*

Covers the Inland navigation rules which apply for vessels operating inland of the United States collision regulation boundaries. Part 89 specifies that every self-propelled vessel over 12 metres or more in length has to retain a copy of these rules if they are operating in waters within the demarcation lines. These vessels must also retain the necessary crew training, and maintenance and test records which complement these rules.

#### *4.3.5.1.8 Part 96*

Outlines the rules for the safe operation of all waterborne vessels and the safety management systems which need to be implemented in order to achieve this. The level of detail within each safety management system will depend on the type and capacity of the vessels being employed.

#### *4.3.5.1.9 Part 126*

Details the requirements for the safe handling of packaged and bulk-solid dangerous cargos at waterfront facilities as they are being both loaded and unloaded from marine vessels. Part 126 applies specifically to the handling, storing, loading and unloading of hazardous materials at waterfront facilities but equally applies to the same activities undertaken on the vessels docked at these facilities. A typical waterfront facility would include all piers and wharves which marine vessels can be secured to. It would also include the land and the water in the immediate vicinity of these structures and all associated buildings, equipment and materials used about the waterfront. It details what fire extinguisher equipment and fire appliances must be provided, along with the types of warning signs necessary to give notice of the hazardous materials. As many facilities run their operations continuously throughout the day, it also specifies the minimum lighting requirements which need to be provided in order for the transfer of hazardous materials to take place during the night. Likewise, these facilities need to have appropriate security measures enacted to deter unlawful entry to both the waterfront facility and the marine vessels.

#### *4.3.5.1.10 Part 127*

Focuses specifically on the requirements for waterfront facilities handling liquefied natural gas and liquefied hazardous gases, detailing aspects such as the design, construction and layout of marine transfer areas for handling these products. It covers the provision of warning signs giving notice of the hazardous materials and the gas detection and alarm systems that must be employed during all transfer and storage processes. It also qualifies the operations procedures required to ensure safety when handling liquefied natural gas and liquefied hazardous gases, along with the emergency measures, including firefighting provision, which need to be in place should an incident ever occur. In addition, it details the maintenance, repairs, inspection and testing requirements to ensure that the facilities can remain fully operational. Due to the nature of the hazardous materials being handled, Part 127 places an important emphasis on the security measures relating to the waterfront facilities in order to deter unlawful access to the marine transfer area where the liquid gases are being stored and transferred.

#### *4.3.5.1.11 Part 142*

Covers the general workplace health and safety provision which needs to be in place for all crews working on-board marine vessels. It details the specific personal protective equipment requirements for each particular marine vessel, along with the general workplace conditions which must be in place to ensure the health and safety of all crew. Specific activities are addressed in relation to specific vessels, as for example, cargo vessels would require a completely different set of procedures compared with passenger ships.

#### *4.3.5.1.12 Part 144*

Details the minimum requirements for lifesaving appliances on-board any vessels navigating the waterways in and around the United States of America. It specifies the minimum quantities and types of life floats and life preservers for each specific vessel. It also details the level of first-aid kit and emergency communication equipment provision which complements these lifesaving appliances.

#### *4.3.5.1.13 Part 145*

Details the minimum provision of firefighting equipment which have to be fitted to all oceangoing vessels and those operating on inland waterways within the United States of America. It outlines the type, number and location of portable and semi-portable fire extinguishers on-board these vessels, with broad firefighting provision being applied to general occupancy vessels and more specific for those transporting dangerous substances and materials.

#### *4.3.5.1.14 Part 153*

Are the necessary controls which need to be implemented in order to prevent pollution of the United States of America waterways by oil or other hazardous substances from marine vessels. It details the general requirements needed to avoid accidental releases in the first place, along with the process of notification if any discharge of oil or a hazardous substance has actually taken place. Clearly, the cargo capacity of some marine vessels can be enormous and accidental releases of their shipment can result in a significant pollution hazard. Likewise, the marine vessel's own fuel tanks can themselves be of sufficient volume that should a rupture of any fuel storage containers occur, this in itself could be the source of pollution.

#### *4.3.5.1.15 Part 157*

Defines the rules which serve to protect the marine environment relating specifically to tanker vessels carrying oil in bulk. It details the general design requirements any such vessel must achieve and the installation of equipment on-board, aimed at preventing accidental release of any oil. Vessels carrying oil in bulk would normally be of a double hull type design, with segregated ballast tanks which provides protection against oil outflow in the event of grounding, ramming or collision.

### *4.3.5.2 Chapter 2—Army Corps of Engineers [33]*

#### *4.3.5.2.1 Part 207*

Covers the waterways navigation regulations specifically applicable to the emergency employment of the Army and other military and civil resources during a natural disaster. It details specific waterways throughout the United States of America, highlighting the navigation requirements for each individual example. This allows for the mass transport of personnel and equipment during natural disaster situations, whilst maintaining the necessary levels of health and safety practices during any such priority deployment.

#### *4.3.5.2.2 Part 245*

Details the requirements for the removal of wrecks and other obstructions from the waterways by the Army Corps of Engineers in order to maintain commercial and military access to the network of navigable waters in and around the United States of America. It details the measures used to determine what constitutes a hazard to navigation in these waters. In addition, it lists what remedial actions need to be implemented in order to reduce risks to marine vessels by the removal of wrecks and other obstructions by either the responsible party or the Army Corps of Engineers.

Part 245 also covers the safety equipment and procedures needed in order to undertake the high-risk activities associated with any such removals to maintain clear waters, taking into consideration factors like the weather and waterway conditions. It also explains the permit process requirements which must be in place in order to undertake these tasks in a safe manner.

*4.3.5.3 Chapter 3—Saint Lawrence Seaway Development Corporation [34]*

The Saint Lawrence Seaway [35] is a system of canals, locks and channels passing through parts of Canada and the United States of America which allows oceangoing vessels to travel from the Atlantic Ocean to the Great Lakes of North America. The Seaway is jointly owned by both countries, with the Saint Lawrence Seaway Development Corporation (SLSDC) in the United States of America being set up to operate and maintain 2 of the 15 locks of the seaway under their direct jurisdiction. The Corporation is an agency within the United States Department of Transportation [36]. The Canadian counterpart is the Saint Lawrence Seaway Management Corporation (SLSMC) [37], who are responsible for the remaining 13 locks situated on Canadian soil.

***4.3.5.3.1 Part 401***

Details the general regulations and rules for the operation and use of the Seaway. It explains how the SLSDC, working with the United States Coast Guards [38], Transport Canada [39] and the SLSMC, provide inspection of commercial vessels using the Seaway. This ensures that their operational safety and environmental performance reaches the standards required to maintain safe use of the seaway and that accidents do not occur which may impede further use of the associated canals, locks and channels. Failure to abide by these rules can result in violation penalties.

### 4.3.6 Code of Federal Regulations—Title 49—Transportation [40]

CFR Title 49—Transportation, sets out the rules and regulations defining all methods of transportation and all types of goods being transported throughout the United States of America and their related security implications. The Regulations are issued jointly by the Departments of Transportation [41] and Homeland Security [42] in consultation with other agencies such as the Hazardous Material Safety Administration [43], the Federal Railroad Administration [44] and the National Transportation Safety Board [45].

CFR Title 49 has specific sections applicable to the regulation of the domestic transportation of hazardous materials by any form of transportation for movements to, from and within the United States of America. Parts 100 to 185 of CFR Title 49 relate directly to the transporting of hazardous materials, controlled by the Pipeline and Hazardous Materials Safety Administration [46]. It covers all aspects of the preparing, transporting and handling of dangerous goods, initially setting out the hazardous material classifications, which in turn determines the packaging requirements for the safe transport of each specific class. It also provides guidelines to the correct use of Hazmat hazard labels [47] for any of the hazardous materials covered by these classifications. In addition, the Regulations detail the limiting amounts of hazardous materials which are permitted in specific containers, along with the required documentation which must be in place for the shipping of any hazardous materials. It also outlines what training and safety plans have to be established before any hazardous materials can be transported.

The Regulations are mandatory for anyone moving goods of any nature in the United States of America. At its core it includes the rules governing all domestic transportation, and organisations failing to comply with it can be penalised. The more stringent rules for transporting hazardous materials will only be applicable for those product listed as such in the Regulations. For transportation of goods over land, CFR 49 Parts 100 to 185 outline the rules, with Part 177 specifically focusing on the carriage of hazardous materials by public highways and Part 174 covering the carriage of hazardous materials by rail. For vessels using the internal waterways of the United States of America, Part 176 covers the movement of hazardous materials by sea and waterways. Accordingly, United States law also stipulates that CFR 49 takes precedence over international regulations covering the movement of hazardous materials by ship. Finally, for the movement of dangerous goods by aircraft, Part 175 covers the loading, unloading and handling requirements of cargos when they are being carried by aircraft.

For all methods of transport, the emphasis is on mandatory safety training which needs to be undertaken before anyone can carry out any movement of dangerous goods. Much of the training is centred on the correct identification, marking and labelling of any product classified on the CFR 49 list of hazardous substances. Part 172 of the Regulations includes all the necessary standards for the marking and labelling of such goods. It also details any emergency response information which the emergency services would need to know before tackling an incident involving these products.

The Environmental Protection Agency (EPA) [48] regulates products identified as hazardous materials, looking specifically how they may impact on local communities and the environment. There are also specific regulations which cover the issues related to environmental clean-up, along with those associated with the handling and disposal of any defined waste hazardous materials. The transport of hazardous materials is regulated by the Hazardous Materials Transportation Act [49]. In addition, the Resource Conservation and Recovery Act [50] provides the legal framework to protect human health and the environment from those listed hazardous materials. At a more general level, the Consumer Product Safety Commission [51] regulates any hazardous materials which may be used in products intended for sale to the general public.

## 4.4 General product safety in the United States of America

Whilst industrial applications of components and products are controlled by the numerous Codes of Federal Regulations, general product safety Regulations in the United States of America are controlled by a further range of Federal Agencies. The exact agency depends very much on the nature of the product, along with the hazards associated with them, and the resultant safety and health risks borne by the users of the product. Some examples of these agencies would include; the Consumer

Product Safety Commission (CPSC), the Federal Trade Commission [52], and the National Highway Traffic and Safety Administration [53]. The detailed characteristics of the product in question would determine which of the many Federal Agencies has to be involved. Typically, agencies such as the CPSC have been established to protect any consumer from the unreasonable risk of serious injury or death relating to any products made commercially available. Specifically, they would look in detail at the potential fire, electrical, chemical, or mechanical hazards associated with these products.

These Agencies are ultimately empowered under the Consumer Product Safety Act [54], which fundamentally seeks to protect the general public in the United States of America against the risks of injuries and death from consumer products. Supporting this, the Consumer Product Safety Improvement Act [55] puts into law the testing and documentation requirements which manufactures have to comply with in order to place their products onto the United States of America market place. The Consumer Product Safety Improvement Act also outlines the safe, acceptable levels of several known hazardous substances which are either the product themselves or a component part of a completed, assembled product. To verify compliance with these acceptable levels, manufacturers have to issue a certificate which states that their product complies with all the applicable Standards and Regulations for that product form. Certain categories of this certification needs to be qualified through testing undertaken by an independent, third party organisation.

Products fall into four discrete regulated groups;

- Health and Body
- Vehicles and vehicle related products
- Hazards, Safety and Firearms
- Other products

### 4.4.1 The health and body group

Covers general consumer products such as alcohol, food, cosmetics and tobacco.

### 4.4.2 The vehicle and vehicle related products group

Covers a broad spectrum of transportation methods which includes aircraft, cars and boats. It includes not only the vehicle itself, but also component parts used in the construction of the vehicle.

### 4.4.3 The hazard, safety and firearms group

Covers the work undertaken by the Agencies managing ammunition, industrial and commercial products, and radioactive materials. Some determining elements of cryogenic technology would, by their hazardous nature, be included in this group.

### 4.4.4 The other products group

Covers many chemicals and substances, which by their very make up can cause injuries if incorrectly handled. The use of any liquid cryogens would by definition be included in this group.

## 4.5 Associated safety organisations in the United States of America

### 4.5.1 General

Prior to 1970, there was no national laws in the United States of America which addressed specific health and safety hazards in the workplace. This was significantly changed with the introduction of the directly related Occupational Safety and Health Act (OSH Act) [56] in 1970 and along with it, the formation of the Occupational Safety and Health Administration (OSHA) [57]. At its fundamental core, the OSH Act ensures the safety and health of workers in their place of work, with the OSHA providing the mechanism of implementing this policy in the many different fields of employment throughout the 50 States forming the United States of America. As a result, their contribution to the reduction in work-related injuries is primary to the principles of the OSH Act and why it was established.

In this vital role, the OSHA are directly responsible for enforcing the OSH Act, and any associated Standards and Code of Federal Regulations, on all applicable regulated organisations. They employ Compliance Safety and Health Officers [58] who are directly empowered to undertake inspections and issue fines for any violations found during these reviews. Focusing mainly on hazardous industries, OSHA inspections would also be carried out following workplace fatalities or where multiple hospitalisations have occurred as a result of an incident taking place. Supporting the role of the OSHA, there are various Safety related organisations functioning at different levels, but all with the specific aim of ensuring compliance with the OSH Act is upheld.

### 4.5.2 The American Association of Safety Councils [59]

The American Association of Safety Councils is an advisory body made up entirely of safety professionals who are considered to be contemporary experts in their respective fields. Their aim is to primarily save lives and to reduce workplace injuries throughout the United States of America. This is undertaken through their provision of highly respected educational Safety programmes. Originally established as a forum, the American Association of Safety Councils has developed to the point where they now promote and share the necessary standards of safety and health in the workplace, teaching both employees and employers the methods which can be used to eliminate potential hazards and accidents in their work environments. As valued guidance, these safe practices are adopted in many diverse industries and businesses across the United States of America, with the Safety Council providing the latest information on regulations and procedures as revisions are implemented.

### 4.5.3 The American Society of Safety Professionals (ASSP) [60]

The American Society of Safety Professionals (ASSP) is an organisation comprising occupational safety and health professionals whose agenda is to prevent workplace fatalities, injuries and illnesses in the United States of America. The role of these professionals can be multifaceted, but specifically includes the managing of occupational safety and health issues found in industrial workplace situations. Much of their remit also involves research and consultant work, where they use risk-based approaches and provide safety and health education in order to reduce incidents occurring in the workplace.

As a collective, the ASSP is made up of various speciality communities, each with a focus on a particular industry or field. These groups encapsulate areas such as engineering, the environment, fire protection, risk management, and transportation. They have direct links with various federal agencies and in particular both the Occupational Safety and Health Administration [61] and the research arm of the National Institute for Occupational Safety and Health [62].

### 4.5.4 The National Institute for Occupational Safety and Health (NIOSH)

The National Institute for Occupational Safety and Health (NIOSH) is the Federal Agency responsible for undertaking research into a variety of work-related injuries and illnesses. Their aim, following these detailed studies, is to make recommendations, and to provide products and services which will effectively prevent injuries and illnesses occurring in the first place. As a Federal Agency, they are part of the Centres for Disease Control and Prevention [63], which itself sits within the Department of Health and Human Services [64]. As a research facility, they cover a diverse range of work-related disciplines, which include such topics as, industrial hygiene, safety, engineering and chemistry.

NIOSH is not a regulatory agency, so they are unable to directly issue safety and health standards which would then be enforceable under law. Instead, they can make informed recommendations which feed into these standards and, through their research programmes, develop the information which ultimately defines safe levels of exposure for things like toxic materials and harmful substances. In addition, they undertake the investigations needed into new safety and health problems as they arise.

### 4.5.5 The National Safety Council (NSC) [65]

The National Safety Council (NSC) is an independent body which promotes overall health and safety in the United States of America. Their focus is on areas where the greatest number of preventable injuries and deaths occur, which includes categories such as domestic and road traffic incidents, along with workplace safety. In addition, they provide first-aid training for safety, compliance and wellness, offering a more detailed Emergency Care Service training programme [66] for denoted rescuer teams [67]. An example of their valued work is where they have recognised the relatively high numbers of fleet driver fatalities, and as a result, have developed strategies for protecting workers while using the road networks. In particular they host seminars where this information is brought to the industries' attention.

The NSC works closely with the Occupational Safety and Health Administration [68] and the Department of Labour [69] to improve workplace safety and reduce injuries and deaths in critical working environments. This is done through the provision of training which is based on best-practices found in the respective industries.

### 4.5.6 The Chemical Safety Board (CSB) [70]

The Chemical Safety Board (CSB) is an independent Federal Agency who have the task of investigating industrial chemical accidents at fixed industrial facilities. Their function is to determine the conditions and circumstances which led to any specific accident occurring, and thereby identify the causes so that similar incidents in the future might be prevented. They are authorised under the 1990 amendment of the Clean Air Act [71], which was established to control air pollution in the United States of America. As such, the CSB have autonomy in their undertakings and cannot be directed by other agencies such as the Environmental Protection Agency [72] and the Occupational Safety and Health Administration [73]. In this way, the identification of chemical hazards not already included by these two agencies would be addressed directly by the CSB.

### 4.5.7 Health protection in the United States of America

Health protection in the United States of America is provided by a variety of national departments who are specialists in their own field. The United States Department of Health and Human Services [74] is the body responsible for the health protection of all citizens of the United States of America. Their primary aim is to enhance and protect the health and wellbeing of all Americans in the home, at work and during recreational activities. The main Division of the Department is the United States Public Health Service [75], who oversee environmental and public health through a further set of Divisions. These include the National Institutes for Health [76], the Centres for Disease Control and Prevention [77], the Agency for Toxic Substances and Disease Registry [78], and the Agency for Healthcare Research and Quality [79].

Principally, there are three separate levels of health protection management in the United States of America; the Federal Health Department [80], the State Health Department [81] and the Local Health Departments [82]. At the top level, the Federal Government provides the guidance and funding for the various regional States to implement Federal led health programmes. Sitting beneath this, the State Health Departments are in turn responsible for establishing State Regulations and Health Policies which not only comply with these programmes but also considers the level of local resources available. More rural States for example would adopt different policies to match their community needs than the more densely populated urban States. The Local Health Departments consequently implement both the Federal and State health policies and the human services provisions, recognising their specific local resource levels and reflecting things like any industry hazards in their particular regions.

### 4.5.8 The America Society of Mechanical Engineers (ASME) [83]

The American Society of Mechanical Engineers (ASME) is a professional association in the United States of America which establishes the codes and standards for a variety of mechanical engineering topics. These cover technical areas such as;

fasteners, elevators, pipelines, power plant systems, and boiler and pressure vessels. The ASME standards are established by various committees of experts in their respective subject areas and many are referenced by the Government Agencies as a means of organisations attaining compliance with Health and Safety Regulations. In the main, the ASME Standards tend to be non-mandatory unless they have been incorporated into a specific Regulation by the Federal, State or Local Government Agencies.

The main ASME Code which would be associated with cryogenic technology is that covering the design and manufacture of pressure vessels. As there are inherently potential dangers and risks associated with this type of equipment, the Federal rules in the United States of America specify that organisations manufacturing or using pressure vessels need to demonstrate compliance to an applicable product standard. By virtue of its qualified standing, the ASME Boiler and Pressure Vessel Code [84] fulfils this requirement. The Code also covers situations where repairs or alterations to existing pressure vessels are undertaken. It specifically defines the guiding rules for the design, fabrication, installation, inspection, maintenance and use of boilers, pressure vessels and nuclear related pressure components. It includes various standards on the materials of construction, welding and brazing procedures, and non-destructive examination of the eventual manufactured pressure vessels.

The ASME Boiler and Pressure vessel Code is a comprehensive document structured to include a variety of pressure related equipment. It is divided into sections, with each covering a different aspect of pressure equipment design, manufacture and testing. Section VIII is the general section covering pressure vessels [85].

- Section I Rules for the construction of power boilers
- Section II Materials
- Section III Rules for the construction of Nuclear facility components
- Section IV Rules for the construction of heating boilers
- Section V Non-destructive examination
- Section VI Rules for the care and operation of heating boilers
- Section VII Guidelines for the care of power boilers
- Section VIII Rules for the construction of pressure vessels
- Section IX Welding, brazing and fusing qualifications
- Section X Fibre-reinforced plastic pressure vessels
- Section XI Rules for the in-service inspection of Nuclear power plant components
- Section XII Rule for the construction and continued service of transport tanks
- Code case Boilers and pressure vessels

In addition, ASME provide Standards for a variety of other engineering related areas which would be applicable for the use of cryogenic technology. These include Codes for overhead and mobile cranes, lifting and rigging equipment, piping and pipelines, valves, flanges, fittings, and gaskets. As a regulatory organisation, they also function as an accreditation body for agencies who provide independent third party verification, testing, certification and inspection of pressure vessels as a means of ensuring compliance with the ASME boiler and pressure vessel code.

### 4.5.9 The National Board of Boiler and Pressure Vessel Inspectors (NBBI) [86]

The National Board of Boiler and Pressure Vessel Inspectors (NBBI) is the professional body which enforces the Pressure equipment laws and regulations in the United States of America. They are represented at City, State and Province levels within the Country. The NBBI was originally created in order to prevent deaths, injury and the destruction of buildings and their infrastructure in situations where pressure equipment was being used. In order to implement this strategy, the NBBI provides qualified inspectors who check that pressure equipment is being properly maintained and used correctly. These commissioned inspectors undertake the task of inspecting pressure equipment to ensure that they retain ASME Code integrity, especially if repairs or alterations have been made to the equipment.

The NBBI also functions as a registering authority for the certification of approved items of pressure equipment. To gain registration, the pressure equipment needs to attain quality criteria which certify that they have been manufactured, tested and are subsequently inspected to the required standards. The pressure items are then stamped with a National Board registration number [87]. This certification is required by most jurisdictions in the United Stated of America and is recognised by the various public safety authorities as a qualification that the pressure equipment has been built and inspected to these defined values. Part of this registration is the creation of a data report by the manufacturer of any particular pressure item. The data report includes; the date of manufacture, materials of construction, design details and certification statements by both the manufacturer and the inspector.

The NBBI also have an administration role in the accreditation of repairs and alterations to pressure equipment. As this type of maintenance is essential for retaining the integrity of pressure items, they are inclusive of all types of changes, from minor weld repairs, to the complete overhaul of items such as safety relief valves. Organisations accredited by the NBBI to undertake this work stamp a further mark on the modified pressure items on completion of the work. This denotes the integrity of the work carried out and that the equipment is safe to use. There are three certificates of authorisation which can be issued by this process;

*R* Certificate of authorisation [88] Issued to accredited organisations who can perform repairs and alterations to pressure-retaining items.

*VR* Certificate of authorisation [89] Issued to accredited organisations who can perform repairs and modifications to pressure relief devices.

*NR* Certificate of authorisation [90] Issued to accredited organisations who can perform repairs and replacement of nuclear power plant pressure components

### 4.5.10 Planning Permission or Development Approval [91]

Planning Permission and Development Approvals are the endorsements required for the construction, or even expansion, as well as demolition, of buildings in many of the States within the United States of America. These are normally provided in the form of a building or construction permit [92]. Many of these permits are also subject to the relevant Building Codes [93], and will also need to pass a Plan Check [94] in order to check compliance with local plans. In the United States of America, the main Building

Codes used for this purpose are the International Building Code [95], the International Residential Code [96], along with the associated electrical, plumbing and mechanical Codes [97]. Planning Permissions are part of the State urban planning process and construction laws. They are generally managed by town planners who work directly for local government agencies.

New constructions would typically require inspection during the construction phase, as well as at project completion. This ensures compliance with national, regional and local Building Codes at all stages of the construction. Failure to obtain the relevant permits can result in fines and penalties. In certain circumstances, the demolition of unauthorised buildings takes place if they do not meet the requirements of the salient Building Codes.

## Relevant cryogenic safety legislation in the United States of America—summary

**Key learning points**

**General**

*The Occupational Safety and Health Act (OSH Act)—1970*—The Federal law controlling health and safety within the private sector and Federal Government organisation

*The Occupational Safety and Health Administration (OSHA)*—Oversee the OSH Act through both Federal legislation and State specific plans

**Code of Federal Regulations**

*The Code of Federal Regulations (CFR)—2019*—The general and permanent rules and regulations published by the Federal Government

*CFR Title 14—Aeronautics and Space*—Covers all aviation activities taking place in the United States of America, including the transport of dangerous substances by air

*CFR Title 16—Commercial Practices*—Are the rules and regulations regarding all commercial practices in the United States of America and includes the strict labelling requirements for the packaging and transport of hazardous substances and mixtures

*CFR Title 23—Highways*—Are the rules and regulations regarding the use of the highways in the United States of America, covering the highways, road-bridge design and motor vehicle safety standards

*CFR Title 29—Labour*—Are the rules and regulations regarding the provision of safe places of employment which are free from hazards which are likely to cause death or serious harm. Part 1910 of CRF Title 29 covers the general occupational safety and health standards

*CRF Title 33—Navigation and navigable waters*—Are the rules and regulations for the use of navigable waters within and around the United States of America, covering areas such as the transport of hazardous material by waterborne vessels. CRF Title 33 is divided into three separate chapters, denoting specific roles of responsibility, namely; The United States Coast Guard, The Army Corps of Engineers and the Saint Lawrence Seaway Development Corporation

*CRF Title 49—Transportation*—Are the rules and regulations for all transportation throughout the United States of America, covering in particular the domestic transportation of hazardous material by any form of transport

**Associated Safety Organisations**

*The American Association of Safety Councils*—An advisory body made up of safety professional, providing educational safety programmes

*The American Society of Safety Professionals (ASSP)*—An organisation comprising occupational safety and health professionals, undertaking research to reduce incidents occurring in the workplace

*The National Institute for Occupational Safety and Health (NIOSH)*—Is a Federal Agency undertaking research into work-related injuries and illnesses

*The National Safety Council (NSC)*—Is an independent body which promotes Health and Safety in the United States of America

*The Chemical Safety Board (CSB)*—Is an independent Federal Agency tasked with investigating industrial chemical accidents at fixed industrial facilities

*The United States Department of Health and Human Services*—Is the body responsible for the health protection of all citizens of the United States of America

*The American Society of Mechanical Engineers (ASME)*—Is a professional Association which establishes codes and standards for a variety of mechanical engineering topics, such as; fasteners, pipelines, and boiler and pressure vessels

*The National Board of Boiler and Pressure Vessel Inspectors (NBBI)*—Is a professional body which enforces the Pressure Equipment Laws and Regulations in the United States of America by providing qualified inspectors to check that pressure equipment is being properly maintained. They also function as a Registration Authority for the certification of approved items of pressure equipment

*Planning Permission and Development Approvals*—Are the endorsements required for the construction, expansion and demolition of buildings in the United States of America

## References

[1] The United States Department for Labor *The Occupational Safety and Health Act*—1970 *(OSH Act)* (www.osha.gov/laws-regs/oshact/completeoshact) (Accessed 2020)

[2] The United States Department for Labor *The Occupational Safety and Health Administration (OSHA)* (www.osha.gov/aboutosha) (Accessed 2020)

[3] USAGOV *Federal Laws and Regulations* (www.usa.gov/laws-and-regs) (Accessed 2020)

[4] The United States Department for Labor—OSHA *State Specific Plans* www.osha.gov/stateplans) (Accessed 2020)

[5] The United States Department for Labor—OSHA *OSHA States* (www.osha.gov/stateplans) (Accessed 2020)

[6] The United States Department for Labor *The Occupational Safety and Health Administration (OSHA)* (www.osha.gov/aboutosha) (Accessed 2020)

[7] The American National Standards Institute (ANSI) *General Public Participation in the American National Standards Process* (www.ansi.org/american-national-standards/info-for-the-general-public/general-public) (Accessed 2020)

[8] National Council for Occupational Safety and Health *The OSHA Inspection: A Step by Step Guide* (www.osha.gov/sites/default/files/2018-12/fy10_sh-20853-10_osha_inspections.pdf) (Accessed 2020)
[9] Impact Safety *Top 5 OSHA Violations and Penalties Affecting Safety* (www.impactsafetyinc.com/top-5-osha-violations-and-penalties-affecting-safety/) (Accessed 2020)
[10] United States Department of Labor *Permissible Exposure Limits (PELs)—Annotated Tables* (www.osha.gov/annotated-pels) (Accessed 2020)
[11] GOVINFO *The Code of Federal Regulations (CFR)*—2019 (www.govinfo.gov/help/cfr) (Accessed 2020)
[12] USAGOV *Branches of the U. S. Government* (www.usa.gov/branches-of-government) (Accessed 2020)
[13] GOVREGS.COM *Code of Federal Regulations—Title 14—Aeronautics and Space* (www.govregs.com/regulations/title14) (Accessed 2020)
[14] The Federal Aviation Administration (FAA) (www.faa.gov/) (Accessed 2020)
[15] Legal Information Institute (LII) *Code of Federal Regulations—Title 16—Commercial Practices* (www.law.cornell.edu/cfr/text/16) (Accessed 2020)
[16] USAGOV *The Federal Trade Commission* (www.usa.gov/federal-agencies/federal-trade-commission) (Accessed 2020)
[17] *The Consumer Product Safety Commission (CPSC)* (www.cpsc.gov/) (Accessed 2020)
[18] GOVREGS *Code of Federal Regulations—Title 23—Highways* (www.govregs.com/regulations/title23) (Accessed 2020)
[19] USAGOV *The Federal Highway Administration (FHWA)* (www.usa.gov/federal-agencies/federal-highway-administration) (Accessed 2020)
[20] History.com *The United States of America National Highway System* (www.history.com/topics/us-states/interstate-highway-system) (Accessed 2020)
[21] *The United States Department of Transportation* (www.transportation.gov/) (Accessed 2020)
[22] U. S. Department of Transportation *Understanding the National Highway Traffic Safety Administration (NHTSA)* (www.transportation.gov/transition/understanding-national-highway-traffic-safety-administration-nhtsa) (Accessed 2020)
[23] UNECE *The United Nations Economic Commission for Europe (UNECE) World Forum for Harmonisation of Vehicle Regulations (WP.*29) (www.unece.org/trans/main/wp29/introduction.html) (Accessed 2020)
[24] GOVREGS *Code of Federal Regulations—Title 29—Labor* (www.govregs.com/regulations/title29) (Accessed 2020)
[25] Bartleby Research *The Role of the Senate and House of Representatives Essay* (www.bartleby.com/essay/The-Role-of-the-Senate-and-House) (Accessed 2020)
[26] United States Department of Labor *The Hazardous Waste Operations and Emergency Response (HAZWOPER) Regulations* (www.osha.gov/SLTC/emergencypreparedness/hazwoper/index.html) (Accessed 2020)
[27] United States Department of Labor *CFR Title 29 Part 1910—Occupational Safety and Health Standards* (www.osha.gov/laws-regs/federalregister/1999-03-22) (Accessed 2020)
[28] United States Department of Labor *Standard Interpretations: Material Safety Data Sheets (MSDSs)* (www.osha.gov/laws-regs/standardinterpretations/1997-11-19-0) (Accessed 2020)
[29] United States Department of Labor *Permit-required Confined Spaces* (www.osha.gov/Publications/osha3138.html) (Accessed 2020)

[30] GOVREGS.COM *Code of Federal Regulations—Title 33—Navigation and Navigable Waters* (www.govregs.com/regulations/title33) (Accessed 2020)

[31] Koordinates.com *The Collision Regulation Boundaries of the United States of America* (https://koordinates.com/layer/20698-us-colregs-demarcation-lines/ (Accessed 2020)

[32] International Maritime Organisation (IMO) *The International Regulations for Preventing Collision at Sea*—1972 *(COLREGS)* (www.imo.org/en/About/Conventions/Pages/COLREG.aspx) (Accessed 2020)

[33] *U.S. Army Corps of Engineers* (www.usace.army.mil/Contact.aspx) (Accessed 2020)

[34] The United States Department of Transportation *Saint Lawrence Seaway Development Corporation* (www.seaway.dot.gov/) (Accessed 2020)

[35] *The Great Lakes Saint Lawrence Seaway System* (https://greatlakes-seaway.com/en/) (Accessed 2020)

[36] *The United States Department of Transportation* (https://transportation.gov/) (Accessed 2020)

[37] The United States Department of Transportation *Saint Lawrence Seaway Development Corporation* (www.seaway.dot.gov/)(Accessed 2020)

[38] U. S. Department of Homeland Security *The United States Coast Guards* (www.uscg.mil/) (Accessed 2020)

[39] *Transport Canada—Government of Canada* (https://tc.canada.ca/en) (Accessed 2020)

[40] GOVREGS *Code of Federal Regulations—Title 49—Transportation* (www.govregs.com/regulations/title49) (Accessed 2020)

[41] *The United States Department of Transportation* (www.transportation.gov/) (Accessed 2020)

[42] The Department of Homeland Security *Homeland Security* (www.dhs.gov/) (Accessed 2020)

[43] USAGOV *The Pipeline and Hazardous Material Safety Administration* (www.usa.gov/federal-agencies/pipeline-and-hazardous-materials-safety-administration) (Accessed 2020)

[44] United States Department of Transportation *The Federal Railroad Administration* (https://railroads.dot.gov/) (Accessed 2020)

[45] The National Transportation Safety Board (www.ntsb.gov/Pages/default.aspx) (Accessed 2020)

[46] USAGOV *The Pipeline and Hazardous Material Safety Administration* (www.usa.gov/federal-agencies/pipeline-and-hazardous-materials-safety-administration) (Accessed 2020)

[47] Legal Information Institute (LII) *Hazmat Hazard Labels* (www.law.cornell.edu/cfr/text/49/part-172) (Accessed 2020)

[48] The U.S. Environmental Protection Agency (EPA) (www.epa.gov/) (Accessed 2020)

[49] The United States Environmental Protection Agency (EPA) *The Hazardous Materials Transportation Act* (https://archive.epa.gov/emergencies/content/lawsregs/web/html/hmta-over.html) (Accessed 2020)

[50] The United States Environmental Protection Agency (EPA) *The Resource Conservation and Recovery Act* (www.epa.gov/laws-regulations/summary-resource-conservation-and-recovery-ac) (Accessed 2020)

[51] *The Consumer Product Safety Commission (CPSC)* (www.cpsc.gov/) (Accessed 2020)

[52] USAGOV *The Federal Trade Commission* (www.usa.gov/federal-agencies/federal-trade-commission) (Accessed 2020)

[53] U. S. Department of Transportation *Understanding the National Highway Traffic Safety Administration (NHTSA)* (www.transportation.gov/transition/understanding-national-highway-traffic-safety-administration-nhtsa) (Accessed 2020)

[54] The United States Consumer Product Safety Commission (CPSC) *The Consumer Product Safety Act (CPSA)* (www.cpsc.gov/Regulations-Laws--Standards/Statutes/Summary-List/Consumer-Product-Safety-Act/) (Accessed 2020)

[55] The United States Consumer Product Safety Commission (CPSC) *The Consumer Product Safety Improvement Act (CPSIA)* (www.cpsc.gov/Regulations-Laws--Standards/Statutes/The-Consumer-Product-Safety-Improvement-Act/) (Accessed 2020)

[56] The United States Department for Labor *The Occupational Safety and Health Act—1970 (OSH Act)* (www.osha.gov/laws-regs/oshact/completeoshact) (Accessed 2020)

[57] The United States Department for Labor *The Occupational Safety and Health Administration (OSHA)* (www.osha.gov/aboutosha) (Accessed 2020)

[58] United States Department for Labor *The OSHA Compliance Safety and Health Officers—Authority for Inspection* (www.osha.gov/laws-regs/regulations/standardnumber/1903/1903.3) (Accessed 2020)

[59] *The American Association of Safety Councils* (https://safetycouncils.org/) (Accessed 2020)

[60] *The American Society of Safety Professionals (ASSP)* (https://assp.org/) (Accessed 2020)

[61] The United States Department for Labor *The Occupational Safety and Health Administration (OSHA)* (www.osha.gov/aboutosha) (Accessed 2020)

[62] Centers for Disease Control and Prevention (CDC) *The National Institute for Occupational Safety and Health (NIOSH)* (www.cdc.gov/niosh/) (Accessed 2020)

[63] *The Centers for Disease Control and Prevention (CDC)* (www.cdc.gov/) (Accessed 2020)

[64] USAGOV *The U.S. Department of Health and Human Services* (www.usa.gov/federal-agencies/u-s-department-of-health-and-human-services) (Accessed 2020)

[65] The National Safety Council (NSC) (www.nsc.org/) (Accessed 2020)

[66] The National Safety Council (NSC) *Emergency Care Service Training Programme* (www.nsc.org/safety-training) (Accessed 2020)

[67] The National Safety Council (NSC) *Workplace First Aid Training* (www.nsc.org/safety-training/first-aid) (Accessed 2020)

[68] The United States Department for Labor *The Occupational Safety and Health Administration (OSHA)* (www.osha.gov/aboutosha) (Accessed 2020)

[69] The U.S. Department of Labor (www.dol.gov/) (Accessed 2020)

[70] The U.S. Chemical Safety and Hazard Investigation Board (CSB) (www.csb.gov/) (Accessed 2020)

[71] United States Environmental Protection Agency (EPA) *The U.S. Clean Air Act* (www.epa.gov/laws-regulations/summary-clean-air-act) (Accessed 2020)

[72] The U.S. Environmental Protection Agency (EPA) (www.epa.gov/) (Accessed 2020)

[73] The United States Department for Labor *The Occupational Safety and Health Administration (OSHA)* (www.osha.gov/aboutosha) (Accessed 2020)

[74] The United States Department of Health and Human Services (HHS) (www.hhs.gov/) (Accessed 2020)

[75] U.S. Department of Health and Human Services (HHS) *The Commissioned Corps of the United States Public Health Service (USPHS)* (www.usphs.gov/) (Accessed 2020)

[76] U.S. Department of Health and Human Services (HHS) *The National Institutes for Health* (www.nih.gov/) (Accessed 2020)

[77] The Centers for Disease Control and Prevention (CDC) (www.cdc.gov/) (Accessed 2020)

[78] U.S. Department of Health and Human Services (HHS) *The Agency for Toxic Substances and Disease Registry (ATSDR)* (www.atsdr.cdc.gov/) (Accessed 2020)

[79] The Department of Health and Human Services *The Agency for Healthcare Research and Quality* (www.ahrq.gov/) (Accessed 2020)

[80] The Federal Health Department *The U.S. Department of Health and Human Services* (www.hhs.gov/) (Accessed 2020)

[81] USAGOV *The State Health Department* (www.usa.gov/state-health) (Accessed 2020)

[82] Wikipedia *The Local Health Departments in the United States* (https://en.wikipedia.org/wiki/Local_health_departments_in_the_United_States) (Accessed 2020)

[83] The America Society of Mechanical Engineers (ASME) (www.asme.org/) (Accessed 2020)

[84] The American Society of Mechanical Engineers (ASME) *The ASME Boiler and Pressure Vessel Code (BPVC)* (www.asme.org/codes-standards/find-codes-standards/bpvc-complete-code-boiler-pressure-vessel-code) (Accessed 2020)

[85] The American Society of Mechanical Engineers (ASME) *The ASME Boiler and Pressure Vessel Code Section VIII* (www.asme.org/codes-standards/find-codes-standards/bpvc-viii-1-bpvc-section-viii-rules-construction-pressure-vessels-division-1) (Accessed 2020)

[86] The National Board of Boiler and Pressure Vessel Inspectors (NBBI) (www.nationalboard.org/) (Accessed 2020)

[87] The National Board of Boiler and Pressure Vessel Inspectors (NBBI) *National Board Registration Numbers* (www.nationalboard.org/SiteDocuments/Registration/NB-264.pdf) (Accessed 2020)

[88] The National Board of Boiler and Pressure Vessel Inspectors (NBBI) *R Certificate of Authorisation* (www.nationalboard.org/index.aspx?pageID=115&ID=160) (Accessed 2020)

[89] The National Board of Boiler and Pressure Vessel Inspectors (NBBI) *VR Certificate of Authorisation* (www.nationalboard.org/index.aspx?pageID=115&ID=161) (Accessed 2020)

[90] The National Board of Boiler and Pressure Vessel Inspectors (NBBI) *NR Certificate of Authorisation* (www.nationalboard.org/Index.aspx?pageID=115&ID=162) (Accessed 2020)

[91] *Planning Permission or Development Approval—Planning Regulations in the USA and Their Implications on Urban Design in the Central City Zone* (www.doiserbia.nb.rs/img/doi/0354–4605/2011/0354–46051102289D.pdf) (Accessed 2020)

[92] Ycharts *Building or Construction Permits* (https://ycharts.com/indicators/us_building_permits) (Accessed 2020)

[93] Buildingsguide *The U.S. Building Codes* (www.buildingsguide.com/blog/resources-building-codes-state/) (Accessed 2020)

[94] LA Department of Building and Safety *Plan Check* (https://ladbs.org/services/core-services/plan-check-permit) (Accessed 2020)

[95] Wikipedia *The International Building Code* (https://en.wikipedia.org/wiki/International_Building_Code) (Accessed 2020)

[96] The International Code Council (ICC) *The International Residential Code* (www.iccsafe.org/products-and-services/i-codes/2018-i-codes/irc/) (Accessed 2020)

[97] Maryland Department of Labor *Electrical, Plumbing and Mechanical Codes* (www.dllr.state.md.us/labor/build/buildmech.shtml) (Accessed 2020)

**IOP** Publishing

The Safe Use of Cryogenic Technologies
A handbook for best practice and training
**Robert Done**

# Chapter 5

# Relevant cryogenic safety legislation in the European Union

## 5.1 The European Union safety legislation

Article 153 of the Treaty on the Functioning of the European Union [1] provided the European Union with the authority to adopt Directives related to workplace Safety and Health. At the topmost level, the European Framework Directive on Safety and Health at Work (Directive 89/391 EEC) [2] covers a wide scope of applications with the aim of improving safety and health at work. By description, it guarantees a minimum level of safety and health requirements throughout the Member States [3] for all work activities undertaken in Europe. In addition, the individual Member States, under the remit of Directive 89/391 EEC, are permitted to adopt more stringent measures as they require. The resultant effect of this process is that European laws are unequivocally harmonised at the same level for all Member States.

Within this functioning framework Directive, additional Directives focus on more specific aspects of safety and health at work, which taken together, form the basis of European safety and health legislation. These European Union Directives are legally binding and have to be adopted as national laws by all the Member States. In many cases, these individual Directives contain more rigorous provisions which override those of the Framework Directive. Member States have to convert these European Union Directives into their own national laws, but in doing so, they are allowed to implement stricter rules in order to further protect their employees in the workplace. It is common therefore for legislative requirements encompassing safety and health at work to vary significantly between European Union Member States. An example of this would be where, as the architects of these Directives, the European Commission [4] is empowered with the design of product safety requirements. It is however the individual national authorities who undertake the inspection of these products and check that compliance with the Directives has been achieved.

 

The individual European Union Directives detail the minimum requirements and fundamental safety and health principles of their specific topic heading. These cover;

- *Specific tasks*—such as the manual handling of loads or the use of work equipment.
- *Specific hazards*—such as the exposure to dangerous substances or the use of pressurised equipment.
- *Specific workplaces and sectors*—such as laboratory facilities, temporary work sites or fishing vessels.
- *Specific groups of workers*—such as pregnant woman, young workers or workers with a fixed-duration employment contract.
- *Certain work-related aspects*—such as the organisation of working time or lone working.

These individual Directives set out risk preventative measures, along with details of how risk assessments are to be undertaken where specific hazards have been identified. They also outline the responsibilities which both employers and employees have in the application of the Directives, and in some cases, detail limiting values for things like certain hazardous substances or agents. To assist with this process, a series of European Guidelines [5] have been written which support the implementation of the Directives and any associated European Standards [6].

A further aspect of the conformity process, openly linked to these Directives, is the application of a CE mark [7] on products which are to be traded on the Single Market in the European Economic Area [8]. By the authorised application of the CE mark, it demonstrates that the manufacturer of the item has checked that it meets the European Union safety, health or environmental requirements and that it complies with all relevant European Union legislation. This in turn allows the free movement of CE marked products within the European market.

## 5.2 Directive 2009/104/EC—Use of Work Equipment [9]

The European Union Directive associated with the Use of Work Equipment defines the minimum safety and health requirements for the use of items of work equipment by employees at their place of work. It outlines the measures which employers need to implement to ensure the safety of all work equipment when it is made available to their workers. In particular, they need to consider the existence of any specific working conditions which may increase the risks to the safety and health of their workers when using particular work equipment. If any such risks cannot be eliminated, the employers needs to take appropriate measure to minimise their consequences. For the use of work equipment related to cryogenic technology, this would include the extremely low temperature regime and the fact that cryogen vapours may be present throughout the period of use of the work equipment.

The definition of work equipment within the text of the Directive includes any machine, apparatus, tool or installation which is used at work. The usage of this equipment can involve any activity, such as; starting or stopping the equipment, its principle use, transport, repair, modification, maintenance and servicing. The Directive also defines a so-called danger zone, which is the area within which an exposed worker is at a greater risk to their health and safety when they are using the work equipment. The Directive covers a range of general purpose work equipment, but also includes specifics, which include the use of mobile equipment, such as fork-lift trucks and the general provisions required for their use. It also covers work equipment used for lifting loads, which includes certain aspect of cranes and lifting attachments, but also takes account of personnel lifts. Equally so, the use of ladders and the specific provisions for the use of scaffolding are addressed in the body of the Directive.

For work equipment specifically employed for the lifting of loads, they must be used in such a way as to ensure the equipment remains stable at all times. The Directive provides guidance to make sure that personnel are not underneath any suspended loads during the use of lifting equipment such as cranes. Any lifting accessories selected as an integral part of the lifting operation must be appropriate for the load being handled and must be marked with its characteristics, such as the maximum load capacity of the part. Storage of lifting accessories should guarantee that they are never damaged or are able to degrade when not in use.

Any work equipment being used must comply with the minimum requirements of the relevant European Union Directives in addition to the Use of Work Equipment Directive, and must be installed correctly. Furthermore, the work equipment must continue to remain compliant with these Directives throughout its subsequent usage, with the provision of adequate maintenance and repair if necessary. In many cases, this process would be further supported by periodic inspections and testing of specific types of work equipment. These inspections have to be undertaken by a competent person as defined within the Directive.

In certain situations, specific risks to employees from the use of work equipment may be such that restricted access to the work equipment is essential. This would be particular to situations where ergonomics and occupational health aspects, such as hearing loss risks, would be manifest. Employers have to take full account of these requirements and must provide their employees with written instructions on how to safely use these particular types of work equipment, along with providing applicable training. Similarly, employees must familiarise themselves with any dangers relevant to the work equipment they are being asked to use.

To this end, and where appropriate, work equipment must be fitted with a suitable emergency stop device which can be operated to prevent dangerous use of the equipment from taking place. In addition, where work equipment is being used which results in the production of hazards, such as the emission of gases, vapours, liquids or dusts, they must be fitted with containment or extraction devices close to the source of the hazard. Where a risk due to the contact with moving parts of the work equipment exists, these must have guards or other devices fitted to prevent access to the danger zones. These guards and protection devices must themselves be of a robust construction, so they themselves do not pose any additional hazard.

Directive 2009/104/EC also defines the safety and health requirements of work equipment used to facilitate work at height. Unlike other Directives, it specifies that collective protection measures for work at height must be given priority over personal protection measures and that whatever measures are employed must not themselves provide additional risks of falling from height. Examples for this would include the use of handrails on scaffolding or fall arrest equipment anchored to a common lanyard when working on roof areas. For work at height undertaken outdoors, all prevailing weather conditions need to be considered for the safety and health of any employees undertaking the work before the work commences. Provision must be made in all work at height situations to prevent, or at least arrest falls, from height.

Included in the Directive is the use of ladders for short-duration, low risk work at height. Specific provisions apply, such as the safe positioning of ladders so that they cannot be displaced during use. Many accidents relating to the use of ladders occur due to the feet of the ladders slipping during use, and therefore the Directive details the use of anti-slip devices or the securing of the uprights to prevent this happening.

For longer-term, higher risk work at height activities, the use of scaffolding is the preferred method. Here however, the structural stability of any erected scaffold assembly needs to be paramount and there are recognised standard configurations outlined in the Directive to enable this. It is important to also consider the initial build and dismantling phase of any scaffold structure, as these can generate additional risks not directly linked to the actual work at height task being undertaken. Any part-built scaffolding must therefore be clearly marked with warning signs and the danger area around it suitably delimited to prevent casual access to it. The specific size of scaffolding decks must be appropriate to the work being performed at height, as well as being suitable for the loads expected to be carried by the structure. This is important, as situations may arise where something like a piece of heavy plant equipment being installed at height may need to be temporarily supported on the scaffold before final installation takes place. By definition within the Directive, only competent persons can assemble and dismantle scaffolding, with this competency being generated by specific training for this type of activity.

Any European Union organisations supplying cryogenic equipment must by definition of the Directive, comply with its key values. All cryogenic work equipment must be intrinsically safe to use and must have the necessary control measures in place, which effectively reduce any risks associated with their operation. Any work equipment parts specifically operating at low temperatures must in particular be protected in such a way to prevent the risk of users of the equipment coming into contact with these extremely cold items.

## 5.3 Directive 2014/68/EU—Pressure Equipment [10]

The European Union Directive 2014/68/EU, implements the harmonisation of Member State laws relating to pressure equipment which are manufactured and made available through the European market. They allow the free movement of all applicable pressure equipment and assemblies within the market, and through the common ground of the Directive, fulfil all Member State safety and health protection requirements for such equipment. This guarantees that all pressure equipment complying with the Directive will not be a safety and health risk while they are being installed, used or maintained. To achieve this, the equipment must meet the safety requirements of the Directive which specifically covers their design, manufacture and eventual testing. They will also need to satisfy the appropriate conformity assessment procedures in order to qualify for a CE mark identifier.

The Directive applies to all pressure equipment and assemblies with an allowable pressure above 0.5 bar. This includes equipment, such as; containment vessels, piping, safety accessories and pressure accessories. It outlines a set of essential requirements which any piece of pressure equipment must satisfy at the time it is manufactured. All pressure equipment must be conformity assessed against these essential requirements before they can be marketed. These requirements are divided into general and specific and will depend on the category of the pressure equipment. Certain types of pressure equipment, for example, those in categories III and IV, need to have a final assessment, which includes the prerequisite of a full inspection and a proof test.

All pressure equipment covered by the Directive are attributed a category number which is based on the level of hazard associated with their operational use. This level of hazard is defined by the operating pressure of the equipment, the volumetric capacity of the equipment and the further consideration of whether the fluid contained in the equipment is considered dangerous or not. Figure 5.1 demonstrates how individual pieces of pressure equipment are initially defined in a particular fluid group and then subsequently referenced to a series of charts within the Directive. The categories themselves range from Sound Engineering Practice (SEP) through to category IV. The higher the category the more stringent the conformity assessment is, much of which has to be verified by an independent, notified body, especially for higher category equipment. As such, once registered in one category, pressure equipment cannot be upgraded to a higher category, as the supporting conformity assessment cannot be retrospectively produced, but must already exist when the item was originally manufactured and tested.

The compliance responsibility associated with any piece of pressure equipment is directly with the manufacturers, importers and distributors. All manufacturing information must be supplied with the equipment to qualify compliance with the Directive. For example, the manufacturer must qualify that the pressure equipment has been made from eligible safe materials as defined within the Directive. Member

| | Vessels | | | | Steam generators | Piping | | | |
|---|---|---|---|---|---|---|---|---|---|
| State of contents | Gas | | Liquid | | | Gas | | Liquid | |
| Fluid group | 1 | 2 | 1 | 2 | | 1 | 2 | 1 | 2 |
| Chart | 1 | 2 | 3 | 4 | 5 | 6 | 7 | 8 | 9 |

**Figure 5.1.** Showing the pressure equipment classifications.

States have so-called 'supervision rights' [11] which allows them to withdraw from the market any imported pressure equipment considered unsafe. They are also given the responsibility to authorise user inspectorates who will undertake certain conformity assessment tasks, such as third-party checking of the appropriate manufacturing conformity assessment modules. These user inspectorates will often also function as the notified body, empowered with the final inspection and testing of the pressure equipment.

## 5.4 Directive 2014/29/EU—Simple Pressure Vessels [12]

Running parallel with Directive 2014/68/EU, which applies to all types of category level Pressure Equipment, Directive 2014/29/EU relates more explicitly to the marketing and putting into service of simple pressure vessels. It establishes the safety standards and essential safety requirements which need to be in place in order to provide the health and safety protection of anyone in the immediate vicinity of the simple pressure vessel, due to any serious hazards resulting from the leakage or bursting of them.

In definition, simple pressure vessels are containment vessels which have been manufactured as a batch and have a working pressure greater than 0.5 bar. They will also contain either air or nitrogen, and are classed as unfired. Pressure vessels of this type are not intended to be fired, so they are not open to burners and combustion gases and are therefore not prone to overheating as part of their use. In addition, included in Directive 2014/29/EU are the parts and assemblies which contribute to the strength of these vessels, which are specifically made of steel or aluminium. Pressure vessels designed for nuclear use, fire extinguishers and those installed on ships and aircraft are specifically excluded from the Directive, but will sit within their own detailed Directives associated with their relevant technology.

Traceability of simple pressure vessels throughout the whole supply chain is a key principle of the Directive and this is managed through conformity assessments, which includes items such as; material sources, manufacture, inspection and testing. The Directive obliges the Member States to establish the relevant competent authorities who can both facilitate these notification procedures and act as a conformity assessment body. For this purpose, the Directive references Regulation (EC) No. 765/2008 [13], which outlines the accreditation requirements of these conformity assessment bodies, along with the overall principles of CE marking requirements associated with the simple pressure vessels [14].

The level of conformity assessments applied to each individual pressure vessel is based on its operating pressure and volume, with procedures ranging from the least to the more stringent as the pressures and volumes increase. These conformity assessments are directly related to the level of risk associated with the use of the vessels and the level of safety required for their marketing and operation. Vessels where the product of the operating pressure and volume exceeds 50 barL have to comply with the essential safety requirements outlined in the Directive. In comparison, those vessels where this product is less than 50 barL, need only to be designed and manufactured according to Sound Engineering Practice. By association, manufacturers, importers and distributors of simple pressure vessels are obliged to conform to all notification process requirements within the Directive.

## 5.5 Directive 2010/35/EU—Transportable Pressure Equipment [15]

The European Union Directive 2010/35/EU—Transportable Pressure Equipment, supports both the Pressure Equipment Directive 2014/29/EU and the Simple Pressure Vessel Directive 2014/68/EU, in detailing the associated rules for the specific category of Transportable Pressure Equipment. The type of equipment this covers includes all pressure containment receptacles and any associated valves and pressure accessories used for the inland transport of dangerous goods as defined in Directive 2008/68/EC [16]. This also includes gas cartridges, open cryogenic receptacles, gas cylinders for breathing apparatus and fire extinguishers. The main aim of Directive 2010/35/EU is to improve the transport safety of Transportable Pressure Equipment and at the same time allowing the free movement of this type of equipment throughout the European Union market. In doing so, it provides for the national transportation of goods in Transportable Pressure Equipment by road, rail and inland waterways throughout the European Union.

To implement the Directive, a set of conformity requirements, along with conformity assessments and their procedures, are detailed. In addition, obligations are stipulated which owners and operators of transportable pressure equipment need to comply with in order to ensure the safe transport of the equipment. An essential part of the overall process is the conformity assessments which provide qualifying evidence that the transportable pressure equipment is safe. These in turn, would be supported by periodic inspections of the equipment to ensure continued compliance, once they have been initially commissioned.

Transportable pressure equipment which has achieved the necessary conformity requirements and have passed all the compulsory tests and inspections are identified with a European Union ‘Pi’ marking [17], denoting that this compliance has been gained. This is a recognisable symbol, adopted by all Member States and ensures the free movement and use of the equipment across national boundaries. To achieve this ‘Pi’ marking, manufacturers of the equipment have to demonstrate that it has been designed, manufactured and suitably documented in accordance with the conformity requirements of the Directive. This evidence requires validation by notified bodies, who are appointed by individual Member States and work to a common framework of compliance criteria.

The Directive also sets out the requirements for the defined authorising organisations who are responsible for the assessment, notification and monitoring of notified bodies. These notified bodies in turn, have the duty of endorsing the conformity requirements which manufacturers and operators of the equipment have to put in place. As this particular Directive is a European Union ordinance, any certificates of conformity assessments, reports of periodic inspections, intermediate inspections and exceptional checks issued by notified bodies are valid in all Member States. However, Member States retain the right to prohibit the use of any transportable pressure equipment, even if they conform to the Directive, which they consider to still pose a risk to safety under specified conditions. This clause forms part of the so-called safeguarding procedures as described in the Directive, which ensures only safe equipment is ever used within the European Union affiliation.

## 5.6 Directive 2008/68/EC—Inland Transport of Dangerous Goods [18]

The transport of dangerous goods throughout the world is governed by agreements made between the relevant international bodies. In turn, the European Union has adopted and integrated these rules into specific Directives which applies to the transport of dangerous goods within, and across the European Union national borders. Directive 2008/68/EC—Inland Transport of Dangerous Goods, encapsulates all aspects of the inland transport of dangerous goods by; road, rail and inland waterways within the European Union Member States, through the use of; vehicles, wagons and inland waterway vessels. This includes any associated activities, such as the loading and unloading of dangerous goods, or the transfer of these goods from one mode of transport onto another. Any transport of dangerous goods between Member States must comply with the requirements of the Directive, although these rules specifically do not apply to the movement of dangerous goods by the armed forces, by seagoing vessels on maritime waterways, or by ferries crossing inland waterways.

Directive 2008/68/EC recognises that the transport of dangerous goods by any means presents a considerable risk of accidents occurring. To this end, measures are detailed in the Directive, which if correctly applied, ensures that the transport will be carried out as safely as possible. In addition, most Member States work to European Agreements for the transport of dangerous goods, which include; the International Carriage of Dangerous Goods by Road (ADR) [19], the International Carriage of Dangerous Goods by Rail (RID) [20] and the International Carriage of Dangerous Goods by Inland Waterways (ADN) [21]. These stipulate the rules for the safe international transport of dangerous goods which can be subsequently adopted for any form of transport at a national level as well. Where Member States do not have a particular mode of transport as defined in the Directive, they are exempt from the implementation of those detailed rules. An example of this would be where a Member State does not currently have a railway network, or as in the United Kingdom, where their inland waterways are not linked to any other European Union waterways.

The Directive functions with a clause mechanism which allows Member States to notify the European Union Commission if there are concerns with the safety provisions applied to any transport operations entering their territories. If these safety provisions are considered insufficient to limit the hazards involved in the movement of any dangerous goods, the individual Member States can propose additional measures which would effectively reduce the associated risks. Where the quantities of certain dangerous goods are relatively small, or where the transport of dangerous goods is over a very short distance, Member States can apply for an exemption to the provisions within the Directive.

Furthermore, the European Union laws distinguish a clear divide between the risks relating to dangerous goods and those for hazardous materials. The specific laws relating to dangerous goods are mainly focused on the transport of those goods, along with any interim storage which may be required during this transport process. In contrast, the laws covering hazardous materials are mainly concerned with the requirements for their long-term storage, such as in warehouses, along with the actual proposed usage of the hazardous materials once they have been held in storage. Appreciating the distinction is critical, as discretely different Directives apply for the two separate situations.

Whatever the method of transportation of dangerous goods, all consignments must be accompanied by a transport document which defines what the goods are and details their associated nature. Each mode of transport has their own specific set of requirements and the documentation must be in accordance with the relevant transport method. Where goods are being transported by multiple methods during the course of their journey, such as marine vessel to rail and then eventually on to road, transport documents must be prepared for each separate method. Preparation of these documents is the responsibility of the organisation sending the goods out for shipment. To transport dangerous goods by air, a dedicated air transport document must be used. Examples of this would be the International Air Transport Association (IATA) Shipper's Declaration of Dangerous Goods [22].

All dangerous goods being transported are required by the International laws to be suitably labelled. Suppliers of the goods have to mark them with the appropriate hazard warning signs and provide safety advice on any external packaging. To facilitate this, a range of internationally recognised hazard symbols have been developed so that anyone, wherever in the world, should be able to instantly recognise the type of hazard the dangerous goods present. Manufacturers of the goods must also provide instructions for the use of the products, either on the label or as a separate leaflet. For any dangerous substances which are intended for use in the workplace, suppliers must also provide Material Safety Data Sheets (MSDS) [23] which provide information relating to the occupational safety and health for their use. These would typically include such things as; instructions for their safe use, any potential hazards they pose and spill-handling procedures. For the transport of hazardous chemicals, Directive 1272/2008—Classification, Labelling and Packaging of Substances and Mixtures [24], explains the principles which have to be adopted, along with the documentation requirements to achieve compliance. For products moving into the European Union from external countries, there is often a requirement for relabelling to be made in order for them to comply with the European Union requirements.

As with the labelling requirements, any trade in dangerous goods must comply with the packaging requirements contained in the relevant standards for a particular mode of transport, in order for them to be moved safely throughout the European Union. All packaging has to be designed and constructed to UN specification standards. In order to qualify for this standard, the packaging must undergo a series of practical, transport related tests, such as being dropped or being splashed by water. More importantly however, the packaging must also meet the demands and effects of the substance it is to contain. Packaging must be certified by national competent authorities established by the individual Member States. All UN approved packaging is marked with a prefix 'UN' and followed by specific codes which are listed in the relevant regulations governing the international carriage of dangerous goods by road, rail, air and sea. The specific codes include:

*Type of container* Drum/wooden barrel/box/bag

*Container material* Steel/aluminium/wood/plastic/glass

*Packaging group—Danger* X—highest degree/Y—Medium degree/Z—Lowest degree

*Mass* Gross mass for solids/Specific gravity for liquids

*Hydrostatic test pressure* Solids/Liquids

Different authorities in each of the Member States are responsible for enforcing the various Directives for the transport of dangerous goods by road, air, rail and sea. For international movement of dangerous goods, each national authority along the route may well enforce their own specific requirements. Any such route must therefore be assessed in advance to ensure compliance is achieved as the respective national borders are traversed.

### 5.6.1 Transport of dangerous goods by road

In addition to the main Directive, Directive 95/50/EC—Checks on road transport of dangerous goods [25], stipulates a standard list of points which organisations undertaking the transport of dangerous goods by road may be checked against to confirm compliance with the main Directive. These checks can be carried out by the individual Member States on any vehicles travelling in their territory or entering it from a third country. In order to check compliance with the international laws governing the transport of dangerous goods by road, a representative proportion of consignments would be checked against these standard points. Each Member State has to submit an annual report to the European Union Commission relating to these checks, outlining the number of checks which have been carried out, the types of infringements recorded and the number of penalties imposed.

Supporting this process, a set of Industry Guidelines for the secure road transport of dangerous goods have been developed by the various international committees in conjunction with the Agreement concerning the International carriage of dangerous goods by road (ADR) [26]. The Guidelines reinforce the need for transport security and vigilance whenever dangerous goods are being moved. They provide organisations undertaking the transport of dangerous goods by road, with a set of options on how they can essentially meet the requirements of the ADR. However, these options are not necessarily binding requirements of each Member State's national legislation and tend to be used as a basis for further consultation between the authorising authorities and the respective industries. The Guidelines include aspects such as, the securing of temporary storage areas during transit and the technical options for preventing the theft of vehicles or loads during transport operations.

As such, overriding security measures should be an integral part of the safety and quality management system of any company involved in the transport of dangerous goods.

The general requirements of ADR are mandatory, but the specific ways in which they are addressed will depend on the individual circumstances of things like the transport chain, the risks relating to the dangerous goods being transported and the possible outcomes of potential incidents. The classification of dangerous goods however is controlled by international standards for any organisation involved in not only the transporting of dangerous goods, but also those involved in the processing and packaging of these goods. Under the remit of Directive 2008/68/EC—Inland transport of dangerous goods, it is therefore important that the dangerous goods are classified correctly as this informs anyone belonging to the supply chain, along with the emergency services, the exact information of what the hazard is. To implement this, dangerous goods are defined within certain classes depending on their predominant hazard and they must be packaged and transported according to the requirements of that classification. The packing group in turn further classifies the level of danger associated with the dangerous goods and will determine the packaging material requirements and the identification marks and labels they must display.

Running parallel with the actual transport of dangerous goods, the training and qualification requirements for drivers and safety advisors [27] are defined within the Directive. Any organisation handling, processing or transporting dangerous goods on a routine basis must appoint a DGSA. Their role is to monitor compliance with the general rules governing the transport of dangerous goods and to ensure that an organisation's procedures and safety measures are sufficiently robust. They must obtain a vocational training certificate and pass a written exam in order to act as a DGSA. This training is normally organised by independent providers or trade associations for the various methods of transport.

### 5.6.2 Transport of dangerous goods by rail

The carriage of dangerous goods by rail is controlled by the Convention Covering International Carriage of Dangerous Goods by Rail (RID) [28] and implemented through Directive 2008/68/EC covering the inland transport of dangerous goods. However, where a European Union Member State has no railway network, they are exempt from the obligations within the Directive. As can happen with many of these Directives, where existing equipment and infrastructure is in place prior to the inception of the Directive, concessions are applied which allows these items to be authorised for use, providing they were compliant to the appropriate national standards applicable during that period. This would cover such things as existing railway wagons and tank wagons, the proviso being that they subsequently have to be maintained to the required levels of safety described in the Directive.

In addition, where the ambient temperature conditions in certain Member States are regularly below −20 °C, more stringent standards are required regarding the operating temperature of materials used for the transport of dangerous goods by rail. The Directive outlines reference temperatures for certain materials which may typically be used for things like packaging and in the construction of containment tanks. Consideration needs to also be made by any rail operator where the movement of dangerous goods is into, or through, territories where these common low temperatures are experienced. The Directive requires the implementation of additional provisions based on defined climatic zones incorporated into the body of the Directive. Likewise, each Member State is allowed to maintain their existing national provision, as well as the minimum requirements set out in the Directive, for the specific transport of liquefied gases.

For the transport of dangerous goods by rail via the Channel Tunnel between the United Kingdom and France, more stringent provisions are defined within the Directive due to the increased risks associated with the movement of these goods through the confines of the tunnels. Albeit the Channel Tunnel has been designed in such a way that there is a level of assurance associated with its safe use, an additional dangerous goods policy has been adopted which further ensures the safety of customers and staff using the facility, along with the integrity of the Tunnels. The policy has been produced in collaboration with industry, trade and professional bodies. The normal International Carriage of Dangerous Goods by Rail (RID) rules still apply and any freight being transported will only be accepted on the basis of these defining rules. The additional Channel Tunnel Dangerous Goods Policy [29] is decreed as a By-Law in the United Kingdom and as an Arrêté Ministériel [30] in France and forms part of the Channel Tunnel Conditions of Carriage [31] which outlines more stringent measures than RID to reflect the very specific characteristics of the Channel Tunnel construction. All carriers are responsible for ensuring that any dangerous goods being transported comply with all the necessary Directives, Regulations and Policy.

### 5.6.3 Transport of dangerous goods by inland waterways

The carriage of dangerous goods by inland waterways is controlled by the Convention Covering the International Carriage of Dangerous Goods by Inland Waterways (ADN) [32] and implemented through Directive 2008/68/EC covering the inland transport of dangerous goods. However, where a European Union Member State has no Inland Waterways, they are exempt from the obligations within the Directive. Where Inland Waterways of Member States, such as the United Kingdom, are not directly connected to the European Union Inland Waterways, the Directive only applies for the training and examination system for associated safety advisors.

The requirements and procedures defined within the Directive cover key topic areas such as the necessity for inspections of transport vessels and containers, the certificates of approval required before transport can be undertaken and the monitoring of both equipment used on the inland waterways, and the implementation of the certificates. Where inland waterways are linked to the sea by maritime waterways and used by seagoing vessels, additional requirements are applied as though the transport of dangerous goods were in fact undertaken by sea. The ADN also allows for existing certificates of approval, which were put in place prior to the

Directive coming into force under the requirements of existing Member State Regulations, to remain valid. Example of these national Regulations which have been incorporated into the Directive are the Regulations for the Carriage of Dangerous Goods on the Rhine (ADNR) [33] and the Regulations for the Carriage of Dangerous Goods on the Danube (AND-D) [34].

Recognising that the Directive over time may well require amendments to its content, a safety committee exists as part of the Economic Commission for Europe [35]. Their task is to review the requirements of the Directive, in particular, with regard to the safety of navigation along the European Union inland waterways. Their remit includes regulating; the construction of inland waterway vessels, the equipment used in the transport of dangerous goods and the competency and training of the crews employed on the vessels used.

### 5.6.4 Transport of dangerous goods by sea

The carriage of dangerous goods by sea for Member States in the European Union is controlled by the Convention Covering the International Maritime Dangerous Goods Code (IMDG) [36]. Its provision is chiefly defined in order to prevent injuries occurring or damage to be sustained to ships and their cargoes. Where the dangerous goods represents a potential marine pollutant, the Code also sets out the minimum requirements in order to prevent harm to the marine environment. The Code includes not only the transporting of dangerous goods on journeys across international seas, but also ferry services between nations.

The IMDG Code details the types of dangerous cargoes which can be transported by sea. These include solids, liquids as well as gaseous substances. Additionally, explosives, flammables, oxidising substances and radioactive materials are also covered by the Code. The Code defines the packaging or containment requirements for each of these substances, and outlines the technical information which must accompany any such cargo. The guidance provided in the IMDG Code has been internationally agreed in order to provide the safe transport of dangerous goods by sea, typically in freight containers and tank containers.

Many dangerous goods being transported by sea will inevitably, on reaching their destination, also require movement by road, rail or inland waterways as well. In these situation, where full compliance with the IMDG Code is implemented, the labelling of dangerous goods will be as per the IMDG Code requirements, overriding the labelling requirements of the International Carriage of Dangerous Goods by Road, (ADR) [37], the International Carriage of Dangerous Goods by Rail (RID) [38] and the International Carriage of Dangerous Goods by Inland Waterways (ADN) [39]. All training of crews and firefighting equipment requirements are also detailed in the Code.

Individual Member States are responsible for implementing the IMDG Code under the format of their own national legislation, along with establishing an enforcement agency whose role it is to regulated the correct execution of the Code. This is a mandatory obligation for those members of the United Nations Government [40], under both the International Convention for the Safety of Life at Sea (SOLAS) [41] and the International Convention for the Prevention of Pollution from Ships (MARPOL 73/78) [42]. Their combined remit covers not only mariners directly involved in the transport of hazardous materials, but also any related industries and services connected to the shipping industry and is endorsed by the International Maritime Organisation (IMO) [43].

As with similar Regulations and Codes, in order to avoid hazardous situations arising during the transport of dangerous goods by sea, a set of classifications for the individual goods have been established. These classifications link directly to the control measures which need to be implemented for the safe movement of the goods by sea, with the level of associated danger recognised by the discrete classification. These are:

- *Classification 1* High-risk & Low-risk explosives
- *Classification 2* Highly inflammable & non-inflammable gases & non-toxic gases
- *Classification 3* Liquids
- *Classification 4* Highly combustible solids & self-reacting solids & solids which when they interact with water could emit toxic gases
- *Classification 5* Oxidising substances
- *Classification 6* Toxic & infective substances
- *Classification 7* Radioactive materials
- *Classification 8* Material which may suffer corrosion & erosion
- *Classification 9* All remaining dangerous goods

As part of the implementation of the Code, all crew members engaged on ships involved in the transport of dangerous goods have to undergo dangerous goods training so that they are able to classify them correctly and provide the precise packaging where necessary. As marking and labels play an important role in the identification of the pertinent hazards, this training needs to provide an understanding of what marking and labelling particular dangerous goods need to have in place. Finally, the training must also include the safe practices associated with the loading and unloading of cargo containing any of the IMDG products.

### 5.6.5 Transport of dangerous goods by air

The carriage of dangerous goods by air for Member States in the European Union is controlled at a governmental level by the International Civil Aviation Organisation's (ICAO) Technical Instructions [44]. These inform the Dangerous Goods Regulations which are published by the International Air Transport Association (IATA) [45] and are applied at an aviation industry level. In combination, these two conventions detail the instructions required for the safe international transport of dangerous goods by air. In addition, the European Union rules regarding the transport of dangerous goods by air are defined in the Commission Regulation 965/2012 [46] which sets out the technical requirements and administrative procedures relating to all air operations. This Regulation is overseen by the European Union Aviation Safety Agency (EASA) [47] and covers commercial air transport operations with aeroplanes and helicopters, along with non-commercial operations with aeroplanes, helicopters, balloons and sailplanes.

Providing certain compliances are met, the majority of dangerous goods can be carried safely by air transport without placing the aircraft or its occupants at risk. The fundamental principle contained in both the Technical Instructions and the Regulations is that should an incident occur involving dangerous goods, that this alone cannot lead to an accident happening on-board an aircraft. As such, dangerous goods training is therefore mandatory for anyone involved in the transport of dangerous goods by air, from the actually flight crew, all the way to the ground crews loading and unloading the aircraft.

Any aircraft designed for the transport of dangerous goods must have an air operator certificate [48] in place, which are issued by the authorising agencies in the respective Member States. In addition, specific limitations would be place on the aircraft operators, such as flight and duty times for the transport of dangerous goods. It is the responsibility of the Member States to also conduct investigations and inspections of any aircraft being used for this purpose. Some specific airlines and countries are allowed a relaxing of the rules under special circumstances, these are often known as State and Operator Variations [49].

## 5.7 Directive 2006/42/EC—New Machinery Directive [50]

Directive 2006/42/EC—New Machinery Directive, applies to all machinery sold within the European Union free market and is concerned with protecting workers and consumers when using this machinery. It details the essential health and safety requirements for all general applications, with additional specific requirements for certain categories of machinery. The Directive applies to machinery, as well as interchangeable equipment on assemblies, safety components, lifting accessories, chains, ropes and webbing, and removable mechanical transmission devices.

Member States have an obligation under the Directive to put in place measures to ensure that any machinery they produce satisfies the relevant safety requirements of the Directive. To apply this, manufactures of machinery have to undertake a risk assessment of the equipment to determine the necessary health and safety requirements and risk reductions, so that the machinery can be designed and constructed in compliance with the Directive. Having done this, operational limits of the machinery can be defined, clearly stating their intended use and any remaining hazards or necessary protective measures needed to be applied during their use. The overriding principle in the Directive is that machinery is designed and constructed so that it is fit for its intended application, and can be operated and maintained without putting individuals at risk. This remit includes the various phases of the machinery's life which includes; initial transportation from its place of manufacture, assembly, use, dismantling and eventual scrapping. Any control systems employed at any of these stages must also be designed and constructed so that they do not cause hazardous situations to arise as they are applied. Some machinery within the Directive have more detailed safety requirements associated with them, specific to their application. These would include machinery used in the food, cosmetics or pharmaceutical industry, wood-working machinery, portable hand-held machinery and machinery for pesticide applications.

Individual Member States are also obliged to provide market surveillance of the Machinery they both produce and import. In this way, a uniform application of the Directive is established throughout the European Union. These Member States also need to appoint competent authorities whose specific role is to monitor the conformity of the Machinery covered under the Directive. However, the responsibility for certifying this conformity remains with the manufacturers. As a guarantee of this conformity, the CE marking [51] is used on all applicable machinery, supported by a technical construction file which includes all essential health and safety requirements. This technical file will include details such as; the materials used to construct the machinery, all inspection reports, information relating to the mandatory fitting of guards and protective devices, maintenance schedules, instructions on how the machinery is to be safely used, and any specific hazards pertaining to the use of the machinery such as; radiation and lasers.

As with similar Directives, a safeguard clause is included which provides individual Member States the power to reject machinery which they consider would still be likely to endanger the health and safety of persons using the equipment, even if the Machinery is compliant with the Directive and carries a CE mark. This may be because the Machinery fails to satisfy the essential requirements of the Directive, or that the standards within the Directive have been incorrectly applied. By enacting this clause, the Member States would be obliged to inform the European Union Commission [52] of the measures they have taken and the reasons for making the rejection decision so that further investigation can be made. The European Union Commission in turn provides a judgement on whether the safeguarding clause was correctly applied and ensures that all Member States are informed of the outcome of their judgement.

## 5.8 Directive 99/92/EC—Risks from Explosive Atmospheres [53]

Directive 99/92/EC—Risks from Explosive Atmospheres, sets out the minimum requirements for the safety and health of any worker who is potentially at risk from explosive atmospheres. This would include any flammable substance in the form of a gas, vapour, mist or a dust, which after ignition would cause combustion to spread to the remaining unburned mixture. Employers have an obligation within the Directive of providing the appropriate technical and organisational control measures which will prevent the formation of explosive atmospheres in the first place. Where the nature of the process activity cannot avoid this, employers must ensure that provision is made to avoid the ignition of explosive atmospheres and to reduce the effect of an explosion should one occur. The key aim throughout the Directive is that working with explosive atmospheres should not put the health of any employee at risk.

The process of enacting any control measures to mitigate the risks of explosions requires employers to undertake a risk assessment of their activities. This must take into account the likelihood that explosive atmospheres will be present or generated by the activity being undertaken by their organisation. It must also consider the likelihood that ignition sources, including potential electrostatic discharges, may be present or become active, which could cause combustion of these explosive atmospheres to occur. Employers must therefore consider the installation set-up, what substances are being used, their processes and the scale of any anticipated effects should an explosion occur.

If during this risk assessment, places are identified where explosive atmospheres may be present, they must be classified in zones according to the type of explosive atmosphere and the level of risk they possess:

*Hazard—Gases, vapours and mists*

*Zone 0* A place where an explosive atmosphere in the form of gas, vapour or mist is present continuously or for long periods.

*Zone 1* A place where an explosive atmosphere in the form of gas, vapour or mist is likely to only occasionally occur in normal operation.

*Zone 2* A place where an explosive atmosphere in the form of gas, vapour or mist is unlikely to occur in normal operation, but if it does occur, will only be present for a very short period.

*Hazard—Powders and dusts*

*Zone 20* A place where an explosive atmosphere in the form of a cloud of combustible dust is present continuously or for long periods.

*Zone 21* A place where an explosive atmosphere in the form of a cloud of combustible dust is likely to only occasionally occur in normal operation.

*Zone 22* A place where an explosive atmosphere in the form of a cloud of combustible dust is unlikely to occur in normal operation, but if it does occur, will only be present for a very short period.

Any area where explosive atmospheres may occur, must have warning signs providing notification of this fact. In addition, an explosion protection document has to be produced for these areas. This document provides evidence that explosion risks have been determined and assessed and that measures have been put in place to allow compliance with the Directive. It should also detail the various locations which have been classified in the respective zones and what equipment is being used in each of these zones.

Applicable precautions will be required for each of the individual zones in order to protect the health and safety of employees working in the vicinity, with the greater hazards commanding more significant safeguards. These would include the necessary training of staff working near to these explosive atmospheres, so that they are aware of the necessary explosion protection which they need to implement. This may well be supported by the application of permits to work for staff undertaking activities in the respective zones. However, employers are obliged to install explosion protection measures which would effectively divert or remove to a safe place, any explosion hazards which may escape from systems, as well as preventing ignition sources from becoming apparent in the respective zones. Electrostatic discharges in particular can be an issue, and workers must be provided with the appropriate clothing made from materials which would not give rise to any such dangerous discharges.

For any equipment being used in the various zones, they must be rated as safe to be used in those explosive atmospheres. They must be selected on the basis of their category rating and have to be compatible with gases, vapours or mists, or dusts. The lower the category number, the more hazardous the atmosphere they can be used in.

*Zone 0* or *Zone 20* Category *1* equipment.

*Zone 1* or *Zone 21* Category *1* or *2* equipment.

*Zone 2* or *Zone 22* Category *1*, *2* or *3* equipment.

These ratings are often referred to as the ATEX rating and are covered by a separate Directive, 2014/34/EU—Equipment and Protective Systems intended for use in potentially explosive atmospheres [54].

## 5.9 Directive 2014/34/EU—Equipment and Protective Systems Intended for Use in Potentially Explosive Atmospheres

Directive 2014/34/EU—Equipment and Protective Systems Intended for Use in Potentially Explosive Atmospheres, applies to any electrical and non-electrical equipment, protective devices and systems used in potentially explosive atmospheres. It establishes the safety and health requirements and standards which need to be met in order for this equipment to comply with the Directive. These requirements exceed the general requirements which all equipment have to conform to, thereby allowing them to be used specifically in these potentially explosive atmospheres.

In broad terms, the Directive aims to minimise the risks of an explosive atmosphere being accidentally ignited. In many workplace situations, effective ignition sources can be varied, and if combined with sufficient oxygen and fuel, they can rapidly cause an explosion. In explosive atmospheres, this fuel can be a gas, mist, vapour or even the fine particles forming a dust. Some ignition sources are obvious, such as open flames and sparks produced during a grinding process. Others are less apparent and would include; lightning strikes, high surface temperatures created during a machining operation and sparks produced through electrostatic discharges. Where electrical equipment is being used in a potentially explosive atmosphere, any bad electrical connection could in effect cause an electric spark and therefore become a source of ignition.

For equipment to be safely used in explosive atmosphere, a set of conformity requirements are defined within the Directive. Equipment meeting these requirements are marked with the CE marking [55] to qualify their compliance. In addition, electrical equipment rated safe for the various classification zones as defined in Directive 99/92/EC—Risks from Explosive Atmospheres [56], are marked with the εx marking [57] to qualify their compliance. All types of equipment have to undergo a conformity assessment and must be supplied with safety instructions and all associated technical documentation by the manufacturer. The conformity assessment normally includes the testing and certification of the equipment by a national notified body, although manufacturers and suppliers are permitted to self-certify Category 3 type equipment. This certification ensures that the equipment is fit for its intended purpose and that is can be used safely in the intended explosive atmospheres.

The equipment covered by the Directive is divided into two main groups:

*Group I—Mining*

*Category M 1* Equipment used in underground mines, ensuring a *very high* level of protection

*Category M 2* Equipment used in underground mines, ensuring a *high* level of protection

*Group II—Surface industries*

*Category 1* Equipment used in areas where explosive atmospheres are present continuously, ensuring a *very high* level of protection

*Category 2* Equipment used in areas where explosive atmospheres are present occasionally, ensuring a *high* level of protection

*Category 3* Equipment used in areas where explosive atmospheres are present infrequently, ensuring a *normal* level of protection

## 5.10 Directive 92/58/EEC—Safety and/or Health Signs [58]

In workplace situations where hazards exist, which due to the processes involved, cannot be avoided or reduced, employers must provide signs warning of these hazards. Directive 92/58/EEC—Safety and/or Health Signs, sets out the minimum requirements and standards for the provision of safety and health signs in these situations. It defines the type of signs required which include; safety signs, health signs, warning signs, emergency escape signs, along with first-aid signs. In addition, acoustic signals, verbal communication and the use of hand signals are also included as part of the Directive.

A key obligation within the Directive is that which is placed on organisations to ensure that not only are safety and health signs installed within their establishments, but that their employees are provided with suitable instruction about these signs. This includes explaining the meaning of the signs and what specific requirements need to be enacted in those work areas where they are displayed. An example of this would be where the presence of hazards identified by a warning sign would necessitate the wearing of specific personnel protective equipment in that work area. All employees would need to be aware of these requirements and respond accordingly.

The Directive covers the display of both permanent and temporary signs and defines the intrinsic features required in order for them to comply with the Directive. These include:

*Prohibiting signs* Such as, 'No smoking' and 'Do not touch'

*Warning signs* Such as, 'Toxic material' and 'Strong magnetic field'

*Mandatory signs* Such as, 'Eye protection must be worn' and 'Safety gloves must be worn'

*Emergency escape* Such as, 'Emergency exit route' and 'Emergency escape route'

*First-aid* Such as, 'First-aid post' and 'Eyewash station'

*Firefighting* Such as, 'Fire extinguisher post' and 'Emergency telephone'

The Directive also covers situations where containers are used for the storage or transport of chemical substances or hazardous mixtures. In order to comply with the minimum requirements of the Directive, these containers have to be labelled with the relevant hazard pictogram which defines the contents and also serves as the warning sign. In addition, supplementary information pertaining to the hazardous substance can be attached to this sign.

Where acoustic signals are being employed to give warning of an imminent hazard, employers need to ensure that these signals are maintained in a good working condition and that staff working in locations where they are fitted are aware of what the signal is for. These sound signals could be devices such as those fitted to vehicles warning that they are reversing, or an alarm fitted to a piece of equipment warning of certain levels having been exceeded. Any audible sound level of these signals must be considerably higher than that of the ambient noise in a working environment. That being said, it is easy to imagine however, an employee in all innocence misunderstanding the importance of an oxygen level alarm sounding in a cryogenic environment if they have never heard the sound before. It is common in many organisations to practice fire emergency evacuation drills in response to audible fire alarms being sounded, but there are many more acoustic signals used in the workplace which would benefit from such a rehearsal. Any signal for the evacuation of premises, must by definition in the Directive, be a continuous sound in order to distinguish between a warning and an evacuation alarm.

The Directive also encapsulates the minimum safety requirements for both verbal and hand signals which are used as part of an organisation's working practices. These would include examples where emergency information is being broadcast within an establishment or where hand signals are employed to convey instructions to a crane operator who may not have a direct line of sight of the load being lifted. The simplicity, clarity and precision of both methods of communication must be overriding in their comprehension, providing an obvious meaning to all employees involved in their use.

## 5.11 Directive 89/656/EEC—Use of Personal Protective Equipment [59]

In the majority of workplace situations, the control of hazards is managed at an individual level by employees wearing appropriate personal protective equipment. Directive 89/656/EEC—Use of Personal Protective Equipment, defines the minimum health and safety requirements of the equipment itself, along with how they should be correctly used. Personal protective equipment can be any type of equipment worn or held by an employee, which serves to protect them against all types of hazards likely to endanger their health and safety. This would include protection from injury as well as from infections. Typically, the hazards addressed by the use of personal protective equipment would include physical, electrical, heat, chemical and biohazards.

There is an obligation within the Directive stipulating that personal protective equipment must be used when any prevailing risk cannot be avoided or limited by an organisations existing safety and health procedures. The type of personal protective equipment covered by the Directive would include such things as; clothing, gloves, helmets, goggles, boots and cleanroom suits. They are generally categorised by the area of the body which they protect and by the type of hazards they are designed to mitigate.

For situations where the use of cryogenic technology are applied, various prevailing hazards would necessitate the use of personal protective equipment. Skin protection is one of these, where the physical agent of extreme low temperature can be damaging to any exposed skin. The use of personal protective equipment, such as cryogenic compatible gloves, serves to provide a barrier between the skin and the apparent hazard. As many tasks in the workplace are performed with the hands, the use of gloves is an essential prerequisite for a large proportion of cryogenic technology applications. Likewise, eye protection from the same hazards is equally vital, preventing eye injuries from occurring which may need some form of medical treatment. Eye protection requirements can obviously vary depending on occupational activities, but the Directive provides generalised guidance, outlining the specific requirements for eye protection for a variety of employments.

The main focus of the Directive is to ensure that any personal protective equipment within the European Union meets common quality and safety standards in order to carry the CE marking [60]. As such, the Directive is divided into three categories of equipment:

*Category I* Simple design, such as gloves, footwear and goggles

*Category II* Items such as personal flotation devices, and dry and wet suits

*Category III* Complex design, such as respiratory equipment and harnesses

Employers have an obligation within the Directive to not only provide the appropriate personal protective equipment to their staff free of charge, but also to ensure that any equipment provided complies with the provisions on their design and manufacture. All personal protective equipment must therefore be appropriate for the risks involved in what an employee is undertaking and must be continuously maintained in good working order. Employers must also provide training in the correct use of any personal protective equipment provided in the workplace.

## 5.12 Directive 89/654/EEC—Workplace Requirements [61]

For general place of work situations, Directive 89/654/EEC—Workplace Requirements, defines the minimum health and safety requirements for an organisation's workplace. This includes; the provision of safe routes to emergency exits, the upkeep of adequate levels of hygiene for the workplace and any equipment used there, and the regular maintenance and inspection of safety equipment provisioned to avert or eliminate hazards. Electrical installations in particular must be designed and constructed in such a way that they do not present a fire or explosion hazard in the workplace. Personnel must also be protected against the risk of accidents occurring due to any contact with electrical equipment, with the Directive providing guidance on the competency requirements of individuals installing this type of apparatus.

The Directive also defines what firefighting equipment, fire detection systems and fire alarms, need to be installed in the workplace. Specific provisions will obviously depend on factors such as the size and occupancy of buildings, what equipment they contain, and if there are chemical substances present. It also covers the ventilation requirements of enclosed workplaces, so that sufficient fresh air is provided to these locations. If this ventilation is provided by a forced system, any associated equipment must be maintained in a good working order and any breakdown of the equipment controlled in such a way to not endanger the health of any employee working in any enclosed workplace. As there is no European Union legislation specifically relating to work in confined spaces, Directive 89/654/EEC encapsulates the duties placed on employers to take into account the specific characteristics of every workplace, which by definition, would also include confined spaces.

Traffic routes within the workplace and more specifically, the so-called danger areas within the working environment, are addressed by the Directive so that safe access for both pedestrians and vehicles do not themselves endanger any workers in their vicinity. Likewise, any danger areas which may pose a risk due to the nature of work conducted there, must have devices preventing unauthorised access to these areas. In addition, authorised workers entering these areas must be directly protected from the hazards present with the application of appropriate measures and any remaining dangers clearly indicated as such.

## 5.13 Directive 98/24/EC—Risks Related to Chemical Agents at Work [62]

As a means of protecting the health and safety of workers from the associated risks of chemical agents, Directive 98/24/EC—Risks Related to Chemical Agents at Work, was established and defines the minimum safety requirements an organisation needs to put in place when working with defined chemical agents. The Directive covers not only chemical agents already present in the workplace, but also those created as a result from work activities involving other chemical agents. One of the main factors within the Directive is the defining of occupational exposure limits [63], as well as biological limit values [64], which inform organisations when they need to implement hazard control measures in relation to the use of chemical agents. The Directive is directly linked to Directive 2019/1831—Indicative Occupational Exposure Limit Values [65], where these limiting values are specifically defined.

Where a binding occupational exposure or biological limit values is defined in the Directive for specific chemical agents, Member States are obliged to establish their own corresponding national exposure and biological limits. As such, these national limits cannot exceed the limiting values stipulated within the Directive.

At an organisation level, the Directive stipulates that employers need to initially determine if they have any hazardous chemical agents in their workplace, and if they do, accordingly assess the risks arising from their use. This would be undertaken by using the safety data sheets [66] for individual chemical agents, which contain all relevant safety information and presents the risks associated with their use. These risk assessments need to be regularly reviewed and amended if there has been any significant change in the processes involving the chemical agents. A further amendment would also be required if any health surveillance results of their employees indicated the presence of a greater risk than had been originally estimated. In certain situations, it is conceivable that process activities in an organisation may involve the use of several hazardous chemical agents simultaneously. Under these circumstances, the exposure risks need to be assessed based on the risks presented by the combination of these chemical agents.

Once the risks have been assessed, employers need to then implement the necessary preventative measures to either eliminate the risks, or at least reduce them to as low a value as possible. This would normally take the form of a combination of specific protection for employees, prevention strategies and monitoring measures in order to protect the safety and health of these workers involved in the use of chemical agents. The best form of defence when using chemical agents, is that of elimination, which by default immediately removes the risk. However, elimination is not always possible, as the very process an organisation is undertaken may be solely dependent on the use of a particular chemical agent. In these situations, the Directive defines how employers should preferably replace any hazardous chemical agent they are using with ones which possess less of a risk.

In order to fulfil the obligations of the Directive, organisations must routinely measure any exposure to chemical agents and compare these findings to the respective national occupational exposure limits. Clearly, if they have at any point exceeded these limiting values, actions need to be introduced in order to remedy the situation. This may be in the form of processes being halted or employees having stricter limits imposed on their working with chemical agents. Linking in with this, employees must be suitably trained, so that they are able to apply any protective measures and precautions in association with the work they are undertaken with chemical agents.

Employers must also comply with any defining measures relating to fire safety arising from the use of chemical agents on their establishments. These, along with action plan procedures, form the basis of a strategy which should be immediately enacted in the event of an accident, incident or emergency related to the presence of the hazardous chemical agents in their workplace. A key element of any such strategy is the correct marking of containers holding chemical agents, with the hazards they present being clearly highlighted.

Where any risk assessments made in relation to the use of chemical agents indicates the requirement of the provisioning of health surveillance, under the stipulations of the Directive, Member States must establish the means for these health surveillances to be carried out. More importantly, they are compulsory wherever there is a binding biological limit value set as defined by the Directive. With certain chemical agents, there are even defined specific prohibitive limits which Member States must ensure are never exceeded.

## 5.14 Directive 2019/1831—Indicative Occupational Exposure Limit Values [67]

For any work undertaken where the use of chemical agents are involved, protection of the health and safety of workers from the associated risks is paramount. To assist with this, non-binding, indicative occupational exposure limit values have been established, which form the body of Directive 2019/1831. These exposure limit values have been derived from the most up-to-date scientific data relating to chemical agents and take into account the availability of measuring equipment to qualify these values. Directive 2019/1831 has a direct correlation with Directive 98/24/EC—Risks Related to Chemical Agents at Work [68], presenting the exposure limit values for each respective chemical agent listed there.

For all chemical agents where an indicative occupational exposure limit value has been defined within the Directive, Member States are required to then establish their own national exposure limit value. They must take into account the exposure limit values defined in the Directive, but should also determine their own national exposure limit values to reflect current national legislation and practice in regard to the use of chemical agents.

The majority of the indicative occupational exposure limit values described within the Directive are in relation to a reference period of 8 hours, which would in principle be an average working day for an employee. These are referred to in the Directive as long-term exposure limit values. For certain chemical agents, a shorter reference period of 15 minutes is defined within the Directive. These are referred to as short-term exposure limit values, and take into account the more direct effects of the chemical agents from short-term exposure. For a particular group of chemical agents, the skin absorption limits are also defined as part of the Directive.

## 5.15 Regulation (EC) No 1272/2008—Classification, Labelling and Packaging of Substances and Mixtures [69]

Regulation (EC) No 1272/2008, defines the harmonised criteria for the classification, labelling and packaging of substances and mixtures, and establishes the subsequent rules controlling the labelling and packaging for these products. To implement this, the hazardous substances and mixtures are initially classified with respect to the physical, health or environmental hazards which they embody. The Regulation specifically excludes radioactive substances and mixtures from this classification, which are included in their own dedicated Directive 2013/59/Euratom [70].

The Regulation details the hazard classification process, which in its basic form consists of an evaluation of available information and the correct identification of the substance and mixtures involved. Specific concentration limits are detailed as distinct thresholds within the Regulation, above which substances and mixtures are considered to be hazardous. Following this classification, the minimum labelling requirements are described, which needs to provide as much information as possible about the risks associated with the substances and mixtures. Labelling would typically include technical data relating to the hazards associated with the substance and mixtures, such as; their toxicity, their potential to cause eye damage or if they are considered to be a carcinogen. These are generalised with a set of hazard pictograms as defined within the Regulation.

Packaging requirements for substances and mixtures are set out in the Regulation to ensure that they are strong enough to endure so-called general handling. It is important that none of the packaging can adversely react with the contents, but it must be solid enough to prevent leakage or its contents from escaping.

Member States must observe the Regulation and establish their own national competent authorities and enforcement agencies who will undertake the role of managing the compliance for the classification, labelling and packaging of substances and mixtures. They must also appoint bodies responsible for receiving European Union information relating to any emergency health response in respect of the substances and mixtures in general use.

Explosives are included as part of this Regulation, with the associated fire risks and resultant projectile hazards forming major considerations. Likewise, flammable and chemically unstable gases have their own specific requirements because of their higher-risk potentials. For general gases under pressure, these fall into four separate classifications; compressed gases, liquefied gases, refrigerated liquefied gases and dissolved gases. Each has its own specific packaging and labelling requirements which reflect the varying nature of the associated hazards.

## 5.16 Directive 2013/35/EU—Electromagnetic Fields [71]

Where risks exist within organisations arising from electromagnetic fields, the minimum health and safety requirements of the exposure of their employees to this physical agent is described in Directive 2013/35/EU—Electromagnetic Fields. The Directive addresses all known short-term biophysical effects caused by the presence of electromagnetic fields. In its present form however, the Directive does not quantify the possible long-term effects of working with electromagnetic fields, since there is currently very little scientific evidence to support any definitive conclusions relating to this. At a fundamental level, the detrimental effects on the human body will depend on the frequency of the electromagnetic field and the duration of the exposure. Exposure limiting systems which form part of the Directive are therefore based on these two variables.

For all possible exposure to electromagnetic fields in the workplace, a risk assessment needs to be undertaken which should be proportional to the situations encountered in specific organisations. This assessment can be supported by the inclusion of magnetic field measurements and any associated calculations, such as exposure patterns of specific employees. The risk assessment must be referenced to action levels (ALs) [72] and exposure limit values (ELVs) which are detailed in the Directive, but should also take into consideration any changes in sensory or neural symptoms experienced by employees working with electromagnetic fields. Health surveillance of employees should be adopted as a norm and the findings used to better inform the production of subsequent risk assessments.

Prevention measures must be established in order to comply with the defined ALs and ELVs, and where possible risks eliminated or reduced to a sensible minimum. Additional measures are necessary for employees at particular risks from electromagnetic fields, such as pregnant women and anyone with implanted medical devices in their bodies. These measures can be implemented through the provision of signs warning of the risks and restricting access to the locations where electromagnetic fields are present. As with many of the other European Union Directives, anyone working with electromagnetic fields must be suitably trained and made aware of the associated hazards.

## 5.17 Directive 90/269/EEC—Manual Handling of Loads [73]

Directive 90/269/EEC establishes the minimum health and safety requirements for the manual handling of loads which could result in a back injury for employees. It outlines the obligations an organisation has in providing items, such as mechanical equipment, in order for manual handling tasks to be avoided. In situations where manual handling cannot be avoided however, additional measures need to be implemented which reduce the overall risk of injury from the manual handling of loads. These could for example be through the assessment of individual work patterns, looking critically at the health and safety conditions of the work involved.

The characteristics of the loads being handled is a key risk factor and the Directive outlines ways in which back injuries can be prevented by scrutiny of this aspect. The load for example may be too heavy or too large to be handled safely, or may be unstable, in that its contents are likely to shift when being moved. Likewise, the characteristics of the working environment can have an effect on how safe a manual handling activity would be. Typically, there is a greater risk of injury occurring where there is insufficient room for manual handling to be safely undertaken, or where the floor is uneven and presents a tripping hazard. By addressing each of these concerns, control measures can be implemented, such as simply reducing the load being handled, which can have a significant benefit to the safety and health of individuals undertaking manual handling activities.

The Directive supports these measure by detailing the common risks encountered during manual handling. It is critical therefore, that whatever measures are implemented, they are communicated to the employee undertaking the manual handling task. This could be as simple as marking the weight of a load on it, or at least providing this information before manual handling is attempted. All employees must be suitably trained to undertake manual handling and also made aware of the risk they may be subject to if they do not perform manual handling correctly.

## 5.18 Directive 2012/18/EU—The Control of Major-accident Hazards Involving Dangerous Substances [74]

Directive 2012/18/EU—The Control of Major-accident Hazards Involving Dangerous Substances, establishes the rules for the prevention of major accidents which might result from specific industrial activities. It attempts to limit the consequences on human health and the environment following any such accident. It is sometimes known as the Sevesco III Directive [75].

As major accidents can have serious consequences, with their impact even having the potential to extend beyond national borders, the Directive ensures that both appropriate precautionary and protective actions are taken by organisation involved in the use of dangerous substances. The Directive is therefore implemented in national legislation of the Member States and enforced by national chemical safety authorities [76].

The main requirement of the Directive is for organisations to make an assessment of their so-called major-accident hazards for any particular dangerous substances they are using. If these substances can cause a release of matter or energy under both normal and abnormal conditions, and in doing so, create a major accident, then a major-accident prevention policy (MAPP) [77] needs to be put in place. The Directive outlines how the MAPP needs to be designed to ensure a high level of protection to human health, as well as the environment, and must be submitted to the respective national competent authority. Implementation of the MAPP can be done through safety management systems and structures within an organisation. It must also include the provision of emergency plans should an accident ever occur. It is the responsibility of the national competent authorities to also undertake inspections at facilities which have registered a MAPP with them. This is to ensure that organisations can demonstrate that appropriate measures have been established in order to not only prevent major accidents occurring but also to mitigate the consequences if such an event did occur.

## 5.19 European Union safety guidelines

In addition to the range of safety Directives established by the European Union, European guidelines relating to health and safety topics are also published. They form non-binding documents which Member States are not obliged to comply with, but have the function of providing best-practice guidance on how the specific European Union Directives are to be implemented. Often put in place with reference to particular Directives, they can be in a variety of guidance formats. Generally, they are published as practical guidelines directly from the European Union Commission [78], with the aim of outlining best practices which organisations can adopt in order to prevent risks associated with their undertakings. They can also be issued as European Council Recommendations [79], European Commission Communications [80] and European Union Social Partners Agreements [81]. Examples relating directly to the use of cryogenic technology would include:

### 5.19.1 Health and safety at work is everybody's business—practical guidance for employers [82]

A publication which provides employers with information about their duties relating to general occupational safety and health. It details why establishing a good safety and health culture in the workplace is beneficial for business, and guides employers through the many basic principles which are essential for organisations to achieve this. Based on the European Framework Directive on Safety and Health at Work (Directive 89/391 EEC) [83], particular national legal obligations apply and these are defined within the document. It covers the specific topics of risk assessments, risk prevention methods and occupational safety and health training. Examples of how risk management can be applied in the workplace are included, covering a range of topics, such as; work-related stress and risks relating to musculoskeletal disorders. It provides a common sense approach so that organisations can identify any potential risks and decide the best way of addressing them.

### 5.19.2 Guidance on risk assessment at work (Directive 89/391/EEC) [84]

A general purpose set of guidance notes providing help and advice to employers, their staff, safety experts and enforcement agencies who are likely to be involved in the preparation and management of workplace risk assessments. As these risk assessments often form the core of many safety and health processes in the workplace, its aim is to provide all stakeholders with the necessary information which allows them to fulfil their duties as described in the European Framework Directive on Safety and Health at Work (Directive 89/391/EEC) [85]. The guidance covers; hazard identification and risk control requirements, which need to be applied by all workplace employees. As a European Union document, it does not take into account how the Framework Directive is implemented in individual Member States, where more stringent requirements may apply for the undertaking of risk assessments.

### 5.19.3 Guidance on the application of the essential health and safety requirements on ergonomics [86]

Ergonomics is the interaction of the human body with the workplace environment and in general terms can be maximised in order to increase productivity by reducing so-called operator fatigue and discomfort. This guidance document supports the Machinery Directive 2006/42/EC [87] and provides examples of key ergonomic factors and their possible negative consequences to both productivity and the health of workers if not implemented correctly. It provides guidance on topics such as; work-rates, human/machine interface, physical stress, psychological stress, and operator discomfort and fatigue. By presenting direction on the application of the essential health and safety requirements set out within the Directive, compliance can be achieved and work-related ergonomic issues should by default be reduced.

### 5.19.4 Guidance on the application of the Machinery Directive 2006/42/EC [88]

A general guidance document aimed at how organisations can provide the health and safety requirements associated with the use of any type of machinery in the workplace, as covered by the Machinery Directive 2006/42/EC [89]. In simple terms, it aims to clarify the concepts and requirements of the Directive so that individual Member States can apply a common interpretation. The guidance can be used by particular groups, such as; machinery manufacturers, importers, distributors, notified bodies, and occupational health and safety agencies. The document can be employed in parallel with the Guidance on the application of the essential health and safety requirements on ergonomics [90].

### 5.19.5 Non-binding guide to good practice for implementing Directive 2001/45/EC (work at height) [91]

As with many of the European Union Directives, how organisations actually implement them and assure compliance is not always the easiest of processes to define. This non-binding guide therefore provides the means by which organisations can choose the most appropriate work equipment for undertaking the specific task of work at height. Employees working at height are by definition exposed to the risks of falling and the guide provides practical examples of so-called good practice on how to prevent this from occurring. It correlates the minimum health and safety requirements set out in the Directive with the practicalities of controlling the risks associated with this type of activity. Its main focus is on the use of temporary equipment for working at height, such as; scaffolds, ladders and ropes, as they are the most commonly employed equipment to perform this work.

The guide assists employers in both small and medium-sized companies, explaining what the best methods are for them to select and use work equipment for all types of work at height. This selection process is based on; risk assessments, the duration of the work and any ergonomic constraints associated with the work, with the guide providing examples throughout. These illustrations are themselves based on good practice undertaken in the European Union Member States and can be used alongside the various European standards covering this discipline. The guide provides suggestions and recommendations prepared by experienced experts from the various Member States. It is not an absolute guide however, and risk assessments still have to be used in conjunction with its findings before a specific method of working is selected. Laws, regulations and standards may also differ slightly between Member States in order to reflect local conditions. Compliance with these specific rules has to be respected, using the guide as an initial reference.

### 5.19.6 Guidelines to Directive 2014/34/EU (ATEX) [92]

This document provides guidance to any organisation involved in the use of equipment and protective systems operating in potentially explosive atmospheres. It supports Directive 2014/34/EU—Equipment and Protective Systems intended for use in potentially explosive atmospheres [93] and clarifies the procedures needed when using such equipment. It attempts to bring a commonality to the safe use of the equipment, providing transparency to the specific roles played by the individual Member State Governments, stakeholders, manufacturers, trade associations and the standards agencies. As a link to Directive 2014/34/EU, it provides guidance on how the essential health and safety requirements can be implemented, and defines the technical details of how this can be achieved. Compiled in collaboration with national and industry experts, it provides an overview of the related European Union product rules which are needed to ensure that only safe equipment can be used in potentially explosive atmospheres.

### 5.19.7 Non-binding guide to good practice for implementing Directive 99/92/EC—Risks from Explosive Atmospheres (ATEX) [94]

Explosions and flash-fire accidents have the potential of producing dramatic consequences, not only in terms of injuries and the possible loss of life, but also in respect of the economic costs such an event would demand. Wherever explosion hazards are identified as a result of work being undertaken with flammable substances, controls must be established to not only prevent an explosion occurring, but also to mitigate the consequences if it did. It is important for organisations to understand that explosions can often take place as a result of a secondary process in what they are undertaking, such as the creation of explosive dusts, and these need to be address in order to comply with the explosive atmospheres Directive 99/92/EC [ATEX]—Risks from Explosive Atmospheres [95].

This non-binding guide has therefore been produced with the aim of assisting Member States as they prepare their national policies for the protection of the health and safety of workers undertaking tasks in explosive atmospheres. It provides guidance on how all salient hazards can be correctly identified and how any associated risks can therefore be assessed. It also outlines specific safeguarding measures, such as surveillance of employees, which can be applied to the working environment, so that explosion protection is provided at all times. One key element of the Directive is the production of an explosion protection document [96] to support these strategies, with the guide providing a framework which can be used to produce this information.

### 5.19.8 Minimising chemical risk to workers' health and safety through substitution [97]

Where possible, whenever chemicals of any nature are being used in the workplace, a practical method of reducing the associated risks is to substitute the chemicals with a lower risk type. Alternatively, it is also practical to consider the possibility of substituting the process incorporating the chemicals, with one which possess a lower overall risk. This guide focuses on how substitution can be used as a risk control measure, which in turn reduces risks to the health and safety of employees involved in the use of chemicals at work. The aim of the guide is for Member States to adopt a common approach in this substitution process.

The document is split into two primary parts, with both sections provide guidance on compliance with national legislation wherever risk management is utilised as part of a chemical substitution process.

*5.19.8.1 Part 1—practical guidance*, is aimed at organisations who have limited experience or knowledge of chemical risk management, although even those with experience can gain from its content.

*5.19.8.2—Report on identifying a viable risk management measure*, is more targeted towards the chemical supply chain and companies who are routinely handling chemicals.

### 5.19.9 ECHA guidance on reach [98]

The European Chemicals Agency (ECHA) [99] manages the technical and administration elements of the European Union Regulation; Registration, Evaluation, Authorisation and Restriction of Chemicals (REACH) [100]. This guidance describes the good practice employed throughout the European Union, which allows organisations to comply with the details of the Regulation. Its aim is to help Member States and individual companies to undertake the safe use of chemicals under the obligations of REACH. As such, companies are obliged to identify and manage any risks linked to the substances they manufacture and market.

These substances have to be formally registered and authorised, a process which involves them demonstrating to the ECHA that the substances can be used safely. The companies must also provide information on how any risks associated with these substances can be controlled.

Overall, the formal registration and authorisation process for chemicals and substances is aimed at the reduction in use of chemicals considered to be substances of very high concern. This can be through the substitution of the substances, or through the replacement of the technologies associated with their use with less hazardous methods. The guidance provides current best-practice approaches of undertaking this risk reduction. It also outlines the safety and health restrictions within the REACH process which are there to protect the health of those working directly with chemicals which would otherwise pose unacceptable risks. There is also guidance on how accurate Safety Data Sheets [101] can be generated for the substances.

Where substances of very high concern are present, and where they cannot be substituted, very often legal obligations entail for not only the producers and suppliers of these chemicals, but also importers of the product as well. The guidance explains the legislation defined within REACH and how it applies to each associated party. There is also an enforcement element to REACH, where national authorities can audit compliance with the Regulation, which often involves verification of items such as the Safety Data Sheet. There are also certain categories of substances which are deemed hazardous to human health and could likewise be detrimental to the environment. The guidance provides the means by which compliance with the Regulation can be achieved to ensure that the risks associated with their use are minimised as much as possible.

### 5.19.10 ECHA guidance on CLP [102]

The European Union Regulation 1272/2008—Classification, Labelling and Packaging of substances and mixtures (CLP) [103] has been put in place to ensure that all hazards presented by chemicals are communicated through the correct classification and labelling of their containers. This European Chemicals Agency (ECHA) guidance provides the means by which suppliers can initially decide the classification of a substance or mixture through a self-classification process. Any substances deemed hazardous by this process, must then be labelled according to the CLP Regulation.

In many cases, the classification of substances has been harmonised throughout the Member States and a common approach to labelling and packaging will have to be adopted for these materials. The guide provides clarity on this issue and it follows therefore that suppliers are obliged to ensure that an adequate risk management process is in place in order for these particular substances to be safely transported through all territories of the European Union. Any such harmonised information is conveyed to the CLP appointed bodies in the Member States so that it can be used for emergency health response in the various Poison Centres [104] throughout the European Union. As National Authorities have an enforcement role in verifying compliance with the CLP Regulations and REACH, the guide outlines the range of actions these agencies will enact to judge such compliance. Organisations can therefore use these provisions to establish conformity assessments of their own in order to pre-judge their ability to pass the tests undertaken by the National Authorities.

### 5.19.11 Practical guidelines of a non-binding nature on the protection of the health and safety of workers from the risk related to chemical agents at work [105]

Directive 98/24/EC—Risks Related to Chemical Agents at Work [106], defines the minimum standards on health and safety which organisations have to put in place where their employees are working with chemical agents. In order for Member States and their associated organisations to achieve compliance with the Directive, the practical guide outlines the methods which allows implementation of the rules within the Directive. In particular, it provides the applied tools on how to measure and evaluate air concentrations in the workplace where chemical agents are being used. It also details how specific risk assessments can be undertaken for each of the chemical agents listed in the Directive and provides ideas on the general requirements of protection and prevention which can be implemented in order to lower overall risks.

Working in parallel with other guidance documents, the guide contains information on the correct labelling requirements, along with a series of Safety Data Sheets [107] for many chemical products currently on the market. There are also many industrial examples which can be used as a template for a particular product or process making use of chemical agents. It also sets out occupational limit values [108] and biological limit values [64] applicable for the use of these chemicals and the measurement methods for a range of chemical agents which are listed in Directive 2000/39/EC—Indicative Occupational Exposure Limit Values [109].

### 5.19.12 Non-binding guide to good practice for implementing Directive 2013/35/EU—Electromagnetic Fields [110]

In order to comply with Directive 2013/35/EU—Electromagnetic Fields [111], organisations need to understand the minimum safety requirements regarding the exposure of their employees to the risks arising from the use of electromagnetic fields. The non-binding guide assists with this process and also provides Regulatory Authorities in the various Member States with a framework on how they can check that any such compliance is in place. The document is divided into two volumes;

#### *5.19.12.1 Volume 1*

This part of the guide provides advice on how organisations can conduct risk assessments for situations where working with electromagnetic fields cannot be avoided. It also outlines the options available to these organisations where it is necessary as a result of this risk assessment for them to implement additional protective or preventative measures. It details the likely health effects and safety risks associated with electromagnetic fields and what the typical sources are likely to be.

It explains how organisations can use the defined Exposure Limit Values [112] and Action Levels [113] within the Directive and how to calculate or measure exposure. It presents examples of protective and preventative measures which can be employed and what emergency procedures should be in place should an employee be exposed to electromagnetic fields. Finally, it highlights the typical symptoms exhibited by someone who has been exposed and the type of health surveillance that should be established for all employees working with electromagnetic fields.

*5.19.12.2 Volume 2*

This part of the guide presents a series of case studies, which show organisations how to correctly approach the undertaking of risk assessments, with illustrations of typical preventative and protective measures that could be adopted in the workplace. These were compiled from genuine work situations and are therefore practical examples of how the Directive needs to be implemented. They cover work areas such as; offices, Nuclear Magnetic Resonance (NMR) spectrometers, electrolysis, medical, engineering workshops, automotive, welding, metallurgical manufacturing, Radiofrequency (RF) plasma devices, rooftop antennas, walkie-talkies and Airports.

## 5.20 European Union planning permission [114]

The planning requirements for the construction of buildings in Member States is not a direct European Union regulatory decree. It is generally the individual Member States which establish standards and permission processes in order to ensure the safe design and construction of buildings in their territories. However, other related European Union legislation and case law often impacts on these individual planning rules, such as those relating to buildings fire-safety standards, environmental impact assessments and conservation, and pollution control. As a result, the European Union has, by association, an indirect impact on individual national planning systems, without the normal mechanism of Directives or Regulations being in place for such.

## 5.21 European Standards

Organisations, who through their undertakings, comply with the various European Union Directives, ensure that a high level of health and safety and environmental protection is provided in what they are doing. Corresponding European Union Harmonised Standards [115] are often associated with the Directives and there is the presumption that products complying with the Standards are by definition meeting the minimum requirements of the Directives. This is acknowledged by the marking of products with the CE mark [116] and the issuing of conformity certificates [117], both of which are recognised by all Member States. The use of these European Union Harmonised Standards by organisations is voluntary and alternative standards can be used instead, although if they are, there is then an obligation to prove that any product made under their guidance still meets the essential requirements of the Directives.

The compilation of the European Union Harmonised Standards is undertaken by one of the European Standardisation Organisations; the European Committee for Standardisation (CEN) [118], the European Committee for Electrotechnical Standardisation (CENELEC) [119] and the European Telecommunications Standards Institute (ETSI) [120]. It is their responsibility to ensure that any corresponding European Union Harmonised Standard meets the essential requirements of the respective European Union Directives they relate to. In addition, these Standardisation Organisations work closely with International Standardisation Bodies such as; the International Organisation for Standardisation (ISO) [121], the International Electrotechnical Commission (IEC) [122] and the International Telecommunication Union (ITU) [123]. The primary aim throughout for all of these organisations it to provide the means by which products can be manufactured and used with safety, health and environmental protection at the heart of the standards they publish.

## 5.22 National legislation on safety and health at work

The various European Union Directives set out the minimum standards for safety and health in the workplace. However, by virtue of all the Member States endorsing Article 153 of the Treaty on the Functioning of the European Union [124], they are required to comply with the minimum standards within these Directives by implementing their own Regulations and Laws through their respective national legislation process. Member States have the option of adopting more stringent rules in order to protect workers, but whatever legislation they statute, it must comply with at least the minimum standards of the Directives. Unfortunately, this results in national safety and health legislation varying in different degrees across the European Union.

To assist organisations in preparing products which need to move across national borders, a list of the national implementation measures is detailed at the end of each Directive. This records the national implementing laws for any particular Directive, which can be used where more detailed scrutiny of the legislation is required. In turn, the European Union Commission [125] carries out evaluations of the practical implementation of the European Union Directives as a means of ensuring full compliance is being managed by all Member States.

## 5.23 CE marking [126]

The CE marking in essence represents a certificate of conformity [127] with the European Union health, safety and environmental protection legislation for any product marketed within the European Economic Area [128]. It is also found on products which are sold outside of the European Economic Area but have been manufactured to European Union standards. It allows the free movement of products within the European market, but is also recognised throughout the World as an indicator of this assured compliance.

Not all products have to display the CE marking, but those which do are defined within the associated product categories within the respective Directives. An example would be where items of pressure equipment which have been designed, manufacture and tested to the Sound Engineering Practice [129] category, are self-assessed by the manufacturer and as such do not need to display a CE mark. For pressure equipment being designed, manufacture and tested to categories I, II, III or IV however, a full compliance assessment is required and these would require a CE marking.

The CE marking does not automatically mean that the product has been made in the European Economic Area, but simply qualifies that it has been correctly assessed before being placed on the market. This assessment must in turn be a check that the finished product complies with all the essential requirements of the relevant Directives and that all health and safety requirements within the Directives have been met. Manufactures are therefore responsible for;

- Undertaking the conformity assessment
- Produce a Technical File associated with the product
- Issue a Declaration of Conformity
- Place the CE marking on the product

### 5.23.1 Undertaking the conformity assessment [130]

This will involve initially identifying the Directives and harmonised standards which are applicable to any particular product. The Directives set out the product categories which require a CE marking, along with the essential safety, health and environmental requirements needed to comply with the Directives. Likewise, the harmonised standards provide more detailed technical provisions which allow compliance to be gained. As the use of these harmonised standards is in principle voluntary, manufactures are permitted to select alternative methods of achieving compliance with the Directive. These substitutes however need to still demonstrate that the product is safe and adequately meets the product-specific requirements of the respective Directives.

It is important to recognise in this assessment process that certain categories of products require an independent conformity assessment to be made by a nationally authorised Notified Body [131]. In many circumstances, the Notified Body would need to undertake various phases of the assessment to ratify conformity at the key design, manufacture and testing stages of a products life. This may include surveillance visits to the manufacturing facilities in order to confirm that things like quality plans [132] are being prepared correctly. The respective Directives define which product categories require this level of assessment. In turn, manufacturers are by definition of the Directives responsible for testing the product and checking that it conforms to the relevant European Union legislation before it is placed on the market. Some of these tests may require conformation granted by the Notified Bodies.

### 5.23.2 Produce a Technical File associated with the product [133]

For any manufactured product bearing the CE marking, a Technical File needs to be established in order to confirm conformity to the relevant requirements within the Directives. Often this is also referred to as the ‘Technical Documentation’ or the ‘Technical Construction File’. It must contain all information which adequately demonstrates this conformity and it must be available as soon as the product has been placed on the European market. Typically, the Technical File has to be retained for ten years from manufacture. It is the responsibility of the manufacturer to produce this Technical File.

The content of the Technical File must consist of all the relevant documents listed in the essential requirements of the associated product Directives. In general this will include:

- A general description of the product
- Information relating to the product's design, e.g. design drawings and component lists
- Installation instructions and the user manual
- A list of all applicable Directives and the essential requirements
- A list of all the Standards applied to verify compliance with the respective Directives
- The product safety analysis, which should take the form of an assessment of the risk probability and severity of the manufactured product and the consequential measures taken to reduce these risks.
- Certificates of CE compliance of the product as issued by the Notified Bodies
- Test reports

The actual structure and format of the Technical File is left to the manufacturer to decide and in most cases it will be retained in an electronic format. The Technical File must remain active and be updated to recognise any future modifications to the product, as well as reflecting any amendments to the Directives. Many of the Directives have defined stipulations that a full copy of the Technical File is retained for possible inspection by the European Union Surveillance Bodies.

### 5.23.3 Issue a Declaration of Conformity [134]

It is the responsibility of the manufacturer of any product to generate and sign a Declaration of Conformity certificate. This forms a legal document proving that the item meets the requirements of the Directives and can therefore have the CE marking [135] attached and can be sold in the European market [136]. The Declaration of conformity document needs to include;

- The name and address of the manufacturer
- The responsible person, where this is a requirement of the Directives
- A list of the relevant Directives applicable to the product
- A list of the standards used in order to fulfil the essential requirements of the Directives
- A declaration statement qualifying that the product has been manufactured to these essential requirements
- The details of any relevant Notified Bodies
- The name and organisational position of the person signing the Declaration of Conformity
- The signature of that person
- The date

### 5.23.4 Place the CE marking [137] on the product

It is the responsibility of the manufacturer or their authorised representatives, to place the CE marking on the product where this is a requirement of the Directives. It must be placed so that it complies with the legal format associated with the product or the data plate requirements as defined in the Directives. It must be visible, legible and impossible to remove. There are certain very specific rules for using the CE marking for specific products which are defined in the respective product-related Directives. Generally, the CE marking should be placed on the product, but where this isn't possible, provision is made for it to be placed on manuals or any other supporting documentation. If a Notified Body was involved in the manufacturing control phase, their identification number must also be displayed alongside the CE marking.

The general rules for CE marking are as follows:

- The CE marking must only be placed on the product by the manufacturer or their authorised representative
- The CE marking cannot be placed on products which are not covered by the relevant European Union Directives
- By using the CE marking, manufacturers take full responsibility for the product's conformity with the essential requirements of the Directives
- Any other marking placed on the product, such as company logo or identification number, must not obscure the CE marking

Regulation (EC) 765/2008 [138], sets out the requirements for accreditation and market surveillance relating to products which utilise the CE marking. There are also defined rules associated with the CE marking images, such as; image size, visibility, readability and permanence. There are many bodies throughout the European Union who enforce the CE marking legislation in order to prevent misuse of the symbol and to ensure that safety of products is maintained to a high standard. Typically, this would take the form of direct enforcement or market surveillance by nominated Public Authorities in each Member State. Clearly, National Laws on how this enforcement is implemented will differ between the Member States.

## 5.24 European Association for injury prevention and safety promotion (EuroSafe) [139]

EuroSafe is a non-governmental association which represents employment organisations and working individuals to prevent injury in the workplace and to promote safety. Their remit includes policies, such as; those for promoting child safety, consumer safety, safety for the elderly, safety of vulnerable road users, safety in sport and the prevention of violence and self-harm. Its membership is made up of representatives from health and safety agencies throughout the European Union, research bodies, private sector organisations, such as; insurance providers, and those civil society organisations who influence public policies on safety.

The aim of EuroSafe is to provide a working platform where policy makers and injury prevention professional can exchange their experiences and thereby initiate collaborative actions relating to health and safety. As such, they are in a position of influence where policy agendas are being discussed, both at a European level but also within the specifics of Member State requirements. They are able to do this by calling upon the wealth of information at their disposal and the expertise in a whole range of injury preventions.

## Relevant cryogenic safety legislation in the European Union—summary

**Key learning points**

**General**

*Article 153 of the treaty on the functioning of the European Union*—Provides the European Union with the authority to adopt Directives related to workplace safety and health

*Directive 89/39 EEC—Directive on safety and health at work*—Guarantees a minimum level of safety and health requirements throughout the Member States for all work activities undertaken in Europe

**Directives**

*Directive 2009/104/EC—Use of work equipment*—Defines the minimum safety and health requirements for the use of items of work equipment by employees at their place of work

*Directive 2014/68/EU—Pressure equipment*—Implements the harmonisation of Member State laws relating to pressure equipment with an allowable pressure above 0.5 bar and which are manufactured and made available throughout the European Market

*Directive 2014/29/EU—Simple pressure vessels*—Applies to the marketing and putting into service of simple pressure vessels which have been made as a batch, have a working pressure greater than 0.5 bar, are unfired and contain air or nitrogen

*Directive 2010/35/EU—Transportable pressure equipment*—Details the associated rules for the specific category of transportable pressure equipment used for the inland transport of dangerous goods

*Directive 2008/68/EC—Inland transport of dangerous goods*—Applies to the transport of dangerous goods within, and across the European Union national borders, by road, rail and inland waterways, by use of vehicles, wagons and inland waterway vessels

*Transport of dangerous goods by sea for Member States in the European Union*—This is controlled by the conventions covering the international maritime dangerous goods code (IMDG)

*Transport of dangerous goods by air for Member State in the European Union*—This is controlled by the international civil aviation organisation's (ICAO) technical instructions

*Directive 2006/42/EC—New Machinery Directive*—Applies to all machinery sold within the European Union Free Market and is concerned with protecting workers and consumers when using this machinery

*Directive 99/92/EC—Risks from Explosive Atmospheres*—Sets out the minimum requirements for safety and health of any worker who is potentially at risk from explosive atmospheres such as flammable substances in the form of a gas, vapour, mist or dust

*Directive 2014/34/EU—Equipment and Protective Systems Intended for Use in Potentially Explosive Atmospheres*—Applies to any electrical and non-electrical equipment, protective devices and systems used in potentially explosive atmospheres

*Directive 92/58/EEC—Safety and /or Health Signs*—Sets out the minimum requirements and standards for the provision of safety and health signs in the workplace

*Directive 89/656/EEC—Use of Personal Protective Equipment*—Defines the minimum health and safety requirements of personal protective equipment worn in the workplace and how they should be correctly used

*Directive 89/654/EEC—Workplace Requirements*—Defines the minimum health and safety requirements for an organisation's workplace

*Directive 98/24/EC—Risks Related to Chemical Agents at Work*—Defines the minimum safety requirements an organisation needs to put in place to protect the health and safety of workers from the associated risks of chemical agents

*Directive 2019/1831—Indicative Occupational Exposure Limit Values*—A set of non-binding indicative occupational exposure limit values relating to the use of chemical agents at work

*Regulation (EC) No 1272/2008—Classification, Labelling and Packaging of Substances and Mixtures*—Defines the harmonised criteria and rules controlling the classification, labelling and packaging of substances and mixtures

*Directive 2013/35/EU—Electromagnetic Fields*—Defines the minimum health and safety requirements for the prevention of exposure of employees to electromagnetic fields

*Directive 90/269/EEC—Manual Handling of Loads*—Establishes the minimum health and safety requirements for the manual handling of loads which could result in a back injury for employees

*Directive 2012/18/EU—The Control of Major-Accident Hazards Involving Dangerous Substances (Sevesco III)*—Establishes the rules for the prevention of major accidents which might result from specific industrial activities undertaken with dangerous substances and the limiting of their consequences on human health and the environment if they do occur

**European Union Safety Guidelines**

*Health and safety at work is everybody's business—Practical guidance for employers*—A publications providing employers with information about their duties relating to general occupational safety and health

*Guidance on risk assessment at work (Directive 89/391/EEC)*—A general purpose set of guidance notes providing help and advice to employers, their staff, safety experts and enforcement agencies who are likely to be involved in the preparation and management of workplace risk assessments

*Guidance on the application of the essential health and safety requirements on ergonomics*—Guidance providing examples of key ergonomic factors and their possible negative consequences to both productivity and the health of workers

*Guidance on the application of the Machinery Directive 2006/42/EC*—A general guidance document outlining how organisations can provide the health and safety requirements associated with the use of machinery in the workplace

*Non-binding guide to good practice for implementing Directive 2001/45/EC (Work at height)*—A non-binding guide providing the means by which organisations can choose the most appropriate work equipment for undertaking work at height

*Guideline to Directive 2014/34/EU (ATEX)*—A document providing guidance to any organisation involved in the use of equipment and protective systems operating in potentially explosive atmospheres

*Non-binding guide to good practice for implementing Directive 1999/92/EC [ATEX] (Explosive Atmospheres)*—A non-binding guide aimed at assisting Member States as they prepare their national policies for the protection of the health and safety of workers undertaking activities in explosive atmospheres

*Minimising chemical risk to workers' health and safety through substitution*—A guidance document focusing in how substitution can be used as a risk control measure for the use of chemicals at work

*ECHA guidance on REACH*—A guidance document produced by the European Chemical Agency (ECHA) describing the good practice employed throughout the European Union which allows organisations to comply with the details of the Registration, Evaluation, Authorisation and Restriction of Chemicals (REACH) Regulation

*ECHA guidance on CLP*—An ECHA guidance document outlining the means by which suppliers can determine the classification of substances and mixtures in accordance with the Classification, Labelling and Packaging of substances and mixtures (CLP) regulation 1272/2008

*Practical guidelines of a non-binding nature on the protection of the health and safety of workers from the risk related to chemical agents at work*—A practical guide outlining the methods which allow the implementation of Directive 98/24/EC—Risks related to chemical agents at work, setting out non-binding occupational and biological limit values

*Non-binding guide to good practice for implementing Directive 2013/35/EU—Electromagnetic Fields*—A non-binding guide for organisations and Regulatory Authorities in Member States to understand the minimum safety requirements regarding exposure of employees to the risks arising from the use of electromagnetic fields

**Associated safety topics**

*Planning permission*—The planning requirements for the construction of buildings in the European Union is the responsibility of each individual Member State with the proviso that other related European Union legislation and case law can often impact on local planning rules

*European Standards*—A range of European Union Harmonised Standards associated with specific Directives which provide compliance to the minimum safety and health requirements of the Directives

*National legislation on Safety and Health at Work*—Member State compliance with the minimum standards within the various Directives, implemented through their own Regulations and Laws

*CE Marking*—A certificate of conformity mark placed on any product marketed within the European Economic Area

*European Association for Injury Prevention and Safety Promotion (EuroSafe)*—A non-governmental association which represents employment organisations and working individuals to prevent injury in the workplace and to promote safety

## References

[1] EUR-LEX *Article 153 of the Treaty on the Functioning of the European Union* (https://eur-lex.europa.eu/collection/eu-law/treaties/treaties-force.html) (Accessed 2020)

[2] European Agency for Safety and Health at Work (EU-OSHA) *The European Framework Directive on Safety and Health at Work (Directive 89/391 EEC)* (https://osha.europa.eu/en/legislation/directives/the-osh-framework-directive/1) (Accessed 2020)

[3] GOV.UK *The European Union Member States* (www.gov.uk/eu-eea) (Accessed 2020)

[4] *The European Commission* (https://ec.europa.eu/info/index_en) (Accessed 2020)

[5] European Agency for Safety and Health at Work (EU-OSHA) *European Guidelines* (https://osha.europa.eu/en/safety-and-health-legislation/european-guidelines) (Accessed 2020)

[6] The European Commission *European Standards* (https://ec.europa.eu/growth/single-market/european-standards_en) (Accessed 2020)

[7] The European Union *CE Marking* (https://europa.eu/youreurope/business/product-requirements/labels-markings/ce-marking/index_en.htm) (Accessed 2020)

[8] The European External Action Service (EEAS) *The Single Market in the European Economic Area* (https://eeas.europa.eu/headquarters/headquarters-homepage_en) (Accessed 2020)

[9] European Agency for Safety and Health at Work (EU-OSHA) *Directive 2009/104/EC—Use of Work Equipment* (https://osha.europa.eu/en/legislation/directives/3) (Accessed 2020)

[10] European Agency for Safety and Health at Work (EU-OSHA) *Directive 2014/68/EU—Pressure Equipment* (https://osha.europa.eu/en/legislation/directive/directive-201468eu-pressure-equipment (Accessed 2020)

[11] European Agency for Safety and Health at Work (EU-OSHA) *European Union Member States 'Supervision Rights'* (https://osha.europa.eu/en/legislation/directive/directive-201468eu-pressure-equipment) (Accessed 2020)

[12] European Agency for Safety and Health at Work (EU-OSHA) *Directive 2014/29/EU—Simple Pressure Vessels* (https://osha.europa.eu/en/legislation/directive/directive-201429eu-simple-pressure-vessels) (Accessed 2020)

[13] EUR-LEX *Regulation (EC) No. 765/2008* (https://eur-lex.europa.eu/legal-content/EN/TXT/) (Accessed 2020)

[14] CE Marking Association *CE Marking Requirements Associated with the Simple Pressure Vessels* (www.cemarkingassociation.co.uk/simple-pressure-vessels/) (Accessed 2020)

[15] European Agency for Safety and Health at Work (EU-OSHA) *Directive 2010/35/EU—Transportable Pressure Equipment* (https://osha.europa.eu/en/legislation/directives/directive-2010-35-eu-transportable-pressure-equipment) (Accessed 2020)

[16] European Agency for Safety and Health at Work (EU-OSHA) *Directive 2008/68/EC—Inland Transport of Dangerous Goods* (https://osha.europa.eu/en/legislation/directives/directive-2008-68-ec) (Accessed 2020)
[17] ECE Global *European Union 'Pi' Marking* (www.eceglobal.com/services/certification-validation/ce-pi-marking) (Accessed 2020)
[18] European Agency for Safety and Health at Work (EU-OSHA) *Directive 2008/68/EC—Inland Transport of Dangerous Goods* (https://osha.europa.eu/en/legislation/directives/directive-2008-68-ec) (Accessed 2020)
[19] The United Nations Economic Commission for Europe (UNECE) *The International Carriage of Dangerous Goods by Road (ADR)* (www.unece.org/trans/danger/publi/adr/adr_e.html) (Accessed 2020)
[20] The Intergovernmental Organisation for International Carriage by Rail (OTIF) *The International Carriage of Dangerous Goods by Rail (RID)* (www.otif.org/en/) (Accessed 2020)
[21] The United Nations Economic Commission for Europe (UNECE) *The International Carriage of Dangerous Goods by Inland Waterways (ADN)* (www.unece.org/trans/danger/publi/adn/adn_e.html) (Accessed 2020)
[22] The International Air Transport Association (IATA) *The International Air Transport Association (IATA) Shipper's Declaration of Dangerous Goods* (www.iata.org/en/programs/cargo/dgr/shippers-declaration/) (Accessed 2020)
[23] European Chemical Agency (ECHA) *Material Safety Data Sheets (MSDS)* (https://echa.europa.eu/safety-data-sheets) (Accessed 2020)
[24] European Commission *Directive 1272/2008—Classification, Labelling and Packaging of Substances and Mixtures* (https://ec.europa.eu/growth/sectors/chemicals/classification-labelling_en) (Accessed 2020)
[25] European Agency for Safety and Health at Work (EU-OSHA) *Directive 95/50/EC—Checks on Road Transport of Dangerous Goods* (https://osha.europa.eu/en/legislation/directives/66) (Accessed 2020)
[26] The United Nations Economic Commission for Europe (UNECE) *The Agreement Concerning the International Carriage of Dangerous Goods by Road (ADR)* (http://www.unece.org/trans/danger/publi/adr/adr_e.html) (Accessed 2020)
[27] Economic Commission for Europe Inland Transport Committee *Training of Persons Involved in the Carriage of Dangerous Goods—ADR chapter 1.3* (www.unece.org/fileadmin/DAM/trans/publications/ADR_2019_vol1_1818953_E.pdf) (Accessed 2020)
[28] The International Rail Transport Committee (CIT) *The Convention Covering International Carriage of Dangerous Goods by Rail (RID)* (https://cit-rail.org/en/rail-transport-law/cotif/) (Accessed 2020)
[29] Legislation.gov.uk *The United Kingdom Channel Tunnel Dangerous Goods Policy By-law* (www.legislation.gov.uk/uksi/2007/3531/schedule/made) (Accessed 2020)
[30] Eurotunnel *The French Arrêté Ministériel* (www.eurotunnelfreight.com/uploadedfiles/xnt/adr_2019_uk.pdf) (Accessed 2020)
[31] Eurotunnel *The Channel Tunnel Conditions of Carriage* (www.eurotunnelfreight.com/uk/conditions-of-carriage/) (Accessed 2020)
[32] The United Nations Economic Commission for Europe (UNECE) *The Convention Covering the International Carriage of Dangerous Goods by Inland Waterways (ADN)* (www.unece.org/trans/danger/publi/adn/adn_e.html) (Accessed 2020)

[33] Central Commission for the Navigation of the Rhine (CCNR) *The Regulations for the Carriage of Dangerous Goods on the Rhine (ADNR)* (https://ccr-zkr.org/12020400-en.html) (Accessed 2020)

[34] The United Nations Economic Commission for Europe (UNECE) *The Regulations for the Carriage of Dangerous Goods on the Danube (AND-D)* (www.unece.org/fileadmin/DAM/trans/danger/publi/adn/agreement_text.pdf) (Accessed 2020)

[35] The United Nations Economic Commission for Europe (UNECE) *The Economic Commission for Europe* (www.unece.org/info/ece-homepage.html) (Accessed 2020)

[36] International Maritime Organisation (IMO) *The Convention Covering the International Maritime Dangerous Goods Code (IMDG)* (www.imo.org/en/OurWork/Safety/Pages/DangerousGoods) (Accessed 2020)

[37] The United Nations Economic Commission for Europe (UNECE) *The International Carriage of Dangerous Goods by Road (ADR)* (www.unece.org/trans/danger/publi/adr/adr_e.html) (Accessed 2020)

[38] The Intergovernmental Organisation for International Carriage by Rail (OTIF) *The International Carriage of Dangerous Goods by Rail (RID)* (www.otif.org/en/) (Accessed 2020)

[39] The United Nations Economic Commission for Europe (UNECE) *The International Carriage of Dangerous Goods by Inland Waterways (ADN)* (http://www.unece.org/trans/danger/publi/adn/adn_e.html) (Accessed 2020)

[40] The United Nations *The United Nations Government* (www.un.org/en/) (Accessed 2020)

[41] International Maritime Organisation (IMO) *The International Convention for the Safety of Life at Sea (SOLAS)* (www.imo.org/en/About/Conventions/Pages/International-Convention-for-the-Safety-of-Life-at-Sea-(SOLAS),-1974 (Accessed 2020)

[42] MARPOL *The International Convention for the Prevention of Pollution from Ships (MARPOL 73/78)* (https://marpol.site/) (Accessed 2020)

[43] The International Maritime Organisation (IMO) (www.imo.org) (Accessed 2020)

[44] The International Civil Aviation Organisation (ICAO) *The International Civil Aviation Organisation's (ICAO) Technical Instructions* (www.icao.int/safety/DangerousGoods/Pages/technical-instructions.aspx) (Accessed 2020)

[45] IATA *The International Air Transport Association (IATA) Dangerous Goods Regulations* (www.iata.org/en/publications/dgr/) (Accessed 2020)

[46] The European Union Aviation Safety Agency (EASA) *The European Union Rules Regarding the Transport of Dangerous Goods by Air Commission Regulation 965/2012* (www.easa.europa.eu/the-agency/faqs/dangerous-goods) (Accessed 2020)

[47] The European Union Aviation Safety Agency (EASA) (www.easa.europa.eu/) (Accessed 2020)

[48] The International Civil Aviation Organisation (ICAO) *Air Operator Certificate* (www.icao.int/safety/OASIS/Pages/AOC.aspx) (Accessed 2020)

[49] The International Civil Aviation Organisation (ICAO) *State and Operator Variations* (www.icao.int/safety/DangerousGoods/Pages/StateVariationPage.aspx) (Accessed 2020)

[50] European Agency for Safety and Health at Work (EU-OSHA) *Directive 2006/42/EC—New Machinery Directive* (https://osha.europa.eu/en/legislation/directives/directive-2006-42) (Accessed 2020)

[51] The European Union *CE Marking* (https://europa.eu/youreurope/business/product-requirements/labels-markings/ce-marking/index_en.htm) (Accessed 2020)

[52] The European Commission (https://ec.europa.eu/info/index_en) (Accessed 2020)
[53] European Agency for Safety and Health at Work (EU-OSHA) *Directive 99/92/EC—Risks from Explosive Atmospheres* (https://osha.europa.eu/en/legislation/directives/21) (Accessed 2020)
[54] European Agency for Safety and Health at Work (EU-OSHA) *Directive, 2014/34/EU—Equipment and Protective Systems Intended for Use in Potentially Explosive Atmospheres* (https://osha.europa.eu/en/legislation/directive/directive-201434eu-equipment-and-protective-systems-intended-use-potentially) (Accessed 2020)
[55] The European Union *CE Marking* (https://europa.eu/youreurope/business/product-requirements/labels-markings/ce-marking/index_en.htm) (Accessed 2020)
[56] European Agency for Safety and Health at Work (EU-OSHA) *Directive 99/92/EC—risks from explosive atmospheres* (https://osha.europa.eu/en/legislation/directives/21) (Accessed 2020)
[57] The European Commission *Equipment for Potentially Explosive Atmospheres (ATEX)* (https://ec.europa.eu/growth/sectors/mechanical-engineering/atex_en) (Accessed 2020)
[58] European Agency for Safety and Health at Work (EU-OSHA) *Directive 92/58/EEC—Safety and/or Health Signs* (https://osha.europa.eu/en/legislation/directives/9) (Accessed 2020)
[59] European Agency for Safety and Health at Work (EU-OSHA) *Directive 89/656/EEC—Use of Personal Protective Equipment* (https://osha.europa.eu/en/legislation/directives/4) (Accessed 2020)
[60] The European Union *CE Marking* (https://europa.eu/youreurope/business/product-requirements/labels-markings/ce-marking/index_en.htm) (Accessed 2020)
[61] European Agency for Safety and Health at Work (EU-OSHA) *Directive 89/654/EEC—Workplace Requirements* (https://osha.europa.eu/en/legislation/directives/2) (Accessed 2020)
[62] European Agency for Safety and Health at Work (EU-OSHA) *Directive 98/24/EC—Risks Related to Chemical Agents at Work* (https://osha.europa.eu/it/legislation/directives/75) (Accessed 2020)
[63] European Chemicals Agency (ECHA) *Occupational Exposure Limits* (https://echa.europa.eu/oel) (Accessed 2020)
[64] Scientific Committee on Occupational Exposure Limits (SCOEL) 2014 *List of Recommended Health-based Biological Limit Values (BLVs) and Biological Guidance Values (BGVs)* (https://ec.europa.eu/social/main.jsp?catId=148&langId=en&internal_pagesId=684&moreDocuments=yes&tableName=INTERNAL_PAGES ) (Accessed 2021)
[65] European Agency for Safety and Health at Work (EU-OSHA) *Directive 2019/1831—Indicative Occupational Exposure Limit Values* (https://osha.europa.eu/en/legislation/directive/directive20191831-indicative-occupational-exposure-limit-values) (Accessed 2020)
[66] European Chemical Agency (ECHA) *Material Safety Data Sheets (MSDS)* (https://echa.europa.eu/safety-data-sheets) (Accessed 2020)
[67] European Agency for Safety and Health at Work (EU-OSHA) *Directive 2019/1831—Indicative Occupational Exposure Limit Values* (https://osha.europa.eu/en/legislation/directive/directive20191831-indicative-occupational-exposure-limit-values) (Accessed 2020)
[68] European Agency for Safety and Health at Work (EU-OSHA) *Directive 98/24/EC—Risks Related to Chemical Agents At Work* (https://osha.europa.eu/it/legislation/directives/75) (Accessed 2020)
[69] European Agency for Safety and Health at Work (EU-OSHA) *Regulation (EC) No 1272/2008—Classification, Labelling and Packaging of Substances and Mixtures* (https://osha.

europa.eu/en/legislation/directives/regulation-ec-no-1272-2008-classification-labelling-and-packaging-of-substances-and-mixtures) (Accessed 2020)

[70] European Agency for Safety and Health at Work (EU-OSHA) *Directive 2013/59/Euratom* (https://osha.europa.eu/en/legislation/directives/directive-2013-59-euratom-protection-against-ionising-radiation)

[71] European Agency for Safety and Health at Work (EU-OSHA) *Directive 2013/35/EU—Electromagnetic Fields* (https://osha.europa.eu/en/legislation/directives/directive-2013-35-eu-electromagnetic-fields) (Accessed 2020)

[72] The European Union 2013 Electromagnetic field action levels (ALs)—Directive 2013/35/EU of the European Parliament and of the Council of 26 June 2013 on the minimum health and safety requirements regarding the exposure of workers to the risks arising from physical agents (electromagnetic fields) *Official J. Eur. Union*

[73] European Agency for Safety and Health at Work (EU-OSHA) *Directive 90/269/EEC—Manual Handling of Loads* (https://osha.europa.eu/en/legislation/directives/6) (Accessed 2020)

[74] EUR-LEX *Directive 2012/18/EU—The Control of Major-Accident Hazards Involving Dangerous Substances* (https://eur-lex.europa.eu/legal-content/EN) (Accessed 2020)

[75] The European Commission *The Sevesco III Directive* (https://ec.europa.eu/environment/seveso/legislation.htm) (Accessed 2020)

[76] European Chemical Agency (ECHA) *National Chemical Safety Authorities* (https://echa.europa.eu/contacts-of-the-member-state-competent-authorities) (Accessed 2020)

[77] EUR-LEX *Major-Accident Prevention Policy (MAPP)* (https://eur-lex.europa.eu/LexUriServ/LexUriServ.do?uri=CELEX:31996L0082:EN:HTML) (Accessed 2020)

[78] The European Commission (https://ec.europa.eu/info/index_en) (Accessed 2020)

[79] Council of Europe *European Council Recommendations* (www.coe.int/en/web/common-european-framework-reference-languages/recommendations) (Accessed 2020)

[80] European Commission *European Commission Communications* (https://ec.europa.eu/info/departments/communication_en)(Accessed 2020)

[81] The European Trade Union Confederation (ETUC) *European Union Social Partners Agreements* (www.etuc.org/en/social-partners-framework-agreements) (Accessed 2020)

[82] European Agency for Safety and Health at Work (EU-OSHA) *Health and Safety at Work Is Everybody's Business—Practical Guidance for Employers* (https://osha.europa.eu/en/legislation/guidelines/health-and-safety-work-everybodys-business-practical-guidance-employers) (Accessed 2020)

[83] European Agency for Safety and Health at Work (EU-OSHA) *The European Framework Directive on Safety and Health at Work (Directive 89/391 EEC)* (https://osha.europa.eu/en/legislation/directives/the-osh-framework-directive/1) (Accessed 2020)

[84] European Agency for Safety and Health at Work (EU-OSHA) *Guidance on Risk Assessment at Work (Directive 89/391/EEC)* (https://osha.europa.eu/en/legislation/guidelines/guidance-on-risk-assessment-at-work) (Accessed 2020)

[85] European Agency for Safety and Health at Work (EU-OSHA) *The European Framework Directive on Safety and Health at Work (Directive 89/391 EEC)* (https://osha.europa.eu/en/legislation/directives/the-osh-framework-directive/1) (Accessed 2020)

[86] European Agency for Safety and Health at Work (EU-OSHA) *Guidance on the Application of the Essential Health and Safety Requirements on Ergonomics* (https://osha.europa.eu/en/legislation/guidelines/guidance-application-essential-health-and-safety-requirements-ergonomics) (Accessed 2020)

[87] European Agency for Safety and Health at Work (EU-OSHA) *Directive 2006/42/EC—New Machinery Directive* (https://osha.europa.eu/en/legislation/directives/directive-2006-42) (Accessed 2020)

[88] European Agency for Safety and Health at Work (EU-OSHA) *Guidance on the Application of the Machinery Directive 2006/42/EC Directive* (https://osha.europa.eu/en/legislation/guidelines/guide-application-machinery-directive-200642ec) (Accessed 2020)

[89] European Agency for Safety and Health at Work (EU-OSHA) *Directive 2006/42/EC—New Machinery Directive* (https://osha.europa.eu/en/legislation/directives/directive-2006-42) (Accessed 2020)

[90] European Agency for Safety and Health at Work (EU-OSHA) *Guidance on the Application of the Essential Health and Safety Requirements on Ergonomics* (https://osha.europa.eu/en/legislation/guidelines/guidance-application-essential-health-and-safety-requirements-ergonomics) (Accessed 2020)

[91] European Agency for Safety and Health at Work (EU-OSHA) *Non-binding Guide to Good Practice for Implementing Directive 2001/45/EC (Work at Height)* (https://osha.europa.eu/en/legislation/guidelines/osh_link) (Accessed 2020)

[92] European Agency for Safety and Health at Work (EU-OSHA) *Guidelines to Directive 2014/34/EU (ATEX)* (https://osha.europa.eu/en/legislation/guidelines/guidelines-directive-201434-eu-atex) (Accessed 2020)

[93] European Agency for Safety and Health at Work (EU-OSHA) *Directive 2014/34/EU—Equipment and Protective Systems Intended for Use in Potentially Explosive Atmospheres* (https://osha.europa.eu/en/legislation/directive/directive-201434eu-equipment-and-protective-systems-intended-use-potentially-0) (Accessed 2020)

[94] European Agency for Safety and Health at Work (EU-OSHA) *Non-binding Guide to Good Practice for Implementing Directive 99/92/EC [ATEX]—Risks from Explosive Atmospheres* (https://osha.europa.eu/en/legislation/guidelines/Newoshinfo_3618) (Accessed 2020)

[95] European Agency for Safety and Health at Work (EU-OSHA) *Directive 99/92/EC [ATEX]—Risks from Explosive Atmospheres* (https://osha.europa.eu/en/legislation/directives/21) (Accessed 2020)

[96] EX Machinery *Explosion Protection Document* (https://ex-machinery.com/Blog/explosion-protection-document) (Accessed 2020)

[97] European Agency for Safety and Health at Work (EU-OSHA) *Minimising Chemical Risk to Workers' Health and Safety through Substitution* (https://osha.europa.eu/en/legislation/guidelines/minimising-chemical-risk-to-workers2019-health-and-safety-through-substitution) (Accessed 2020)

[98] European Agency for Safety and Health at Work (EU-OSHA) *ECHA Guidance on REACH* (https://osha.europa.eu/en/legislation/guidelines/echa-guidance) (Accessed 2020)

[99] The European Chemicals Agency (ECHA) (https://echa.europa.eu/) (Accessed 2020)

[100] European Agency for Safety and Health at Work (EU-OSHA) *The European Union Regulation: Registration, Evaluation, Authorisation and Restriction of Chemicals (REACH)* (https://osha.europa.eu/en/legislation/directives/regulation-ec-no-1907-2006-of-the-european-parliament-and-of-the-council) (Accessed 2020)

[101] European Chemical Agency (ECHA) *Material Safety Data Sheets (MSDS)* (https://echa.europa.eu/safety-data-sheets) (Accessed 2020)

[102] European Agency for Safety and Health at Work (EU-OSHA) *ECHA Guidance on CLP* (https://osha.europa.eu/en/legislation/guidelines/labelling-and-safety-data-sheets-clp) (Accessed 2020)

[103] European Agency for Safety and Health at Work (EU-OSHA) *The European Union Regulation 1272/2008—Classification, Labelling and Packaging of Substances and Mixtures (CLP)* (https://osha.europa.eu/en/legislation/directives/regulation-ec-no-1272-2008-classification-labelling-and-packaging-of-substances-and-mixtures) (Accessed 2020)

[104] The European Commission *European Union Poison Centres* (https://ec.europa.eu/growth/sectors/chemicals/poison-centres) (Accessed 2020)

[105] European Agency for Safety and Health at Work (EU-OSHA) *Practical Guidelines of a Non-binding Nature on the Protection of the Health and Safety of Workers from the Risk Related to Chemical Agents at Work* (https://osha.europa.eu/en/legislation/guidelines/practical-guidelines-of-a-non-binding-nature-on-the-protection-of-the-health-and-safety-of-workers-from-the-risks-related-to-chemical-agents-at-work) (Accessed 2020)

[106] European Agency for Safety and Health at Work (EU-OSHA) *Directive 98/24/EC—Risks Related to Chemical Agents at Work* (https://osha.europa.eu/it/legislation/directives/75) (Accessed 2020)

[107] European Chemical Agency (ECHA) *Material Safety Data Sheets (MSDS)* (https://echa.europa.eu/safety-data-sheets) (Accessed 2020)

[108] European Chemicals Agency (ECHA) *Occupational Exposure Limits* (https://echa.europa.eu/oel) (Accessed 2020)

[109] European Agency for Safety and Health at Work (EU-OSHA) *Directive 2000/39/EC—Indicative Occupational Exposure Limit Values* (https://osha.europa.eu/en/legislation/directives/directive-2000-39-ec-indicative-occupational-exposure-limit-values) (Accessed 2020)

[110] European Agency for Safety and Health at Work (EU-OSHA) *Non-binding Guide to Good Practice for Implementing Directive 2013/35/EU—Electromagnetic Fields* (https://osha.europa.eu/en/legislation/guidelines/non-binding-guide-good-practice-implementing-directive-201335eu) (Accessed 2020)

[111] European Agency for Safety and Health at Work (EU-OSHA) *Directive 2013/35/EU—Electromagnetic Fields* (https://osha.europa.eu/en/legislation/directives/directive-2013-35-eu-electromagnetic-fields) (Accessed 2020)

[112] EUR-LEX *Electromagnetic Fields Exposure Limit Values—Directive 2013/35/EU—Electromagnetic Fields—Article 3* (https://eur-lex.europa.eu/legal-content/EN/TXT/?uri=CELEX:32013L0035) (Accessed 2020)

[113] EUR-LEX *Electromagnetic Fields Action Levels—Directive 2013/35/EU—Electromagnetic Fields—Article 3* (https://eur-lex.europa.eu/legal-content/EN/TXT/?uri=CELEX:32013L0035) (Accessed 2020)

[114] www.ceenrg.landecon.cam.ac.uk/research/environmental-law-and-governance/european-union-planning-law (Accessed 2020)

[115] European Agency for Safety and Health at Work (EU-OSHA) *European Union Harmonised Standards* (https://osha.europa.eu/en/european-standards) (Accessed 2020)

[116] The European Union *CE Marking* (https://europa.eu/youreurope/business/product-requirements/labels-markings/ce-marking/index_en.htm) (Accessed 2020)

[117] Your Europe European Union *Technical Documentation and EU Declaration of Conformity* (https://europa.eu/youreurope/business/product-requirements/compliance/technical-documentation-conformity/index_en.htm) (Accessed 2020)

[118] The European Committee for Standardisation (CEN) (https://standards.cen.eu/) (Accessed 2020)

[119] The European Committee for Electrotechnical Standardisation (CENELEC) (www.cenelec.eu/aboutcenelec/whoweare) (Accessed 2020)

[120] The European Telecommunications Standards Institute (ETSI) (www.etsi.org/) (Accessed 2020)

[121] The International Organisation for Standardisation (ISO) (www.iso.org/home.html) (Accessed 2020)

[122] The International Electrotechnical Commission (IEC) (www.iec.ch/homepage) (Accessed 2020)

[123] The International Telecommunication Union (ITU) (www.itu.int/en/Pages/default.aspx) (Accessed 2020)

[124] EUR-LEX *Article 153 of the Treaty on the Functioning of the European Union* (https://eur-lex.europa.eu/collection/eu-law/treaties/treaties-force.html) (Accessed 2020)

[125] The European Commission (https://ec.europa.eu/info/index_en) (Accessed 2020)

[126] The European Union *CE Marking* (https://europa.eu/youreurope/business/product-requirements/labels-markings/ce-marking/index_en.htm) (Accessed 2020)

[127] Your Europe European Union *Technical Documentation and EU Declaration of Conformity* (https://europa.eu/youreurope/business/product-requirements/compliance/technical-documentation-conformity/index_en.htm) (Accessed 2020)

[128] Eurostat *The European Economic Area* (https://ec.europa.eu/eurostat/statistics-explained/index.php/Glossary:European_Economic_Area_(EEA)) (Accessed 2020)

[129] The Engineering Equipment and Materials Users Association (EEMUA) *Sound Engineering Practice* (https://www.eemua.org/Glossary/S/Sound-EP.aspx) (Accessed 2020)

[130] Your Europe European Union *Technical Documentation and EU Declaration of Conformity* (https://europa.eu/youreurope/business/product-requirements/compliance/technical-documentation-conformity/index_en.htm) (Accessed 2020)

[131] The European Commission *Nationally Authorised Notified Body* (https://ec.europa.eu/growth/single-market/goods/building-blocks/notified-bodies_en) (Accessed 2020)

[132] The International Organisation for Standardisation (ISO) *Quality Plans* (www.iso.org/news/ref2310.html) (Accessed 2020)

[133] CEMarking.Net *Producing a Technical File Associated with a Product* (https://cemarking.net/technical-file/) (Accessed 2020)

[134] Your Europe European Union *Technical Documentation and EU Declaration of Conformity* (https://europa.eu/youreurope/business/product-requirements/compliance/technical-documentation-conformity/index_en.htm) (Accessed 2020)

[135] The European Union *CE Marking* (https://europa.eu/youreurope/business/product-requirements/labels-markings/ce-marking/index_en.htm) (Accessed 2020)

[136] The European External Action Service (EEAS) *The Single Market in the European Economic Area* (https://eeas.europa.eu/headquarters/headquarters-homepage_en) (Accessed 2020)

[137] The European Union *CE Marking* (https://europa.eu/youreurope/business/product-requirements/labels-markings/ce-marking/index_en.htm) (Accessed 2020)

[138] EUR-LEX *Regulation (EC) No. 765/2008* (https://eur-lex.europa.eu/legal-content/EN/TXT/) (Accessed 2020)

[139] European Association for Injury Prevention and Safety Promotion (EuroSafe) (www.eurosafe.eu.com/home) (Accessed 2020)

# Chapter 6

# Relevant international safety and health organisations

## 6.1 General

On the whole, the management of health and safety in the workplace is assumed at a national level, with the full remit of divergent laws and regulations being applied, conditional to the country within which the work is being undertaken. Considered independently in this way, these legal standings can vary enormously from nation to nation, and yet they are often influenced by much larger governing bodies or even universal statutes which control any particular aspect of occupational health and safety. However, it is a common understanding that much can be gained in sharing good practice, and many organisations have been established so that international collaboration can take place in order to improve the overall standing of workplace wellbeing, taking it from a provincial standpoint to a more global level of enactment. This is often driven by economics, especially where trade across national borders forms an integral part of the distribution of products and the provisioning of services throughout the entire world. The use of aircraft and maritime vessels between nations alone has grown significantly in more recent times, to a level where many of the goods we receive these days may have travelled huge distances from their point of origin, traversing countless national boundaries in the process. To this end, the need of some form of international standardisation within the remit of organisations in relation to health and safety is often the catalyst to the conception of these mutual associations. Likewise, this in turn is commonly fuelled by the modern way in which vocations can take place at an incredibly rapid pace, especially where this is often divorced from individual national restrictions.

The primary role of many of these transnational organisations is to offer a framework by which international cooperation can routinely take place. In fact, in many cases, it is crucially recognised that certain safety and health issues pertaining to the movement of goods across territorial borders, can only be adequately addressed

doi:10.1088/978-0-7503-3123-4ch6  

at a collectively global level, with an overall shared understanding of the prevailing risks. Developing from the general cooperation between countries and their abundant professional groups, many of these organisations have taken this idea of the exchange of information from its nascent form, all the way to the guiding level of a powerful lobby on the world stage, often with governmental backing and citation.

To this end, these international organisations can be divided into two distinct types; those classed as intergovernmental and those classed as non-governmental. Examples of international intergovernmental organisations would include the United Nations (UN) [1], the Organisation for Economic Co-operation and Development (OECD) [2], and regional entities, such as the European Union (EU) [3] and the North American Free Trade Agreement (NAFTA) [4] between Canada, the United States of America and Mexico. Examples of international non-governmental organisations with a focus on occupational health and safety would include the International Commission on Occupational Health (ICOH) [5], the International Social Security Association (ISSA) [6], the International Association of Labour Inspection (IALI) [7] and the International Organisation for Standardisation (ISO) [8], who covenant the many aspects of standardisation relating to work activities. In addition, some international associations of key professional groups would also have a generic interest in occupational health and safety through their broader undertakings.

The following are examples of some of these transnational organisations which have either direct or indirect impact on the type of work where the use, or transport, of cryogenic technology is undertaken.

## 6.2 Intergovernmental international organisations

### 6.2.1 The United Nations Economic and Social Council [9]

The United Nations Economic and Social Council is a principal organisation within the remit of the United Nations [10] and is responsible for managing the economic and social elements of the United Nations through a series of specialised agencies. These agencies work closely with the various United Nations commissions and Member States in formulating policy recommendations in all aspects of the United Nations' function. Of the 15 specialised agencies currently in existence, several play a key role in the formulation of safety related policies where the use of cryogenic equipment would be relevant. These would include:

### 6.2.2 International Civil Aviation Organisation (ICAO) [11]

The International Civil Aviation Organisation sets out the various codes and principles which determine all types of international air transport throughout the world. They define the required techniques of international air navigation in order to ensure the safety of the industry no matter where the flights begin and end. This is provided by the adoption of standards and recommended practices relating not only to air transport itself, but also the prevention of unlawful interference to the process due to such things as terrorist activities. In addition, they define the necessary border-crossing procedures which are critical for international civil aviation to function as it does with daily flights routinely taking place over international territories. Their remit includes all the main aspects of air navigation, air transport infrastructures, flight inspections and the establishing of protocols in order for air accident investigations to take place following aviation incidents. By default, individual nations are members of the ICAO by virtue of their having signed to the Convention on International Civil Aviation [12].

Directly linked to the function of international civil aviation, the Air Navigation Commission (ANC) [13] is a technical body within the ICAO made up of Commissioners, nominated by the various United Nations Member States. These Commissioners are independent experts and do not serve as Member State representatives, but provide a collective voice in the formation of the ICAO's International Standards and Recommended Practices for air transport [14]. Other similar and associated international organisations for air transport would include; the International Air Transport Association (IATA) [15], the Civil Air Navigation Services Organisation (CANSO) [16], the Airports Council International [17], along with various trade associations such as airport authorities and those representing the various airlines themselves.

### 6.2.3 International Labour Organisation (ILO) [18]

As a United Nations specialist agency, the International Labour Organisation sets out the various International labour standards [19] in order to provide both social justice and promote decent work for employees throughout the world. They also provide technical assistance to developing nations in order to meet the challenges of a changing employment landscape as new technologies impinge on old established values. It is made up of representatives of the United Nations Member State governments, along with employers and their employees. Collectively, they create a set of labour standards applicable to the various international industries, aimed at providing opportunities for both women and men to gain employment, with the assurance of social protection and an entitlement to lifelong learning.

### 6.2.4 International Maritime Organisation (IMO) [20]

The International Maritime Organisation sets out the framework for the international maritime safety and practices. Through cooperation among the United Nations Member States and the numerous worldwide shipping industries, their aim is to improve maritime safety and to prevent marine pollution through the regulation of all shipping activities. Their work is supported by technical sub-committees who work specifically in the areas of marine safety, environmental protection, legal matters, maritime security, efficiency of shipping, and education. The IMO has a broad remit, covering many international shipping requirements, including the safety of life at sea, the prevention of pollution of the sea, along with port facilities security. They undertake these through a series of conventions which also include maritime issues, such as; the safe navigation of ships, search and rescue operations, wreck removals, tonnage measurements, ship recycling, the reduction of smoke emissions from shipping, the training and certification of seafarers, and modern piracy.

One key area which the IMO dedicates their efforts towards is that of controlling the general environmental contamination by ships when they are out at sea through accidental pollution. They implement this through a series of regulations which operate a range of measures designed to prevent large ship accidents and to minimise their consequential effects on marine life. Although shipping accidents are rare, they can occasionally occur and as a result would consequently have a significant impact on the local environment. However, it is the greater environmental threat posed by more routine procedures which are a bigger overall problem associated with the maritime industry and one which the IMO implement measures with the aim of minimising their effects on the environment. Typically, these would include addressing things like the regular cleaning of oil tanks or the disposal of engine room waste. As such, the IMO implement the international convention for the prevention of pollution from ships [21] which covers accidental and operational oil pollutions, along with pollution by chemicals, goods in packaged form, sewage and air pollution.

### 6.2.5 World Health Organisation (WHO) [22]

The World Health Organisation sits as a coordinating authority for all matters relating to international public health, with a broad spectrum of responsibilities covering such things as health, sanitation and diseases. Its mandate in order to implement these strategies includes providing advice on the provisioning of universal healthcare, monitoring public health risks, providing advice on healthy diets and nutritional advice, and coordinating international responses to health emergencies. Its role includes the provisioning of technical assistance to nations when required, the setting of international health standards and guidelines, and the collection of data related to global health issues. They provide health statistics which serves as a forum for the international dialogue on health issues.

### 6.2.6 The International Network of Safety and Health Professional Organisation (INSHPO) [23]

The established remit of the INSHPO is to provide a global forum for the various national occupational safety and health professions throughout the world. In doing so, this encourages the international collaboration between these organisations, with the goal of improving safety and health in the workplace by sharing their common aims and aspirations. By recognising that occupational safety and health issues are not always limited by national margins, and that many associated laws are often derived from an international perspective, the INSHPO provides a route by which safety and health professionals can collate their policies and develop the standards to operate at a global level. This takes into account that the world is an ever changing place in relation to safety and health at work. Moreover, the annually increasing global distribution of goods, the extensive migration of workforces between countries, and the rise in activities of international corporations, all define the fact that safety and health in the workplace must now routinely be recognised within a global, rather than countrywide, scope.

As such, the INSHPO's inception grew from the understanding that to influence policymakers at an international level, safety and health professionals needed a strong platform from which to emphasis their common interests. To this end, the United States of America's Society of Safety Professionals (ASSP) [24], the Canadian Society of

Safety Engineering (CSSE) [25], and the United Kingdom's Institution of Occupational Safety and Health (IOSH) [26] created the INSHPO in 2001. Since then, its membership has grown and now includes a total of 13 professional member organisations from a total of 11 countries.

### 6.2.7 The Organisation for Economic Co-operation and Development (OECD) [27]

As an international intergovernmental economic group, the Organisation for Economic Co-operation and Development (OECD) plays an important role in relation to topics such as, global environmental health issues, safety in the use of chemicals, the methods by which the evaluation of chemical risks are undertaken, and employee protection from ionising radiation. Established in 1961, it is made up of 37 Member Countries who are each represented by a delegation which collectively form the overall OECD Council. For each work area of the OECD, there are further Substantive Committees, generally made up of subject-matter experts from the respective Member Countries. There is also the OECD Secretariat providing the administrative support to the council and its committees, which is constitutionally led by a Secretary-General.

In undertaking its primary work, first and foremost, the OECD provides a means by which individual national policies on health and safety can be compared, with the aim of identifying good practice where possible and implementing this globally where possible. This in turn can then lead to the formation of common international policies by the OECD, which are accordingly adopted by each of the Council Member Countries. Significantly, many of the OECD resolutions are subsequently implemented by other global organisations, who, by fully recognising the integrity of the resolutions and the valuable work undertaken by the OECD, endorse their important work. Funding for the OECD is by contributions from Member Countries.

### 6.2.8 The European Union [28]

The function of the European Union in regard to occupational safety and health is enacted through the creation of a raft of topic-specific Directives by the Commission of the European Union. The adoption of these Directives by Member States is compulsory, wherein they have to be translated into their own national laws as part of their signing of Article 153 of the Treaty on the Functioning of the European Union [29]. As such, the European Union Directives encompass an immense range of occupational health and safety subjects, all of which are aimed at harmonising the various national legislations of the European Union Member States into one common, coherent procedure.

The Directives themselves set out the topic-related minimum safety requirements and fundamental principles which all Member States have to abide by in order to establish trade within the European Union. In addition, countries outside of the European Union wishing to trade with Member States would also have to adopt the values of the Directives for their specific products. The guiding remits within the Directives would include the methods of risk prevention and risk assessment of all associated work activities, as well as defining the specific responsibilities attributed to both employers and their employees in order for them to manage their work activities safely. The Directives are defined within one of three categories; the framework Directives, those that cover the risks which employees can be exposed to, and those that describe the rules governing the design of work equipment. In addition, associated technical standards are developed by the European Committee for Standardisation, also known as the Comité Européen de Normalisation (CEN) [30], the European Committee for Electrotechnical Standardisation (CENELEC) [31] and the European Telecommunications Standards Institute (ETSI) [32]. A series of European guidelines are also published, with the aim of providing guidance to the implementation of any particular European Directives as well as the European standards.

As part of the overall process of creating and managing the various European Union safety and health Directives, the European Agency for Safety and Health at Work (EU-OSHA) [33] has been established. Their role is to collect, analyse and then disseminate any relevant safety and health information to all parties involved in safety and health at work within the European Union. It is staffed by occupational safety and health experts, along with communication and administrative specialists. The information they provide is distributed throughout the European Union in order to provide policymakers with the technical evidence needed to inform any future policies relating to occupational safety and health. In addition to this collating and analysis function, the EU-OSHA are also directly involved in undertaking safety and health research in order to guide specific risk prevention methods. The information they gain from this process is published through a series of safety and health awareness campaigns.

### 6.2.9 The North American Free Trade Agreement (NAFTA) [34] and The United States–Mexico–Canada Agreement (USMCA) [35]

The North American Free Trade Agreement (NAFTA) is a trade agreement brokered between Canada, the United States of America and Mexico. First coming into force in 1994, its inception created a trilateral trade bloc of the North American countries, which as a result, became one of the largest in the world. Its fundamental remit effectively eliminated or reduced existing trade and commercial barriers between the three countries. In addition to the main agreement, two side agreements were also established at the same time; the North American Agreement on Labour Cooperation (NAALC) [36] and the North American Agreement on Environmental Cooperation (NAAEC) [37]. The main objective of NAALC is to improve working conditions and living standards through consultations and the exchange of trade information and provides an oversight mechanism which ensures that labour laws are being enforced in all three countries. Likewise, the NAAEC agreement is concerned with conservation and the protection of the environment throughout the trilateral trade bloc.

NAFTA has since been replaced with the United States–Mexico–Canada Agreement (USMCA) in 2020. Many of the original provisions from NAFTA have been incorporated into USMCA, with included updates covering certain trade tariffs and a greater assurance of the cross-border flow of technical data. USMCA also strengthens existing environmental and working regulations between the three member countries.

## 6.3 Non-governmental international organisations

### 6.3.1 The International Commission on Occupational Health (ICOH) [38]

The International Commission on Occupational Health (ICOH), is an international non-governmental professional society with the aim of fostering the scientific progress, knowledge and development of occupational health and safety, with an emphasis on occupational diseases. Established in 1906, it is now the leading international scientific organisation in the field of occupational health, providing a forum for the scientific and professional communication on all occupational related activities. It routinely issues guidelines and reports on occupational health and establishes scientific committees which cover a variety of fields of occupational health and safety. Every three years the ICOH hosts a World Congress on Occupational Health [39] where current occupational health research findings are discussed. The ICOH is made up of 26 scientific committees and 4 scientific working groups studying a diverse range of topics including; accident prevention, the chemical industry, industrial hygiene, musculoskeletal disorders and work-related thermal factors.

### 6.3.2 The International Social Security Association (ISSA) [40]

The International Social Security Association (ISSA) is a global organisation of agencies responsible for the administration of social security, in particular, concerned with the prevention of occupational risks. Its aim is to encourage, through national laws, the provision of social security protection in the event of an employment accident or occupational disease for any affected workers and their families. At its heart, the ISSA strives to promote and develop social security throughout the world in order to improve both the social and economic conditions of the world's population. It does this by providing high level professional guidance on social security administration matters, along with delivering expert knowledge, services and support to its members

### 6.3.3 The International Association of Labour Inspection (IALI) [41]

The International Association of Labour Inspection (IALI) was created in 1972 as a means of providing a professional forum for labour inspectors, where the exchange of both information and experience can take place between the various interested parties. It is a non-governmental organisation which promotes the cooperation between the various national inspectorates, authorities and related institutes involved in labour inspection, so that a greater consistency of practice between inspectorates in the different countries of the world can be made. Their remit is broad, covering safety and health, along with the working environment topics such as; the recording systems employed for accidents, labour inspections at large construction sites, human factors, risk assessment and occupational health.

As the complexity of many national labour laws evolve, and as technological innovation becomes more prevalent in the working environment, the IALI's role has adapted to embrace these developments in their own continuing programme of work. For example, the introduction of new chemicals and compounds onto a world market, means the IALI needs to provide employers and employees with up-to-date guidance so that any associated preventative strategies for health and safety relating to the use of the chemicals and compounds can be implemented. At its core, the IALI ensures that no one can be harmed in any way by their need to work. It also recognises the concerns that any industrial impact may have on the environment.

### 6.3.4 The International Organisation for Standardisation (ISO) [42]

The International Organisation for Standardisation (ISO) is a global federation of national standards bodies, made up from a variety of representative countries. Its main objective is to promote the development of standards throughout the world, so that the safe international exchange of goods and services can be facilitated. It supports the interests of producers, users, consumers, national governments and the broader scientific community in their activities. The technical work undertaken by ISO, covering a significant range of topics, are formally published as a series of International Standards [43]. The scope of these standards is all inclusive, except for the standards for electrical and electronic engineering, which are the responsibility of the International Electrotechnical Commission (IEC) [44].

Membership of ISO is through one of three routes; member body, correspondent member or subscriber. A member body is a specific national body representative of standardisation in its country. Generally, these tend to be governmental institutions or organisations incorporated at a national level by public law. A correspondent member on the other hand, tends to be an organisation in a developing country which as yet does not have its own national standards body. A subscriber membership is for countries with much smaller-scale economies who wish to embrace and observe the development of any future standards as they are published.

The technical work undertaken by ISO is conducted through their technical committees, with the proposal for the work generally coming from a member body or other international organisations. This work then feeds into the development of International Standards, which are either adopted directly by member bodies or implemented through being incorporated into a country's own national standards. The imperative with all these International Standards is that of occupational safety, fulfilled within a wide range of topic areas including; aerospace, buildings, medical equipment, the environment, personal safety and ergonomics. Each of these International Standards are continuously reviewed and updated accordingly to reflect current research findings and contemporary experience.

Some specific ISO standards applicable to the use of cryogenic equipment would include the following;

#### *6.3.4.1 ISO 21029–1:2018*

Specifies the requirements for the design, fabrication, initial inspection and proof test of transportable vacuum-insulated cryogenic fluids storage vessels of not more than 1000 litres volume.

*6.3.4.2 ISO 20421–1:2006*

Specifies the requirements for the design, fabrication, initial inspection and proof test of large transportable vacuum-insulated cryogenic fluids storage vessels of more than 450 litres volume, which are either permanently or temporarily attached to a means of transport. This includes all fixed tanks, demountable tanks and portable tanks employed for one or more modes of transport. The standard does not however cover the general vehicle requirements, such as the functioning of brakes and lighting, which are included in their own associated relevant standards and regulations.

*6.3.4.3 ISO 21009–1:2008*

Specifies the requirements for the design, fabrication, initial inspection and proof test of static vacuum-insulated cryogenic fluids storage vessels designed for a maximum allowable pressure of more than 0.5 bar. In the same context, the standard can also be used as a guide for storage vessels with a maximum allowable pressure of not more than 0.5 bar.

*6.3.4.4 ISO 13985:2006*

Specifies the construction requirements for refillable fuel tanks for liquid hydrogen which are intended to be permanently attached to land vehicles. It also includes the required testing methods of the fuel tanks to ensure that a reasonable level of protection from loss of life and property resulting from fire and explosion is provided as part of the construction remit.

*6.3.4.5 ISO 21011:2008*

Specifies the requirements for the design, manufacture and testing of valves which are rated to operate at temperatures from ambient to cryogenic. The standard applies to all types of cryogenic valves, including vacuum-jacketed cryogenic valves up to a DN 150 size.

*6.3.4.6 ISO 21013–1:2008*

Specifies the requirements for the design, manufacture and testing of pressure relief valves which are rated to operate at temperatures from ambient to cryogenic and are to be used specifically with cryogenic fluids. The standard applies to all types of relief valve up to a DN 150 size which are designed to relieve single-phase vapours or gases. The relief valves can also be constructed and tested such that they are suitable for use with more than one gas or even with mixtures of gases.

*6.3.4.7 ISO 21013–3:2016*

Specifies the calculation methods for determining the required mass flow of pressure relief valves for a set of specified conditions:

- Vacuum-insulated vessels with an outer jacket and insulating material intact under normal vacuum, outer jacket at ambient temperature, inner vessel at the temperature of the contents and at the specified relief pressure

- Vacuum-insulated vessels with an outer jacket and insulating material intact under normal vacuum, outer jacket at ambient temperature, inner vessel at the temperature of the contents and at the specified relief pressure, with the pressure regulator of the pressure source functioning at full potential
- Vacuum-insulated vessels with their insulation system remaining in place, but with the loss of vacuum, outer jacket at ambient temperature, inner vessel at the temperature of the contents and at the specified relief pressure, under fire engulfment conditions
- Vacuum-insulated vessels containing fluids which have a saturation temperature below 75 K at 1 bar with their insulation system remaining in place, but with a loss of vacuum, with air or nitrogen in the vacuum space
- Vacuum-insulated vessels containing fluids which have a saturation temperature below 75 K at 1 bar with their insulation system remaining in place, but with a loss of vacuum, with air or nitrogen in the vacuum space and under fire engulfment conditions
- Vessels with insulation systems totally lost and under fire engulfment conditions

#### *6.3.4.8 ISO 4126–1:2013*

Specifies the general requirements for pressure relief valves irrespective of the fluid for which they are designed. It applies to pressure relief valves with a flow diameter of 4 mm and above, with a set pressure of 0.1 bar gauge and above. The standard places no limitations on operating temperature of the safety valves.

#### *6.3.4.9 ISO 24490:2016*

Specifies the minimum requirements for the design, manufacture and initial testing of pumps for cryogenic use. This includes both centrifugal and reciprocating type pumps.

#### *6.3.4.10 ISO 21014:2019*

Defines a set of practical methods by which the heat-leak performance of cryogenic storage vessels can be determined. The methods can be applied to both open and closed cryogenic systems, but does not specify the requirement levels for specific insulation performance.

#### *6.3.4.11 ISO 21028–1:2016*

Specifies the toughness requirements of metallic materials used for the construction of cryogenic storage vessels operating in the temperature region below −80 °C. The standard is not applicable however for unalloyed steels or cast materials.

### 6.3.5 The International Electrotechnical Commission (IEC) [44]

The International Electrotechnical Commission (IEC) is the international organisation that prepares international standards for all electrical, electronic and related technologies. The standards they publish cover topics such as; power generation, transmission and distribution, home appliances, office equipment, semiconductors, fibre optics, batteries, electromagnetics, solar energy and marine energy. In addition, there are also IEC standards defining; terminology and symbols used in the industry, electromagnetic compatibility, measurement and performance, and safety and the environment. As part of their remit, the IEC also manages a set of global conformity assessment systems that certify whether electrotechnical equipment conform to its international standards.

## Relevant international safety and health organisations—summary

**General**

*National management of health and safety*—The management of health and safety in the workplace is generally assumed at a national level, with laws and regulations being applied specifically applicable to the country within which the work is being undertaken.

*International management of health and safety*—Many health and safety organisations have been established allowing international collaboration to take place where trade across national borders forms an integral part of the distribution of products and the provisioning of services throughout the World.

*International organisations*—International health and safety related organisations can be divided into two distinct types; those classed as intergovernmental and those classed as non-governmental.

**Intergovernmental International Organisations**

*The United Nations Economic and Social Council*
The United Nations Economic and Social Council is a principal organisation within the remit of the United Nations, responsible for managing the economic and social elements of the United Nations through a series of specialised agencies, such as:

*International Civil Aviation Organisation (ICAO)*
The International Civil Aviation Organisation sets out the various codes and principles which determine all types of international air transport. They define the techniques of international air navigation in order to ensure the safety of the industry.

*International Labour Organisation (ILO)*
The International Labour Organisation sets out the various International labour standards in order to provide both social justice and promote decent work for employees throughout the World.

*International Maritime Organisation (IMO)*
The International Maritime Organisation sets out the framework for the International maritime safety and practices. Their aim is to improve maritime safety and to prevent marine pollution through the regulation of all shipping activities.

*World Health Organisation (WHO)*
The World Health Organisation is a coordinating authority for all matters relating to international public health, with a broad spectrum of responsibilities covering such things as health, sanitation and diseases, and coordinating international responses to health emergencies.

*The International Network of Safety and Health Professional Organisation (INSHPO)*
The INSHPO provide a global forum for the various national occupational safety and health professions throughout the World, with the goal of improving safety and health in the workplace by sharing their common aims and aspirations.

*The Organisation for Economic Co-operation and Development (OECD)*
The OECD plays an important role in relation to global environmental health issues, safety in the use of chemicals, the methods by which the evaluation of chemical risks are undertaken, and employee protection from ionising radiation.

*The European Union*
The European Union provides topic-specific Directives relating to occupational safety and health which have to be adopted by Member States and translated into their own national laws. These European Union Directives cover an immense range of occupational health and safety subjects, all of which are aimed at harmonising the various national legislations of the Member States.

*The North American Free Trade Agreement (NAFTA) and the United States–Mexico–Canada Agreement (USMCA)*
The NAFTA and USMCA are agreements established between Canada, the United States of America and Mexico, effectively eliminating or reducing trade barriers between the three countries and strengthening existing environmental and working regulations between the three member countries.

**Non-Governmental International Organisations**

*The International Commission on Occupational Health (ICOH)*
The ICOH is an international non-governmental professional society with the aim of fostering the scientific progress, knowledge and development of occupational health and safety, with an emphasis on occupational diseases.

**The International Social Security Association (ISSA)**

The ISSA is a global organisation of agencies responsible for the administration of social security, in particular, concerned with the prevention of occupational risks. Its aim is to encourage, through national laws, the provision of social security protection in the event of an employment accident or occupational disease for any affected workers and their families.

*The International Association of Labour Inspection (IALI)*
The IALI provides a professional forum, where the exchange of both information and experience can take place between labour inspectors. It is a non-governmental organisation which promotes the cooperation between the various national inspectorates, authorities and related institutes involved in labour inspection, so that a greater consistency of practice between inspectorates in the different countries of the world can be made.

*The International Organisation for Standardisation (ISO)*
The ISO is a global federation of national standards bodies, made up from a variety of representative countries. Its main objective is to promote the development of standards throughout the world, so that the safe international exchange of goods and services can be facilitated.

*The International Electrotechnical Commission (IEC)*
The International Electrotechnical Commission (IEC) is the international standards organisation that prepares international standards for all electrical, electronic and related technologies.

## References

[1] The United Nations (UN) (www.un.org/en/) (Accessed 2020)
[2] The Organisation for Economic Co-operation and Development (OECD) (www.oecd.org/) (Accessed 2020)
[3] Europa.eu *The European Union (EU)* (https://europa.eu/european-union/index_en) (Accessed 2020)
[4] International Trade Administration (ITA) *The North American Free Trade Agreement (NAFTA)* (www.trade.gov/north-american-free-trade-agreement-nafta) (Accessed 2020)
[5] The International Commission on Occupational Health (ICOH) (www.icohweb.org/site/homepage.asp) (Accessed 2020)
[6] The International Social Security Association (ISSA) (https://ww1.issa.int/) (Accessed 2020)
[7] The International Association of Labour Inspection (IALI) (www.iali-aiit.org/) (Accessed 2020)
[8] The International Organisation for Standardisation (ISO) (www.iso.org/home.html) (Accessed 2020)
[9] The United Nations Economic and Social Council (www.un.org/ecosoc/en/home) (Accessed 2020)
[10] The United Nations (UN) (www.un.org/en/) (Accessed 2020)
[11] International Civil Aviation Organisation (ICAO) (www.icao.int/Pages/default.aspx) (Accessed 2020)
[12] The International Civil Aviation Organisation (ICAO) *The Convention on International Civil Aviation* (www.icao.int/publications/pages/doc7300.aspx) (Accessed 2020)
[13] The International Civil Aviation Organisation (ICAO) *The Air Navigation Commission (ANC)* (www.icao.int/about-icao/AirNavigationCommission/Pages/default.aspx) (Accessed 2020)
[14] The International Civil Aviation Organisation (ICAO) *The ICAO's International Standards and Recommended Practices for Air Transport* (www.icao.int/safety/SafetyManagement/Pages/SARPs.aspx) (Accessed 2020)
[15] The International Air Transport Association (IATA) (www.iata.org/) (Accessed 2020)
[16] The Civil Air Navigation Services Organisation (CANSO) (https://canso.org/) (Accessed 2020)
[17] The Airports Council International (https://aci.aero/) (Accessed 2020)
[18] International Labour Organisation (ILO) (www.ilo.org/global/lang--en/index.htm) (Accessed 2020)
[19] International Labour Organisation (ILO) *International Labour Standards* (www.ilo.org/global/standards/lang--en/index.htm) (Accessed 2020)
[20] International Maritime Organisation (IMO) (www.imo.org/) (Accessed 2020)
[21] International Maritime Organisation (IMO) *The International Convention for the Prevention of Pollution from Ships* (www.imo.org/en/About/Conventions/Pages/International-Convention-for-the-Prevention-of-Pollution-from-Ships-(MARPOL).aspx) (Accessed 2020)

[22] World Health Organisation (WHO) (www.who.int/) (Accessed 2020)
[23] The International Network of Safety and Health Professional Organisation (INSHPO) (https://inshpo.org/) (Accessed 2020)
[24] The United States of America's Society of Safety Professionals (ASSP) (www.assp.org/) (Accessed 2020)
[25] The Canadian Society of Safety Engineering (CSSE) (www.csse.org/) (Accessed 2020)
[26] The United Kingdom's Institution of Occupational Safety and Health (IOSH) (https://iosh.com/) (Accessed 2020)
[27] The Organisation for Economic Co-operation and Development (OECD) (www.oecd.org/) (Accessed 2020)
[28] Europa.eu *The European Union (EU)* (https://europa.eu/european-union/index_en) (Accessed 2020)
[29] EUR-LEX *Article 153 of the Treaty on the Functioning of the European Union* (https://eur-lex.europa.eu/collection/eu-law/treaties/treaties-force.html) (Accessed 2020)
[30] The European Committee for Standardisation/The Comité Européen de Normalisation (CEN) (https://standards.cen.eu/) (Accessed 2020)
[31] The European Committee for Electrotechnical Standardisation (CENELEC) (www.cenelec.eu/aboutcenelec/whoweare) (Accessed 2020)
[32] The European Telecommunications Standards Institute (ETSI) (www.etsi.org/) (Accessed 2020)
[33] The European Agency for Safety and Health at Work (EU-OSHA) (https://osha.europa.eu/) (Accessed 2020)
[34] International Trade Administration (ITA) *The North American Free Trade Agreement (NAFTA)* (www.trade.gov/north-american-free-trade-agreement-nafta) (Accessed 2020)
[35] The Office of the United States Trade Representative *The United States–Mexico–Canada Agreement (USMCA)* (https://ustr.gov/trade-agreements/free-trade-agreements/united-states-mexico-canada-agreement) (Accessed 2020)
[36] The U.S. Department of Labor—Bureau of International Labor Affairs *The North American Agreement on Labour Cooperation (NAALC)* (www.dol.gov/agencies/ilab/trade/agreements/naalcgd) (Accessed 2020)
[37] The Office of the United States Trade Representative *The North American Agreement on Environmental Cooperation (NAAEC)* (https://ustr.gov/sites/default/files/naaec.pdf) (Accessed 2020)
[38] The International Commission on Occupational Health (ICOH) (www.icohweb.org/site/homepage.asp) (Accessed 2020)
[39] International Labour Organisation (ILO) *World Congress on Occupational Health* (www.ilo.org/global/topics/safety-and-health-at-work/events-training/events-meetings/world-congress) (Accessed 2020)
[40] The International Social Security Association (ISSA) (https://ww1.issa.int/) (Accessed 2020)
[41] The International Association of Labour Inspection (IALI) (www.iali-aiit.org/) (Accessed 2020)
[42] The International Organisation for Standardisation (ISO) (www.iso.org/home.html) (Accessed 2020)
[43] The International Organisation for Standardisation (ISO) *International Standards* (www.iso.org/standards.html) (Accessed 2020)
[44] The International Electrotechnical Commission (IEC) (www.iec.ch/homepage) (Accessed 2020)

# Chapter 7

# Cryogenic hazards

## 7.1 General

Hazards exist in everything we undertake, from the simplest of tasks, such as making a hot cup of coffee, to the extremes associated with the handling of significantly dangerous substances. By definition, a hazard is something that has the potential to cause harm [1]. As such, this harm encompasses not only potential injuries which might be caused directly by the hazard, but also other types of detrimental effect, such as diseases, property damage and the irreparable damage caused to the environment by certain specific hazards. Hazard perception however is subjective, and what may appear as an extreme hazard to a cautious individual may be discerned as a low risk to a person with a more adventurous character. The initial reconciliation of hazards and their correct identification is therefore critical in any risk management process. This is especially so where an assessment of the latent hazards in the workplace is key to making the necessary risk reductions in order for safe working practices to take place there.

That being said, hazards can be either natural in characteristic or those created directly by humans. An example of a natural hazard would be the sort of inclement weather experienced during the winter months, where strong stormy winds can be a danger to both life and property. Ironically, some of these natural hazards can be exacerbated by human interaction, such as situations where river flooding can occur due to poor land management, especially where inadequate drainage facilities have been provided in floodplain regions. Largely, hazards in the workplace will be brought about through the direct interaction with human activities and may involve a product hazard, such as the use of chemicals, or a process hazard such as a machining operation. The risks associated with any of these hazards can be further defined as the probability that exposure to the hazards will result in a negative consequence. Significantly for risk-reduction systems, a hazard poses no risk if there is no actual exposure to it.

doi:10.1088/978-0-7503-3123-4ch7  

Hazards are typically classified based on their potential, but they can often be of a dormant nature, thereby representing a lower probability of harm occurring from exposure to them in this state until some secondary process elevates them to a more significant status. Any interaction with a hazard is referred to as an incident, with the severity of the consequences with that interaction ranging from very minor, all the way through to extreme. An extreme consequence would characteristically encompass fatalities and life-changing injuries. All risk assessment should look specifically at the hazard and the associated consequences that exposure to the hazard would produce, and must consider the probability of these consequences occurring. If for example, there is no possibility of a hazard contributing towards an incident, then the overall risk associated with that hazard is deemed low.

In many situations relating to cryogenic technology, hazards are prevalent throughout the workplace. A large proportion of these hazards can be identified based on the presence of stored energy, which on sudden release, could cause damage or harm. Typical forms of this stored energy would be pressurised gases, liquid cryogen storage, reactive chemicals and the introduction of heat into a cryogenic system. Other related hazards would typically involve the presence of hazardous situations rather than a physical product. Examples included in this category would be work in confined spaces, oxygen depleted atmospheres and manual handling activities, in particular where repetitive motions are involved.

Less tangible processes included in the cryogenic technology framework can also provide their own specific hazards. One such category would be the potential biological hazards associated with the cryogenic-freezing of organic samples. Often referred to as biohazards, these can be defined as biological agents that pose a threat to the health of many living organisms, as well as to the environment. Typical biological hazards would include viruses, parasites, bacteria, fungi, toxins and disease-causing pathogens. The effects which biological hazards can have on the human body normally manifest themselves as diseases, typically contracted in the form of infections by bacteria, antigens, viruses or parasites.

Many workplace applications, including these linked to the use of cryogenic technology, will involve the use of chemicals. By virtue of their intrinsic properties, these chemicals can cause harm to humans, property, and the environment. Health hazards associated with the use of chemicals very much depend on the amounts being used and as a consequence, the likely exposure to them. Often, chemicals have specifically defined absolute exposure limits [2] or intake limits [3] associated with them, above which the possibility of death would be high. Others in contrast, can have a cumulative biological effect over time, causing diseases or long-term damage to the human body. Some types of chemicals used in industrial situations can often cause respiratory, digestive or nervous system problems if they are accidentally inhaled, ingested or absorbed through the skin. Fortunately, most chemical hazards these days have been identified and are well defined. However, it is important to recognise that new chemicals are being introduced to the market each year, many of which will have only a limited amount of technical and human reaction data associated with them on their initial release. Harmful chemicals can also occur naturally due to geological irregularities in certain regions. These would include things like the release of radon gas from igneous and metamorphic rocks such as granite, or even natural arsenic contamination of groundwater supplies.

The application of cryogenic technology often involves the movement of large items of equipment. As such, there will be associated ergonomic hazards with these undertakings, where the prevailing physical conditions may pose the risk of injury to the muscles and ligaments of the human frame. Specific ergonomic hazards would include things like awkward or extreme postures due to the nature of the work being undertaken, repetitive motions, and poor lighting of the working environment. Ergonomic hazards are not limited to just the workplace however and very often occur in non-occupational settings as well, with a consequential impact on a person's ability to subsequently undertake work as normal.

In addition to the many primary factors such as mechanical hazards involving the use of machinery or industrial processes, there can also be so-called psychological hazards found in the workplace. These affect the general wellbeing of employees within organisations, which can have a profound consequence on their ability to undertake their duties, especially when they are amongst other people. Recognised as a particular challenge where employees are struggling with mental health problems, these factors are often linked openly to direct psychosocial hazards such as occupational stress and workplace harassment.

Hazards which are brought about by human actions can often result in a range of resultant effects to not only other humans but to infrastructure buildings and the ecosystems making up the immediate environment. Equally so, hazards can also result from the direct inaction by humans, such as the failure to close a venting valve on a chemical processing assembly which could result in the release of excessive quantities of dangerous chemicals. Whatever the initial cause of the hazard, it is important that the ensuing impact they can have is completely understood before procedures are implemented to mitigate their bearing. It is no good for example, spend vast amounts of money for protection against chemicals, when in reality those being used are considered to be passive and pose little overall risk to their users. Therefore, having good processes in place for the correct identification of hazards and their consequences is key to any successful safety management system being established in the workplace.

## 7.2 Identifying hazards

Often, one of the most difficult aspects of assessing the safety of a process, is the initial identification of all the related hazards associated with a particular procedure and then recognising what existing controls are in place which may satisfactorily mitigate their effect. Likewise, understanding these two fundamentals will also inform any assessor as to the level of further risk-reduction controls still needed in order to lower any remaining risks. With cryogenic technology, there are many potential hazards associated with its use, all of which need to be reconciled before any work with the technology can begin. These can range from the more apparent, such as the extremely cold ice formations around shut-off valves which pose a health risk, to the invisible hazards found where oxygen depletion has occurred in the working environment due to an accidental release of inert gas. It is often difficult therefore to simply ascertain all the particular hazards for any specific situation and an overall approach towards hazard identification needs to be adopted. Sometimes they are obvious, such as an exposed blade on a cutting machine, which is clearly extremely sharp. Others are less so, such as a faulty piece of electrical equipment which looks from the exterior to be in a perfect working condition, but would not manifest itself until the equipment is eventually used.

A useful method therefore of proving that all aspects of a process have been considered is to use the following detailed checklist for the working environment. This goes some way towards ensuring that all the relevant hazards in any particular cryogenic technology application have been deliberated.

### 7.2.1 Premises

A check of the physical environment in which any cryogenic technology is being used. This could include considering aspects such as:

- The dimensions of the work area.
- What entry/exist access there is?
- The condition of the flooring.
- The lighting conditions.
- What ventilation is present?
- The general tidiness of the work area.
- Are there any existing accident reports [4] pertaining to the work area?
- If there are accident reports, have the recommendations in the report been applied?

### 7.2.2 Plant and substances

This is to check the type of cryogenic and secondary equipment being used in the work area, ensuring that they are fit for purpose, as well as looking at any chemicals or substances which may form part of the overall process being undertaken. This could include considering:

- What equipment is being used in the work area?
- Has this equipment been suitably maintained or inspected?
- Is the equipment being used for the correct application?
- Are cryogens being stored in the work area?
- What chemicals and substances are being used in the work area?
- Are the safety data sheets [5] available for these chemicals and substances?
- Are the associated mitigation controls being applied for the use of these chemicals and substances?

### 7.2.3 Procedures

The use of any cryogenic equipment should have procedures in place for their safe operation. These would include any relevant safe systems and its associated documentation [6], instructions on how to use the equipment safely and the method of work to be used for specific activities. This could include:

- What safe systems are currently in place?
- What instructions are available for the use of equipment in the work area?
- Have staff been correctly instructed on how to operate the equipment in the work area?
- Are method statements [7] in place for specific tasks?
- Is a permit system [8] employed for specific operations?

### 7.2.4 People

This covers the individuals undertaking tasks involving the use of cryogenic technology. It recognises their training levels as well as their competency and suitability for the activity being employed. Competencies can be a difficult area to quantify, as simply attending training does not necessarily mean that an individual is by default competent to undertake certain tasks. In the truest sense, competencies are not skills which can be simply learned, but rather stand as inherent qualities and characteristics an individual would possess which are desirable for the safe application of cryogenic technology. This could include:

- What specific training have staff undertaken which allows them to operate in the work area?
- Is this training up-to-date?
- What additional experience do staff have which supports this training?
- Are there any health/physical reasons why the staff cannot undertake specific tasks in the work area?
- Are any staff on a health surveillance programme associated with specific tasks they undertake in the work area?
- If they are, are the necessary controls being applied to the work they undertake?

## 7.3 Specific hazards associated with cryogenic technology

### 7.3.1 Pressure build-up

This hazard relates specifically to the storage vessels which are used to keep cryogens in their cold, liquid state. Pressure build-up is a continuous process which is caused directly by the gradual warming over time of the cryogenic liquids within their storage vessels. This in turn results in the partial evaporation of the cold liquid into its warmer gaseous phase. Eventually, this vaporisation process can create a significant rise in vapour pressure above the liquid cryogen surface, which in a sealed system could potentially exceed the design pressure limits of the containment system. This can cause dangerous conditions where the containment vessel may bulge in order to accommodate the increase in internal pressure, possibly causing permanent deformations to the structure. If not properly controlled and released by suitable measures, eventually this excess vapour pressure can result in the catastrophic failure of some critical part of the storage vessel. As pressurised gas has more potential energy than pressurised liquids [9], the rupture of gas systems can be incredibly energetic in release, with devastating consequences, normally consisting of an explosion of the containment vessel. A typical failure of a pressurised metal containment vessel would expel metallic shrapnel over some significant distance, along with producing an expanding pressure-wave which itself can cause further damage and injury if not correctly mitigated. It is therefore important that wherever this pressure build-up is likely, some form of protection system is fitted to prevent the possible failure of the storage vessel.

Typically, this would be a small pressure relief valve (PRV) [10], of suitable specification, incorporated into the storage vessel design so that as any internal pressure is generated due to the evaporative process, the spring-loaded relief valve will operate. In doing so, this will open a small gas exit route, allowing a quantity of the pressurised gas to be released from the vessel into the atmosphere. As the internal pressure is consequentially lowered due to the release of this excess pressure, the balancing forces against the spring will reduce and the PRV will eventually reseal. The process of pressure build-up will then repeat, followed by the PRV operating again when the pressure is high enough to bring it into use once more. As a component part of a pressure system, in many countries throughout the world, this PRV has to be formally registered before it can be implemented and would also need to undergo a statutory inspection [11] for its continued use to be maintained. As most PRVs are sealed by using a spring to generate the required sealing force, it is not uncommon over time that these parts need to be routinely replaced as their efficiency is impaired, resulting in them operating at pressures beyond their design range. PRVs generally come as pre-set rated versions or those which can be adjusted over a pressure range based on the particular spring used in the valve.

Although pressure build-up rates may vary depending on factors such as the size of the storage vessel and how often they are moved, they are typically quoted as being between 1% and 4% of the vessel volume per day [12]. This is very much governed by the age of the storage vessel and the integrity of the thermal insulation around it. As the insulation will inevitably degrade over time, more external heat will as a result be transferred to the inside of the storage vessel, more readily raising the temperature of the liquid cryogen and therefore accelerating the evaporation of it into its gaseous form. Irrespective of the pressure build-up rate, the fitting of a PRV will periodically vent small amounts of gas from any storage vessel into the surrounding atmosphere as a result of this constant thermodynamic process. It is important that this released gas is quantified, as over a period of time, this can accumulatively add to the existing workplace atmosphere which may eventually lead to low oxygen levels as more of the cryogen gas is steadily released. Clearly, the older the storage vessel are, the more gas is likely to be released and the higher the risk of low oxygen levels being prevalent in the workplace.

There are occasions however, when the pressure build-up inside the cryogen storage vessel may exceed that experience through the normal evaporative build-up and could be sufficiently high enough for the PRV to be continuously fully open without a discernible drop in the internal pressure taking place. In these situations, pressure can still dangerously increase inside the vessel as there is a defining limit to the quantity of gas which the PRV can release in its fully open state. This situation would occur for example if there were a major failure of the storage vessel insulation and a rapid boil-off of the contained cryogen occurs as a consequence. It is normal therefore for a pressure-release bursting disc [13] to be included in parallel within the safety release circuit which will operate when a pressure build-up exceeds a value slightly higher than the pressure release setting of the PRV. Generally, burst discs offer a full-bore exit route for any pressurised gas, which taken in conjunction with the gas already being vented through the PRV, should provide enough gas escape for a boil-off situation to be suitably controlled.

It is worth noting though that bursting discs are one-off use safety devices, in that they literally burst open in a controlled way and do not reseal as the PRVs do once the pressure build-up has been reduced. After they have operated, bursting discs will require replacing with a new disc before the pressure system can be sealed again. Also, bursting discs are rated for a specific pressure and cannot be adjusted as some of the PRVs can be. Typically, they have a larger tolerance range compared to PRVs as their bursting operation is based on the failure mechanics of the material they are made from. Typically, these can be as high as ±10% of the operating pressure [14]. It is important therefore that where they are used in parallel with a PRV, that the lower setting value of the burst disc is always rated higher than that of the PRV.

### 7.3.2 Expansion ratio

The cause of any pressure build-up in cryogenic storage vessels is as a direct result of the respective cryogen's expansion ratio. This is the ratio of the volume it will occupy as a liquid at its boiling point temperature, compared to the volume it will occupy in its gaseous form at ambient temperatures and normal atmospheric pressure. Notably, each cryogen has a different expansion ratio and therefore a differing level of hazard severity depending on which is being used. The following chart (figure 7.1) demonstrates the range of these expansion ratios [15, 16]. It is worth noting that these values often vary depending on the primary source of the data and will therefore differ marginally across the literature they are derived from.

| | Gas volume from liquid | Relative Density |
|---|---|---|
| Liquid methane | 1 : 623 | 0.56 |
| Liquid nitrogen | 1 : 710 | 0.9 7 |
| Liquid helium | 1 : 780 | 0.14 |
| Liquid hydrogen | 1 : 865 | 0.07 |
| Liquid oxygen | 1 : 875 | 1.11 |
| Liquid neon | 1 : 1470 | 0.70 |
| Liquid argon | 1 : 860 | 1.39 |
| Liquid $CO_2$ | 1 : 790 | 1.70 |

**Figure 7.1.** Showing the various expansion ratios and relative densities for the most common liquid cryogens.

Having noted this fact, it is important to recognise however, that irrespective of the actual cryogen being used and the definitive value of its expansion ratio, the expansion from liquid into gas is always going to be significantly high and represents a notable hazard for any application of cryogenic technology where liquid cryogens are present.

Considering these expansion ratios in more detail for specific cryogens, liquid nitrogen for example, would produce over 700 litres of gas from 1 litre of liquid if allowed to fully evaporate. For liquid helium, this increases to nearly 800 litres of gas for every 1 litre of liquid. As an illustration and an attempt to put this in context, an average family-sized travel suitcase has an internal capacity of approximately 50 litres, which in turn is the typical quantity of cryogens used in a small capacity cryostat. If this 50 litres capacity suitcase were therefore full of liquid nitrogen which was then released and allowed to warm up and evaporate, it would result in about 35 $m^3$ of gas being produce, expanding into its surroundings. 35 $m^3$ is the internal volume of an average sized small bedroom in a modern house with dimensions of 3.5 m × 3.5 m × 3.0 m high. Any such gas filling a room in this way would have the effect of displacing the breathable oxygen and thereby making the room a dangerous place to occupy. As can be seen, to entirely fill a small room with nitrogen gas only requires a relatively small amount of that gas in its liquid state. It is not uncommon for facilities utilising cryogenic technology to operate from small laboratories where cryogen storage vessels in the hundreds, if not thousands of litres capacity are employed. As a primary hazard, the expansion ratio of the cryogens must therefore always be considered whatever the application. As a quick reference guide, figures 7.2 and 7.3 present the oxygen concentration values resulting from a release of a variety of quantities of liquid nitrogen and liquid helium respectively into a range of room volumes. Safe values of oxygen concentration are represented on the charts in green, with unsafe and dangerous levels marked in orange and red. The normal oxygen

| Room volume $m^3$ | Volume of liquid nitrogen released, litres | | | | | |
|---|---|---|---|---|---|---|
| | **1** | **2** | **5** | **10** | **25** | **50** |
| **10** | 19.3 | 17.8 | 13.4 | 6.0 | 0.0 | 0.0 |
| **25** | 20.2 | 19.6 | 17.8 | 14.8 | 6.0 | 0.0 |
| **50** | 20.5 | 20.2 | 19.3 | 17.8 | 13.4 | 6.0 |
| **75** | 20.6 | 20.4 | 19.8 | 18.8 | 15.8 | 10.9 |
| **100** | 20.6 | 20.5 | 20.0 | 19.3 | 17.1 | 13.4 |

**Figure 7.2.** Showing the oxygen concentration of a set of room sizes for a variety of liquid nitrogen releases.

| Room volume $m^3$ | Volume of liquid helium released, litres | | | | | |
|---|---|---|---|---|---|---|
| | **1** | **2** | **5** | **10** | **25** | **50** |
| **10** | 19.1 | 17.5 | 12.6 | 4.5 | 0.0 | 0.0 |
| **25** | 20.1 | 19.5 | 17.5 | 14.3 | 4.5 | 0.0 |
| **50** | 20.4 | 20.1 | 19.1 | 17.5 | 12.6 | 4.5 |
| **75** | 20.5 | 20.3 | 19.7 | 18.6 | 15.3 | 9.9 |
| **100** | 20.6 | 20.4 | 19.9 | 19.1 | 16.7 | 12.6 |

**Figure 7.3.** Showing the oxygen concentration of a set of room sizes for a variety of liquid helium releases.

concentration in the air is 20.8%, with the minimum safe content for entry into a working area generally recognised at around 19.5%. Anything below this level is therefore deemed unsafe for personnel to be present and entry is generally forbidden.

### 7.3.3 Hot burns

Suffering any type of elevated temperature burn whilst using cryogenic technology seems to be at best an oxymoron. The very contradiction in the fact that the temperatures needed to cause hot burns are far removed from the extreme cold of cryogenics, presents this natural paradox. But hot burns can arise in situations where the leaking of specific cryogens, which by their makeup are also flammable, can ignite and cause any resultant fire. These would typically include liquid and gaseous hydrogen, and natural gas which is composed mostly of methane. The burns can be caused either by direct contact with the flames emitted from any such fire, but can also be the result of radiated heat from the heart of the fire. Most gases burn with a bright luminous flame which are generally easy to see and avoid. Hydrogen flames however give off very little visible radiation and are therefore much harder to detect, especially in well-lit situations.

A typical hot burn will cause immediate damage to any exposed skin, directly caused by the dry heat of the associated fire. Burns of this nature are generally found to be very painful and may result in the reddening or peeling of the skin, blisters, swelling, or even the formation of white or charred skin. Hot burns must be immediately brought down in temperature by applying cool or lukewarm water on the affected area for at least 20 min The burn can then be covered by placing something like cling-film over it to prevent infection into the injury. Depending on the severity of the burn, minor injuries should be kept clean, whereas more serious burns would require medical attention. If the injured person has also inhaled any smoke or fumes being released by the fire, they should also seek medical attention as damage to both the throat and the lungs is also likely under these circumstances.

Burn severity is defined by the nature of the actual damage to the skin and its respective layers. The skin is made up of three discrete layers; the outer layer, known as the epidermis, the layer beneath this which contains blood capillaries and nerve endings, known as the dermis, and the deeper layer of fat and tissue beneath this, known as the subcutaneous fat. Burns are defined as one of four types, each having a different appearance and different associated symptoms [17].

#### *7.3.3.1 Superficial epidermal burn*

This is where the outer epidermis is damaged. The skin will typically appear red and will be swollen and painful, but will not blister.

#### *7.3.3.2 Superficial dermal burn*

This is where both the epidermis and part of the dermis are damaged. In this case the skin will have a pale pink appearance and would be painful. There may be small blisters around the affected area.

*7.3.3.3 Deep dermal burn*
This is where the epidermis and the dermis layers are both damaged. Here the skin will turn red and blotchy and can either be dry or moist. The skin will also become swollen and blistered. Ironically, this type of burn will either be very painful or relatively painless, depending on the amount of damage to any localised nerve endings.

*7.3.3.4 Full thickness burn*
This is where all three layers of the skin are damaged. The skin under these conditions will normally be burnt away and the exposed tissues underneath may appear pale or blackened. The remaining skin will be dry with no blisters and range from white to black in colour and have the texture of leather.

### 7.3.4 Cold contact burns

Having already considered the extremely low temperatures involved in the use of cryogenic technology in earlier chapters, and how they can be a direct danger to the human body, it is self-apparent that any accidental interaction with equipment at these temperatures can be injurious to a varying degree of severity. Contact with cryogens or cold equipment will produce cold-burn effects to the skin, similar in fact to those cause by elevated temperature burns which we are all likely to be more familiar with. Touching uninsulated pipes for example carrying cryogenic liquids may even result in the skin freezing to the pipework as the temperatures involved are sufficiently low enough that they can very rapidly cause permanent damage to the skin and the underlying tissues. Cold contact burns very often leave the damaged area with blisters or even deeper tissue damage that unfortunately is often irreparable.

Sustaining cold contact burns will come about due to the skin being exposed to the extremely cold temperatures of things like liquid cryogens or parts of equipment already operating in the cryogenic region. Exposure rates can obviously vary depending on the operation being undertake, but coming into prolonged contact with any extremely cold item will rapidly decrease the circulation in the exposed skin and the body's natural heat quickly transfers from the skin to the cold object. With this decreased circulation, the skin and the underlying tissues will begin to lose the sense of feeling and the possibilities of blood clots occurring are increased due to the slower movement of blood locally to the injury. As a result, the skin tends to turn blue, with swellings appearing as the tissues themselves react to this rapid loss of circulatory blood. The skin will then turn white as the freezing effects of the underlying tissues progresses until blisters form. At this point, the original lack of sensation is negated by a sudden impulse of pain through the local region of the injury.

Whatever the level of cold contact exposure, it is critical that medical assistance is obtained as quickly as possible in order to mitigate the consequential effects. In fact, if cold contact burns are caught in sufficient time, it is possible for a complete recovery from the injury to be made. It is not uncommon however, that permanent damage to the blood vessels in the affected area of the skin will result and that an enduring loss of sensation in the injured area will be apparent. It is also important to remember that any gases released as a result of the vaporisation of cryogenic liquids will remain extremely cold for some period and can permanently damage the more delicate surface of the eyes if prolonged contact with these gases is made. The wearing of suitable eye-protection is therefore critical where cold vapour clouds are likely to be prevalent in whatever cryogenic operation is being undertaken. For cryogenic burn injuries, local first aiders would normally be trained in the treatment of this type of injury and should have the correct medical supplies to support any such treatment. In an ideal situation, facilities where the use of cryogenic technology is extensive would be expected to have several personnel suitably trained to provide treatment for this type of injury.

One positive aspect relating to cryogenic burns is that tissue damage does not normally occur immediately upon direct contact with small quantities of cryogenic liquids, such as those encountered through an accidental splash or spillage. Initially, the normal blood supply to the skin tissue acts as a robust heat source and creates a very thin, heat-insulating gas film at the liquid-tissue interface. This gas film can remain intact for sufficient time for any spilled cryogens to be safely removed from the skin either by gravity's help or by careful wiping of the affected area. This time delay is usually sufficiently long enough for severe injuries to be avoided, even when larger volumes of cryogenic fluids are involved. This protective gas film process is called the Leidenfrost effect [18]. It is often seen where a cold liquid, in contact with a surface which is significantly hotter than the liquid's boiling point, rapidly produces a very thin insulating vapour layer, effectively preventing the liquid from rapidly boiling. There is an almost repulsive force created by this vapour layer and when observed, droplets of liquid will hover over the hot surface rather than coming into physical contact with it. An example of this can be seen if a droplet of cold water is dropped onto a very hot frying pan. As the water creates the insulating vapour layer between it and the frying pan, it will appear to initially levitate over the hot surface, rapidly flitting about as the dynamics of the vapour film change over time. Eventually the insulating film disappears through the process of evaporation and the water droplet in turn is rapidly boiled away.

As a general policy, if cryogenic fluids are accidentally splashed onto the body, immediately flush the affected area with tepid water and remove any saturated clothing before the cryogen can permeate through the fabric of the clothing. It is especially important to remove gloves or footwear as quickly as possible if the cryogenic liquid is splashed onto them, as large volumes of the cryogen can easily be trapped inside, increasing the risk of damage to the skin of the hands and feet by any prolonged contact.

### 7.3.5 Frostbite

Often associated with the heroic endeavours of famous Arctic and Antarctic explorers, frostbite is in effect an extreme case of a cold contact injury. It directly occurs when the skin or other delicate body tissues are exposed to low temperatures for any prolonged period of time, causing them to freeze. As significantly low temperatures are prevalent in the use of cryogenic technology, frostbite is a noteworthy hazard which needs to be thoroughly considered in order to protect the health and wellbeing of those working directly with any such cold equipment. Having similar symptoms as those experienced through cold contact burns, frostbite would typically manifest itself initially as a numbness in the affected areas of the body, followed eventually by swelling and blistering around the injury. The actual mechanism by which damage is caused to the body is through the formation of ice crystals in the blood vessels of the skin as it freezes, which not only causes direct damage to the surrounding tissues, but can also cause blood clots following thawing of the affected regions. Areas of the body normally susceptible to frostbite would include the cheeks, ears, nose, and more commonly, the fingers and toes.

At a medical level, frostbite is defined as either superficial or a deep injury [19], depending on the severity of the exposure and the cold temperatures involved. In the most sever of cases, the affected skin will become black and so-called mummified, with irreparable, permanent damage having occurred. Clearly, protracted exposure to cold environments runs the greater risk of producing the most damage to the skin and its underlying tissues, as well as causing secondary complications such as hypothermia. An example of this would be the undertaking of activities such as winter sports, where exposure to low temperatures for prolonged periods is highly likely to occur due to the very nature of their required location in cold, snowy regions. Those individuals partaking in the sports are therefore at a much greater risk of suffering the consequences of frostbite. Likewise, anybody working close to cold gases released from liquid cryogens for prolonged periods of time are equally exposed to the sorts of temperature which can eventually bring about this type of injury.

The easiest ways to prevent frostbite is to avoid any contact to the low temperatures in the first place, or at least limit the exposure time. This can be further enhanced with the wearing of proper clothing to suit the ambient conditions. Where frostbite has however occurred, the general treatment for anybody suffering the condition is through the rewarming of the injured area by using things like warm towels, slowly bringing the damaged skin and tissues back to room temperature. Significantly, it is possible that permanent nerve damage can result from even the mildest cases of frostbite, which would manifest itself as a perpetual loss of feeling in the injured area. Unfortunately, where the injury is so severe, amputation of the affected limb is often necessary. At an occupational level, it is generally factors such as inadequate clothing being worn or the lack of appropriate personal protective equipment (PPE) such as the correct cryogenic rated gloves, which contributes to a heightened risk of frostbite.

The actual mechanism by which individuals will suffer differing levels of frostbite are complex, but frostbite is certainly more likely when the human body's ability to either produce, or retain heat, is impaired for some particular reason. These could be due to physical factors, such as an individual suffering an existing illness or being subject to physical stress by the nature of the work they are undertaking. It could also be brought about because of environmental factors, the more common example being the prevalence of existing cold weather, which has already caused the human body to lose a significant amount of its heat.

Treatment of frostbite should be provided as early as possible with the aim of raising the temperature of the damaged skin slowly back to normal. Any tight clothing worn by the injured person should be first loosened and the affected area placed in tepid water. Hot water or direct heating should be avoided as these can cause burning, especially as the frozen tissue at this stage will lack any sensation to temperature. Do not however try to remove any frozen clothing from the injured person until the area has been completely thawed as there remains a risk of skin having been frozen to the clothing until it is evident that this is no longer the case.

It is important to recognise that as the frozen tissues of the skin are thawed, cellular disruption and swelling, such as edema can occur. This is where an abnormal accumulation of fluid in the body causes localised swelling, generally in the ankles, feet and legs [17]. In most cases it can cause severe pain. It is important that the affected area is not rubbed or massaged as this causes the blood vessels locally to widen and as a result decreases the body's ability to retain heat. Also, do not allow the casualty to smoke or drink alcohol as the effect of both of these will thin the blood and interfere with its clotting process. Clotting of the blood is beneficial whenever there is an injury, allowing any laceration to be sealed.

### 7.3.6 Hypothermia

Long exposure to extremely cold environments can lead to cases of hypothermia, where typical symptoms would include the individual's reactions becoming slowed and their general capabilities impaired without them being aware of such. Hypothermia is defined as the state when the human body core temperature drops below 35.0 °C for any extended period of time [16]. A normal body core temperature in comparison is in the range 36.5 °C to 37.0 °C [16] and is maintained at this level by the body's own internal biological reactions. Any prolonged depression of this temperature is therefore considered to be a direct threat to the life of the sufferer. By association, it is possible for employees who are working for extended periods of time in cold conditions, especially where they relate to the use of cryogenic technology, to readily suffer the effects of hypothermia. Management of their tasks and the correct observance of safety practices are therefore critical in order to prevent any exposure which may bring about this potentially life threatening condition.

At the early stage of hypothermia, the human body will involuntarily begin to shiver in an attempt to warm itself up by stimulating movement of the muscles. At the same time, some element of mental confusion starts to occur as the brain's function is impaired, with the sufferer demonstrating uncharacteristic traits, such as making random mistakes in what they are undertaking. In addition, a numbing of their senses could occur, so that eventually they are no longer cognisant of the extremely low temperatures surrounding them and the effects this is having on them. In turn, continuous exposure can lead to localised frostbite to the hands and feet, especially if they are not aware of any contact they are making with cold equipment, such as that used with cryogenic technology, which they may be employing.

Ironically, as the condition worsens, the body's shivering responses gradually stops, giving the false impression at that point that the sufferer has completely recovered.

Quite distinctive though, there is also a further impairment of their cognitive abilities, making them appear drowsy and unresponsive, a contra indication that in fact they are still suffering from the condition. Eventually, the sufferer will fall into an unconscious state, and if not treated at this point, all their vital signs will gradually cease. Hypothermia can therefore be a serious hazard if not treated correctly and quickly.

In addition to the primary cause of prolonged exposure to extreme cold condition, the mechanism by which the body moves into hypothermia can be further exacerbated by pre-existing medical conditions. These could result in a natural decrease of heat production within the body or even an increase in the heat being lost from the body. Typical examples where this occurs would include individuals with low blood sugar levels and those of advanced age. Diagnosis of the specific condition of hypothermia is therefore often determined based on an individual's symptoms with reference to the surrounding risk factors as well as the direct measurement of their core body temperature.

The treatment of those individuals suffering from mild hypothermia includes providing them with warm drinks, warm clothing and encouraging them to undertake some form of light physical activity, such as walking, which will naturally encourage the body to raise its core temperature. Those suffering moderate levels of hypothermia will need to be wrapped in warm, or even heated blankets, in order to prevent further loss of heat from their skin, whilst trying to raise their core body temperature. They should be moved only gently, as at this stage their body will already be shutting down vital organs in an attempt of preserving heat. Individual's suffering severe hypothermia will require specific medical treatment provided by the emergency services, especially if they have become unconscious at this stage. If they are found to have no pulse, cardiopulmonary resuscitation (CPR) [16] will be necessary until the emergency services have arrived.

Prevention of hypothermia can be done by two principal methods. The first is by controlling, and thereby limiting, the period of time an individual can be exposed to cold temperatures. This could be through a permit type system, whereby a task duration time limit is specified and is scrupulously controlled by the permit issuer. The second, is through the provision of suitable personal protective equipment (PPE) and warm clothing, which will provide necessary protection to the body, thereby avoiding its core temperature from dropping. For anyone working with cryogenic technology, a wealth of protective clothing is available for the variety of task likely to be encountered.

### 7.3.7 Toxicity

Toxicity is defined as the level by which particular substances can cause damage to an organism. Its effect can be total, where it damages the entire organism, or focused, where it only effects some substructure of the organism, such as individual cells making up the organism [20]. As such, toxicity is considered a significant hazard which can have damaging effects on the human body at not only a cellular level but also to particular individual organs making up the body. Toxic substances are therefore defined relating to the health hazards they pose, with detailed reference to their lethality to; the entire body, specific organs or their propensity to cause cancers.

Fortunately, most of the gases and liquefied cryogens used in the context of cryogenic technology are known to have relatively low toxicity. However, in certain high concentrations, some of the particular gases employed may cause some form of dizziness or nausea if they are accidentally inhaled. The exact percentage will vary depending on the specific gas being employed and how each will cause very particular health effects if they are breathed in. At an extreme level, liquefied carbon monoxide for example would release large quantities of carbon monoxide gas into the surroundings, which if accidentally inhaled can cause immediate death. Another example would be where the prolonged breathing of pure oxygen can result in the harmful physiological effect known as hyperoxia which occurs when both the tissues and organs of the human body are exposed to an excess supply of oxygen [21]. Damage to lipids, proteins and nucleic acids in the body can result, along with the oxygen reacting with the surrounding biological tissues of the body. As a result, short exposure will impair the central nervous system. Longer exposure creates disorientation, breathing problems and can even result in vision changes. It is always critical therefore to check the safety literature, such as the material safety data sheet [22], supplied with any gases which are being handled. This information will provide any toxic hazard values which can be used to technically inform any associated risk assessments and risk-reduction strategies.

### 7.3.8 Embrittlement of materials

In the world of mechanical engineering, technical strength data for most materials is readily available for room temperature applications. Likewise, there is also a wealth of information available to the design engineer for those same materials used at elevated temperatures. For cold temperature applications however, the information is often sparse and in many cases generally referenced to out-of-date grades of metals and polymers which were subject to testing many years ago. Little contemporary strength information is therefore available for new grades of materials working under extreme cold conditions. This can often make the task of material selection for cryogenic technology a difficult task and may inadvertently hide important characteristics of certain grades and the effects that the cryogenic temperatures may have on their structure.

For all types of materials, the mechanical property known as the Young's modulus [23] is an important characteristic which provides guidance in their suitability for specific applications. Fundamentally, it is a measurement of the stiffness of solid materials in their elastic state, where any deformations caused by the application of loads will return to zero when the load is eventually removed. It is characterised as the relationship between the stress and strain in a material which is subjected to any uniaxial deformation. Young's modulus is generally represented as an upper-case **E** in this relationship.

$$E = \frac{\text{Direct stress}}{\text{Direct strain}}$$

As stress is defined as a load divided by an area, and strain as the extension under load divided by the original length, the following can be determined:

$$E = \frac{F/A}{x/L}$$

$$E = \frac{\text{FL}}{\text{Ax}}$$

Every material possess a unique value of Young's modulus, generally quoted as that determined under room temperature conditions. Where cryogenic technology is being used, which consequentially results in materials being cooled to extremely low temperature, the inherent Young's modulus value will correspondingly increase by a comparative amount. Typically, this increase can be as high as 10% to 20% at liquid helium temperatures (4 K) compared to that at room temperature (300 K).

As a result of this change in Young's modulus, the cooling of any particular material will generally increase its strength and its stiffness. By doing so, this reduces the material's overall ductility, and as a consequence, will increase its brittleness, thereby making it vulnerable to crack propagation and sudden failure. Figure 7.4 shows the clear difference between brittle and ductile materials as plotted on a stress-strain curve and their corresponding changes in strength. In particular, materials such as plastics, rubber, Teflon (PTFE) and carbon steels, all become extremely brittle at cryogenic temperatures. In addition, where cyclical stresses are in existence, this embrittlement may become more manifest, as the mechanisms which promote embrittlement, such as crack propagation and crack-growth rates, are enhanced at the cryogenic temperatures.

Not all combinations of material grades, applied stresses and ambient low temperatures cause embrittlement to take place. In fact, each material, and often each grade of that material, has to be assessed accordingly for their susceptibility as

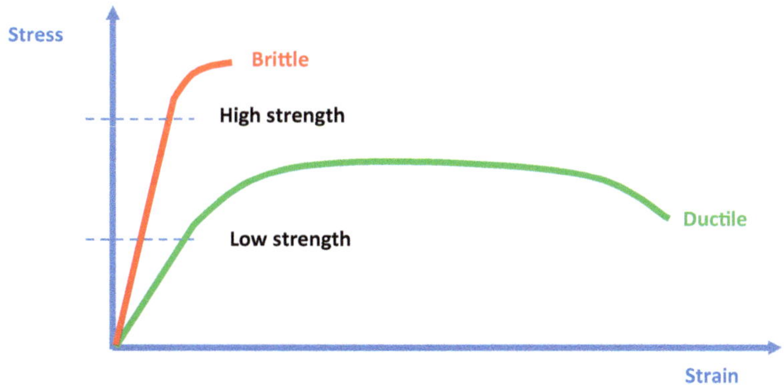


**Figure 7.4.** Showing the different strengths of both ductile and brittle materials.

| Material | Brittle -transition temperature |
|---|---|
| Nylon | 89 K |
| Teflon | 102 K |
| Natural rubber | 102 K |
| Zinc | 144 K |
| Steel | 200 K |

**Figure 7.5.** Showing the defined brittle-transition temperatures for a variety of material.

they do differ greatly. Where temperature is the dominant factor however, very often, a brittle-transition temperature will be defined, below which the loss of ductility of any particular material is so severe, that embrittlement will inevitably take place. Typical values are defined below in figure 7.5 [24]:

### 7.3.9 Material stresses

Mechanical stress is a physical property relating the internal forces within an object, by which each element within an engineering material exerts on each other. This stress can be caused by external forces, such as gravity, or more typically, due to internal created forces within materials, for example, where a structure is used to support a particular mass. By this token, any physical deformation of a solid material causes so-called internal elastic stresses to be generated. These stresses can be imagined as a series of compression springs, whereby on their release, the deformation is negated and the material returns to its original state. Under certain conditions, internal stresses may also exist without the presence of any discernible deformations taking place. These would typically be due to external influences such as changes in temperature in the engineering material, or the presence of strong electromagnetic fields. Where mechanical stresses exceed the defined strength limits of the parent material, permanent deformations will inevitably occur.

This being so, the cooling of materials down to cryogenic temperatures will inevitably cause physical contraction, in some cases, sufficiently large enough to generate high internal stresses in components which have been rigidly fixed as part of an overall design assembly. As a consequence, these mechanical stresses may result in the failure of critical parts if they are not suitably mitigated. Where localised exposure to liquid cryogen spillages can occur for example, the sudden, rapid shrinkage of any affected material can introduce localised high internal stresses in the corresponding parts. Not all materials will respond well to contact with cryogenic technology and adverse reactions can result in the subsequent failure of the component's integrity.

In fact, materials such as plastics become so brittle under these extreme conditions that their own internal stresses cause them to shatter without the effects of any external forces coming into play. Likewise, the rapid cooling to cryogenic temperatures, or rapid heating of materials from cryogenic temperatures, can also introduce undesirable internal stresses. In principal, these excessively fast changes in temperature do not allow the natural mechanisms within the materials to accommodate the equally rapid contraction or expansion taking place. Where possible, components should therefore be both cooled down at a relatively slow rate and then eventually reheated gradually to prevent any such thermal shock effects taking place.

One guiding principle in the selection of metal alloys for cryogenic use should be that they do not undergo a phase transformation on cooling, thereby changing to a more brittle state which would result in an overall loss in ductility. As any such phase transformation in turn would make the metal susceptible to embrittlement, it follows that any internal stresses caused by aspects such as contraction, could be damaging to the internal structure of the metal. Generally, face-centred cubic (FCC) [25] crystal structures are more ductile than body-centred cubic (BCC) [26] crystal structures due to the presence of the additional slip planes making up their structure. In a profound way, certain grades of materials will undergo a transition from the ductile FCC to a more brittle crystal structure when cooled to cryogenic temperatures.

This is not a problem with most structural metals, however, the so-called ferritic and martensitic grades of stainless steel do become brittle as the surrounding temperature is significantly reduced [27]. Typical ferritic grades would include 430 (1.4016) [28] and martensitic grades 410 (1.4408) [29].The austenitic grades however, such as 304 (1.4301) [30] and 316 (1.4401) [31], remain tough at cryogenic temperatures. An important caveat to the selection of austenitic grades however is that cold working of them can affect their cryogenic toughness. This is due to the gradual formation of martensite [32] during the cold working process. Martensite is an extremely hard crystalline structure form of the steel where the FCC austenite has transforms to a body-centred tetragonal (BCT) [33] form which is supersaturated with carbon. Observed under a microscope, martensite resembles the brittle fronds of a fir tree, interlocked with each other in such a complex way to promote the formation of myriad slip planes in the crystal structure. This has the effect of reducing the fatigue strength of the alloy, but can often be controlled through compositional changes in the alloy which appropriately stabilises the austenite. In addition, it is also possible to heat-treat the metal after cold working in order to stress-relieve it and transform the crystal structure from the brittle BCT back to the ductile FCC austenite. Figure 7.6 shows the basic crystal structure of BCC, FCC and BCT materials.

Where the welding of stainless steels has taken place, this too can cause localised brittle failure to occur at cryogenic temperatures due to a change in the crystal structure around the heat affect zone of the weld. Generally, ferrite levels in this weld-region will be higher than in the surrounding metal matrix, leaving the weld zone with poor impact resistance. If internal stresses are sufficiently high enough at

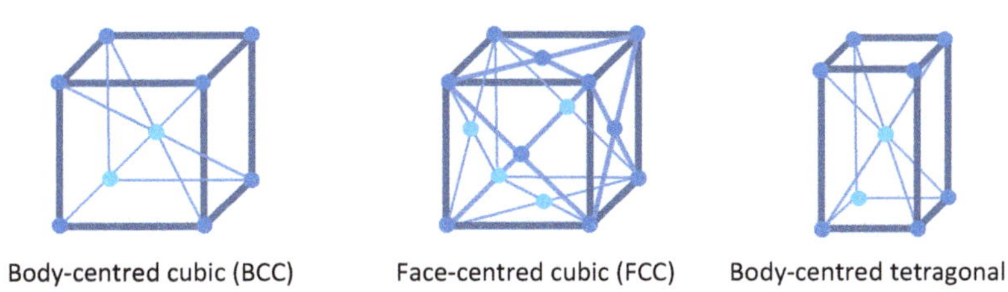


**Figure 7.6.** Showing the basic crystal structures for FCC, BCC and BCT materials.

these weld points, failure can occur even at relatively low stress values. Again, this can be managed by the correct selection of the grade of metal being used and through the inclusion of additions to the alloy which can stabilise the metal and thereby negate the damaging effects around the weld. Typically, for cryogenic work, metal grades with low carbon content are preferred.

### 7.3.10 Contraction of materials

All engineering materials change their shape due to the corresponding effects of an increase in their surrounding temperature. This is defined as the thermal expansion, with the majority of materials experiencing an expansion process when subjected to some form of heating. Likewise, most materials undergo a negative expansion when cooled and this is defined as a thermal contraction (figure 7.7). Typically, contractions of 0.3% of the original length can take place between ambient room temperature down to the liquid nitrogen liquefying temperature of 77 K.

These expansions and contractions are fundamentally due to changes in the kinetic energy of the molecules making up the engineering materials in response to a change in their ambient temperature. If the temperature is higher for example, the molecules will gain energy from their surroundings and therefore their kinetic energy will be higher. Conversely, if the surrounding temperature is lowered, the molecules will liberate some of their energy to the surroundings, resulting in a lower overall kinetic energy. This kinetic energy manifests itself at a molecular level in the form of molecular vibrations. As this vibration is increased, the molecules will separate more to accommodate the heightened movements of each neighbouring molecule. Likewise, if the vibrations are reduced, the molecules will move closer together. Taken singularly, these fractional changes in the average separation of the molecules are too small to reconcile. If taken accumulatively however, they form the defining physical expansion and contraction of that particular material.

In order to determine the exact value of expansion or contraction of materials, the relative expansion is divided by the change in temperature. The resultant is called the material's coefficient of linear thermal expansion/contraction [34], and generally, this value will itself vary over a particular temperature range. In essence, the coefficient describes how the size of any material will change with a corresponding change in ambient temperature, deliberately focusing on changes in linear dimensions.

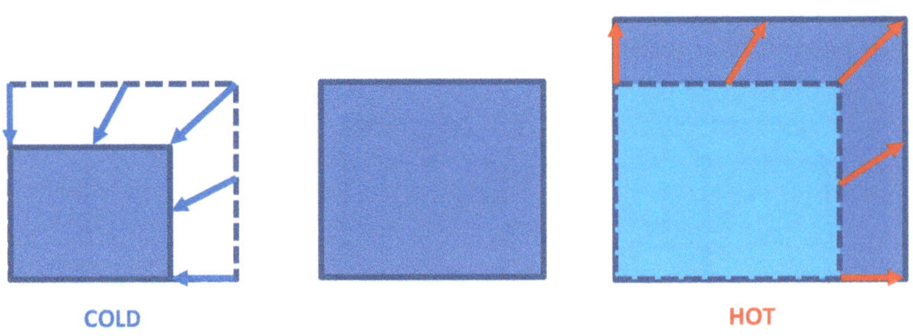


**Figure 7.7.** Showing the basic expansion and contraction of materials due to a change in temperature.

Where liquids are concerned however, it is often more common for the volumetric thermal expansion/contraction coefficient [35] to be considered, as liquids will expand or contract at the same rate in every direction as temperatures are changed. These physical changes in the volumes of liquids form the basis of how a liquid-filled thermometer functions, with a corresponding change in volume being identified on a column scale as temperature changes take place.

In terms of the implications contraction effects have where components are being used in cryogenic environments, an awareness of the freedom of these parts to undergo any shrinkage is critical. If for example, parts are free to contract, the resultant thermal strain in the components can simply be determined by calculating the specific changes in dimensions by the application of the relevant coefficient of linear thermal contraction. However, if parts are not free to contract, as they are being physically constrained as integral parts of an assembly, then internal stresses will be created by the lowering of the temperature. As shrinkage of the parts takes place, these stresses can be calculated by determining the strain that would normally occur if the component were unconstrained and then by applying the elastic stress-strain relationship of the materials Young's modulus [36] value in order to give the resultant stress required to effectively reduce this strain to zero. It is not uncommon for the stresses induced due to contraction of fixed parts to be significant, and in extreme cases, complete failure of assemblies can occur due to this phenomena.

Significantly, the coefficient of linear thermal contraction for all materials changes as the temperature is gradually lowered. In the majority of cases, the coefficient reduces at a non-linear rate as there is a corresponding reduction in temperature. This factor needs to be considered when designing components used with cryogenic technology, as adopting only room-temperature coefficient values will provide errors in the calculations of the actual changes in lengths encountered. On the whole, most general uses of engineering materials do not have to consider any extreme temperature changes, so this is never a considered problem. For cryogenic applications however, significant temperature changes of upwards of 300 K are the norm and real, integrated values are required in order to recognise the changes of the coefficient as the temperature of the component parts are so meaningfully reduced.

An example of where the contraction of materials becomes critical is if there are pressure or vacuum systems employing dissimilar materials as part of their sealing arrangement. As the corresponding parts are made from different materials, each will contract by a conflicting amount. Unless allowance has been made in the design for these variances in dimensional changes, the seals in particular may become over stressed and could in extreme circumstances fail. More likely however, is that leakages will result due to the relative movement of joints as the individual parts of the seal arrangement contract at a different rate.

Another common occurrence where contraction needs to be considered for cryogenic equipment, is where room-temperature limits and fits of interacting parts may become compromised at cryogenic temperatures causing the seizure of parts. This is especially so in moving parts, such as bearing assemblies operating within a cryogenic environment. Once more, this is solely due to the different contractions of mating parts, especially if they are not manufactured from the same material.

A brass bearing bush for example, supporting a stainless steel drive shaft, would not only have a larger coefficient of linear thermal contraction than the stainless steel to start off with, but in addition, this coefficient would change at a different rate to that of the stainless steel equivalent as the temperature is reduced to that of a cryogenic regime. The result would be that any clearances prevalent at room-temperature between the bearing and the shaft, may no longer exist at liquid helium temperatures. In this situation, the function of the bearing mechanism may well be impaired by the brass bush having shrunk so much that it now tightly captures the stainless steel drive shaft, preventing any intended rotation from taking place.

It is common in these situations to enlarge the clearances of parts at room-temperature so that as the contraction of each component part takes place, some level of clearance remains at the cold temperature. This can easily be calculated and included as part of the initial design process where dimensional tolerances are determined. It is important to note however, that it is not uncommon to see such assemblies which are designed specifically for cryogenic applications to appear to have excessive clearances at room-temperatures. This is acceptable logic and must not be corrected, otherwise the result will be seized components at cryogenic temperatures. The only time where this room-temperature clearance may be a problem is where the accuracy of the mechanism is required at both room-temperature and cryogenic temperatures. In fact, alignment at the two temperature extremes will be difficult, as the final limits and fits will be nominally designed to suit the cryogenic application and not the room-temperature one. It is normal under these circumstances for some form of spring compensation to be included as part of the design of the intended parts, ensuring that any potential backlash in room-temperature systems are removed where possible.

### 7.3.11 Liquid cryogens

By definition, liquid cryogens are liquids that boil at temperatures below 173 K (−100 °C) [16]. As such, they have over the years been the main component of many low-temperature applications, and are justifiably used throughout the cryogenic technology industry. They are the principal factor by which cryostats function, used within them as a cryogenic fluid bath where the latent heat of their boil-off can be utilised to provide enormous cooling power and thereby reduce the temperature of any item inserted inside the cryostat. Equally so, their availability forms an ever-growing infrastructure of use, stretching from industrial practices to the more prosaic research applications.

At a practical level there are a range of liquid cryogens, some more readily available than others, with an equally diverse portfolio of applications associated with each type. The common facets they share however are defined by the fact that they exist only at extremely cold temperatures, which in itself has safety implications associated with this temperature regime. Likewise, they also share the mutual fact that they possess large liquid to gas expansion ratios, often above 700 for the more routinely used cryogens. The significance of this being that even a relatively small liquid cryogen spill will result in the release of an extremely large volume of gas, which under certain circumstances can displace air from within the working environment. In addition, some specific liquid cryogens such as; fluorine ($F_2$) and oxygen ($O_2$), are chemically reactive. Others, such as; hydrogen ($H_2$) and methane ($CH_4$) are flammable. As a consequence, the associated hazards for each individual liquid cryogen needs to be recognised before they are handled in any quantity and the corresponding risks minimised to allow their use in a critically safe manner. Part of this risk assessment process therefore requires users of liquid cryogens to fully understand their specific properties, looking in detail at each cryogen's function and the detailed risks they carry.

At the end of their use, the disposal of liquid cryogens must be undertaken in a safe and secure manner. Cryogen storage vessels must never be emptied into any domestic drainage systems or road-side drains as their material of construction are highly likely to be incompatible with the extremely low temperatures at which the cryogens are stored at. Likewise, the disposal of liquid cryogens must be done in a way to prevent any cryogens and vaporised gas from accumulating in trenches, pits or confined spaces where they could bring about an oxygen depleted atmosphere. Any disposal method should allow the gradual vaporisation of the unwanted cryogens, as far away from personnel as possible, preferably in the open air. Where the disposal is undertaken inside the workplace, any vaporised liquid should be vented in a way that it prevents any localised increase in the oxygen concentration of the surrounding air, which could in itself give rise to a greater potential for fire to occur. Also, any vents utilised as part of this process must be directed away from structures and equipment which are more liable to undergo mechanical failure if they are exposed to significantly low temperatures.

### 7.3.12 Liquid nitrogen [16]

Nitrogen gas makes up the majority of the Earth's atmosphere, typically in the order of 78% [37]. It is inert, which means it will not produce combustion if ignited. Although it constitutes a very large proportion of our breathable air, on its own, nitrogen will not support life. Instead, it is the second largest component of that air, oxygen at nominally 21% [38], which provides the vital mechanism to support life.

In its liquid state, nitrogen remains inert and has a specific gravity of approximately 0.8 measured with respect to water. It is non-toxic, colourless, odourless, non-corrosive, non-flammable, and has a relatively low boiling temperature of 77 K. The latent heat of vaporisation of liquid nitrogen is about 160 kJ $L^{-1}$, which means that a heat input of 1 watt will produce a boil-off of about 23 c.c. of liquid per hour. It has a liquid to gas expansion ratio of 710, which means during evaporation, that it can act as an asphyxiant through the displacement of the normal oxygen concentration in the breathable air. Breathing a nitrogen rich atmosphere can cause dizziness, nausea, loss of consciousness and possible death depending on the concentration levels involved. In its gaseous form, it has a specific gravity of 0.967 measured with respect to air. The implication of this being that any boil-off releases of cold nitrogen gas would tend to fill an enclosed room from the floor upwards, significantly depleting the oxygen from the interior to a dangerously low level. In addition, it is possible under these conditions for cold gaseous nitrogen to accumulate at low points in work areas, such as pits and trenches. As these effects are undetectable by the human body, there is far too often an arbitrary complacency about the handling of liquid nitrogen and the risk it poses from asphyxiation, and serious accidents do happen wherever it is used. It demands respect and an overriding safety consciousness if it is to be managed in the workplace without causing injuries and deaths.

The storage and handling of liquid nitrogen is typically managed by the use of vacuum-insulated storage Dewars. These are made in a variety of sizes to suit the specific application and need to be fitted with a pressure safety relief valve and bursting disc to protect the Dewar from the naturally occurring gradual pressure build-up as the liquid nitrogen warms up over time. Under normal conditions, the pressure relief valve will periodically vent any excess pressure as the contained liquid nitrogen continuously releases small amounts of nitrogen vapour above the liquid surface.

Because of the relatively low temperature at which nitrogen exists in its liquid state, combined with its large latent heat of vaporisation, there is a high risk of frostbite occurring if human skin comes into contact with the liquid, even for a matter of a few seconds. The use of peripheral equipment such as transfer lines can also be a potential cold-contact source causing cold-burns and frostbite if the correctly insulated personal protective equipment (PPE) is not worn. Especial care needs to be observed when particular operations, such as the filling of the Dewars or decanting liquid nitrogen from the Dewars, are undertaken as there is the greater potential for spills to occur during these processes.

Liquid nitrogen has many varied applications and is often employed as a secondary cooling mechanism in liquid cryogen cryostats. Here it provides cooling to radiation shields which are designed to reduce heat flow into the inner liquid helium storage bath of the cryostat. Both cryogens need to be replenished as they boil away, the rate at which they do being dependent on the overall size of the cryostat and the heat load of the equipment being taken down to the intended cryogenic temperatures. The object to be cooled is inserted into the cryostat, normally attached to the end of a stick assembly. At an economic level, liquid nitrogen is a relatively inexpensive consumable, costing in the range $0.50 to $1.50 per litre [39].

Liquid nitrogen can also be employed for more fundamental applications and is often used as a refrigerant for trapping water vapour and oil vapour from gas systems where high purity is required. This is normally undertaken by having a U shaped pipe through which the gas is transferred. This pipe is then housed in a small insulated Dewar, into which liquid nitrogen is directly introduced. The liquid nitrogen significantly reduces the temperature of the gas, thereby preferentially condensing any water and oil suspended in it. On condensing, the water and oil collect at the bottom of the U pipe whilst the remaining cleaned gas continuous it route and is transferred to its intended purpose. Often the U pipe is formed from glass, allowing the operator to observe the trapping of the water and oil as this occurs.

At an industrial level, liquid nitrogen can have other uses, one in particular being its application for contracting metal parts which are then inserted into housings. This is a common means by which many bearing assemblies are constructed. Due to the relatively high temperature difference between room temperature and the liquid nitrogen boiling point, relatively large contractions of parts can be achieved by this method. The part to be assembled is often machined to have an interference fit with the part it will eventually be housed in. Immersion in liquid nitrogen reduces the interference fit to a point where clearances are achieved with the mating part. The process of insertion has to be undertaken relatively quickly however, as once removed from the liquid nitrogen, the part begins to warm up and recovers its original dimensions.

Apart from its potential to cause an oxygen depleted atmosphere as it expands from its liquid to gaseous state, the other main risks involved in the use of liquid nitrogen are generally to do with the low temperature it exists at. In addition to causing direct injuries, this cold temperature can result in ice forming on exposed systems where atmospheric water vapour has frozen on contact. More importantly though, its regular and routine use as part of the overall demands of cryogenic technology, means there is generally a higher risk associated with its application. In simple terms, the more it is being used throughout the world, the more likely an accident or injury will inevitably occur due to its application.

### 7.3.13 Liquid helium [16]

Helium gas is the second lightest and second most abundant element in the Universe, with hydrogen gas being both the lightest and most abundant. On Earth however, it is relatively rare in the atmosphere, typically only appearing as little as 5.2 ppm by volume. Fortunately though for the cryogenic industry, helium gas is often found trapped underground within natural gas reserves in concentrations upwards of 7% by volume, from which this can be commercially extracted. Due to its low specific gravity of 0.138 measured with respect to air, helium gas is often used as a lifting gas in lighter-than-air craft, such as aeronautical balloons. In its gaseous form, helium is non-toxic and inert. This makes it useful at an industrial level, where it is regularly utilised as a protective gas for welding processes, as well as employed as a pressurising and purge gas for many pressure assembly applications.

In its natural state, helium gas exists as nine isotopes, of which only helium-3 and helium-4 remain stable. However, for every one million helium-4 atoms found in the atmosphere, there is only a single helium-3 atom detected as a trace element. As such, both helium-3 and helium-4 are used in their cold liquid forms for a variety of cryogenic technology purposes. Liquid helium has the lowest boiling point of all cryogens, with helium-4 having a boiling point of 4.2 K, although it can typically be cooled to about 1.0 K if an evaporative cooling process is employed. Likewise, helium-3 can be cooled down to around 0.2 K using the same method. In addition, mixtures of helium-4 and helium-3 are frequently used in devices such as dilution refrigerators to routinely achieve temperatures below 1.0 K.

Liquid helium is recognised as the other main player alongside liquid nitrogen in the cryogenic technology cooling world, used as a cooling mechanism in cryostats, as well as for the cooling of superconducting materials such as those integrated into high field superconducting magnets. In its liquid state, it is colourless, odourless, non-corrosive, non-flammable and inert. Significantly, as it is non-corrosive, there is no requirement for special materials or coatings to be applied wherever it is used. Critically however, any materials selected as part of a liquid helium assembly, would have to be suitable for use at the extremely low temperatures at which it exists. In particular, materials would need to retain good strength without becoming too brittle under these challenging conditions.

With its extremely low boiling point temperature of 4.2 K, other gases in the vicinity of liquid helium can initially vaporise, but can also under ideal conditions solidify at this cold temperature. This can be a significant issue on poorly insulated transfer lines and especially where these solidified gases can form and collect around safety pressure relief systems, such as pressure release valves. In extreme situations these solidified gases can cause an immovable plug in the exhaust vent of these safety valves, rendering the valve inoperable. Similarly, under certain conditions, these solidified plugs can form in the necks of liquid helium storage Dewars and vessels, thereby isolating the pressure relief system entirely. If such a plug does form, internal pressures inside the storage containers can subsequently be created as their contents continue to evaporate over time. If these pressures become significant, catastrophic failure of the containers can occur.

In addition, liquid helium can be responsible for liquefying oxygen out of the surrounding air due to the same vaporising process, creating small localised pockets of pure oxygen, giving rise to potential fire hazards as a result. This comes about through the initial liquefying of the air, which at its dew point contains approximately 50% of oxygen by volume. Liquid helium must therefore be stored and handled under positive pressures or in a closed system to prevent the infiltration of the immediate air around it, which would be a consequential source of this oxygen. Surfaces on which condensation might occur, thereby further promoting the formation of this oxygen enriched liquid air, should ideally be cleaned to oxygen service standards and deliberately kept free of oil, grease or any other combustible materials. In an ideal situation, all cryogenically cold surfaces should be vacuum-insulated to prevent any such condensation, but this often carries an economic cost. Where this is not possible, any selected materials of construction should be oxygen-compatible grades and certainly non-combustible.

Similar to the main properties associated with liquid nitrogen, liquid helium has a relatively large liquid to gas expansion ratio of 780, which may lead to the unintended build-up of high pressures in closed systems where vaporisation is allowed to happen without the provision of some form of safety venting system. The importance of providing self-operating pressure relief system is paramount in these situations. However, where this expanding helium gas is allowed to escape from any closed system, it can easily cause a rarefied air atmospheres if there is a significant amount of the liquid evaporated into its gaseous form. On its own, helium plays no biological role, and provides no support for any life on Earth. It is therefore considered to be an asphyxiant gas if released in sufficient quantities into the working environment. As a gas, helium is odourless, colourless and tasteless, and therefore provides no obvious warning indicators to anyone working in its vicinity, making them unaware of any dangerous releases of the gas when it occurs.

Apart from the likely sounds associated with a large, sudden release of evaporated gas, any small accumulative build-up of helium gas would be undetectable by the human senses. Inhalation of helium gas in excessive quantities can cause dizziness, nausea, vomiting, loss of consciousness, and at the extreme, death. Significantly, where the accumulation of helium gases causes almost zero oxygen concentration values in the working environment, unconsciousness and death can occur within seconds, and without any warning.

As helium in its gaseous form is lighter than air, any releases of cold helium gas would tend to fill an enclosed room from the ceiling downwards, possibly venting safely through natural vents or openings at elevated levels. Ironically, because of this fact, anyone initially fainting in a helium rich environment, due to the lack of oxygen caused by its release, will in fact fall down into an oxygen rich layer of air, as yet unaffected by the helium. It is possible, providing that they do not sustain significant injuries from their fall, for an escape to be enacted from the workplace, by the individuals crawling along the floor, so that they remain in the normal air until they are safely out of the affected area. Contradicting this concept however, it has been shown in certain studies that helium gas can be more transient and can in fact diffuse through an air column in a direction opposing its natural buoyancy [40]. This would mean that any release of helium gas must be considered as a hazard and this buoyancy effect should be taken into account during any asphyxiation risk assessment for liquid helium releases.

Due to this potential asphyxiation risk, under no circumstances should a liquid helium Dewar be accompanied in a passenger lift where it is being transported between levels of a building. A sudden release of helium gas in such a confined space would rapidly displace any contained oxygen and could be fatal to anybody travelling in the lift with the Dewar. Procedures should therefore be adopted where the Dewar is sent by one person at the starting level outside of the lift and received at its destination by a second person. Some means of preventing passenger entry into the lift on any floors in between should also be adopted, such as deploying personnel at each floor, prohibiting entry.

Notably, the speed of sound in helium is approximately three times of that in air. As a consequence, a common party trick can be performed where helium gas is inhaled in small quantities, resulting in a temporary increase in the resonant frequency of the vocal cords. This has the effect of changing the timbre of the voice, resulting in a duck-like quality to anything being subsequently spoken. The effect is fortunately short-lived, and as soon as normal air replaces the helium in the vocal tract, the tone is returned to normal. Clearly, inhaling too much helium and too often can bring about the first stages of asphyxiation, as it deleteriously prevents oxygen from gaining access to the lungs.

The storage of liquid helium is typically managed by the use of vacuum-insulated storage Dewars. Liquid helium has however a low specific gravity of approximately 0.125 measured with respect to water. Therefore, the design of storage Dewars made specifically for liquid helium will factor this into their construction, and compared to other cryogen storage vessels will probably have thinner sections of material used throughout. As the mass of even the total liquid helium content of these Dewars will be relatively low, related stresses in their structure due to this mass will be equally low.

It is important to recognise however, that Dewars designed for liquid helium use must not be used for the storage of any other cryogens with a higher specific gravity value. Dewars are made in a variety of sizes to suit the specific application and need to be fitted with a pressure safety relief valve and bursting disc to protect the Dewar from gradual pressure build-up. Under normal conditions, the pressure relief valve will periodically vent excess pressure as the contained liquid helium warms up over time.

As liquid helium tends to be used for more specific applications within the cryogenic technology field, and has certain limitations on its overall supply, it is generally transported and stored in smaller Dewars compared to the other generic cryogens used throughout. Depending on its intended application, the stored liquid helium may be drawn directly from its storage vessel, or more commonly, decanted into smaller Dewars for subsequent use. It is worth noting that any decanting process carries with it a heightened risk due to the inadvertent release of cryogens and gas as part of that procedure. It therefore follows, that the more frequent these decanting processes take place, there is a greater likelihood of an accident occurring. For this reason, it is good practice to transfer liquid helium through insulated transfer lines which not only reduces any loss of liquids during the process, but also reduces the likelihood of the condensation of air on any exposed cold surfaces.

Liquid helium has a relatively low latent heat of vaporisation of approximately 2.6 kJ $L^{-1}$, which means that it will evaporate rapidly when heated. This is self-evident when liquid helium is first transferred into any warm or partially cooled cryogenic equipment, with the rapid production of helium vapour resulting from the process. This low latent heat of vaporisation property also means that even with a relatively small deterioration of the insulating vacuum of a liquid helium storage vessel, a large increase in the rate of boiling of the liquid contents will result. Any pressure relief device fitted to such a storage vessel must therefore be of an adequate capacity to release such a large quantity of vapour that may be caused by this type of event. It would not be acceptable for example to simply fit a safety relief valve based purely on its operating pressure. The full-bore venting capacity of the valve would also need to be assessed for a maximum gas release application when required.

The cost of producing liquid helium is significantly higher than that of liquid nitrogen and typical prices range from $4.0 to $15.0 per litre [41]. Much of this cost is due to having to bring the gas down to a temperature of 4.2 K in order to achieve its liquid state. Helium is extracted from the natural gas it is found within by a process called fractional distillation.

However, a more contemporary issue exacerbating the market cost is the steady depletion of global reserves of crude helium, set against an ever increasing global demand for the same commodity. The unit price has for example risen by at least 100% over the last 10 years [42]. New reserves are being established all the time, but the markets are volatile and can often be influenced by geopolitics, which can from time to time bring about global supply shortages of the helium. Many facilities around the world have come to recognise the spiralling costs incurred with the purchase of this resource and have as a result installed helium recovery systems as part of their infrastructure. This process involves the collection of any vented helium from the cryogenic appliances in their facility, as well as the compression, drying, storage and eventual liquefaction of the gas for reuse in its cold, liquid form. Accepting that any such system in itself involves a huge capital expense to set up, and that maintenance costs need to be considered as part of the overall cost, some facilities may find that the economics of installing helium recovery systems do not balance. For larger establishments however, where they are routinely purchasing large quantities of liquid helium, the long-term gains acquired through the overall reduction in the purchase of new liquid helium supplies makes it a viable and economic venture. An added advantage of recovering the helium in this way is that it also provides a much safer way of managing the released helium gas from any cryogenic systems. As the gas is automatically captured through a series of connected recovery lines, only small amounts will be accidentally released into the workplace environment, reducing the risks of oxygen depleted atmospheres from being created.

Any remaining safety issues relating to the use of cryogenic helium are focused on its extremely low temperature which can result in cold burns to exposed skin if accidental spills do occur. It is essential therefore that the correct personal protective equipment (PPE), such as cryogenic rated gloves, faces visors and safety glasses, are worn whenever handling the cryogen. Likewise, only vacuum insulated syphons should be used for the transfer of helium from Dewars, with the actual transfer being undertaken slowly to prevent any thermal shocks occurring in the syphon. Any purging of these transfer syphons, along with the receiving cryogenic equipment must be done using dry, helium gas released ahead of the flow of liquid helium in order to prevent any capture and subsequent freezing of water vapour from the air when placed in direct contact with the cold cryogen.

### 7.3.14 Liquid hydrogen [16]

Hydrogen gas is the lightest element on Earth and also the most abundant chemical substance in the known Universe. It has a specific gravity of 0.0696 measured with respect to air, which originally made it an ideal choice as a lifting gas for early lighter-than-air craft, such as aeronautical balloons, as its natural buoyancy causes it to rapidly travel upwards at ambient temperatures. In 1900, for example, the maiden flight of a brand of hydrogen filled rigid airships called Zeppelins was undertaken [43], with a whole industry of similar flying airship evolving soon after. However, hydrogen is a high-energy fuel which, when mixed with air and ignited, releases a large amount of energy in the form of heat or explosive forces. Sadly because of this, commercial hydrogen airships were brought to an abrupt halt in 1937 with the mid-air fire and ultimate destruction of the infamous Hindenburg airship [44]. As such, hydrogen is now only used as a lifting gas for specialist devices such as unmanned weather balloons.

The commercial production of hydrogen is mainly from the steam-reforming of natural gas supplies, but can also be undertaken through the electrolysis of water. Hydrogen is found naturally in the molecular form $H_2$ and it is colourless, odourless and highly flammable. Liquid hydrogen has a boiling point of 20.28 K, placing it neatly between liquid helium and liquid nitrogen on the cryogenic temperature scale.

In its liquid state it retains its colourless feature and has an extremely low specific gravity of approximately 0.071 measured with respect to water. Liquid hydrogen is non-corrosive, which negates the need for special materials to be applied to prevent corrosion wherever it is used. Any selected materials would however need to be compatible with the extremely low temperature at which it exists. In particular, brittle fractures can occur due to the materials being used below their so-called brittle-transition temperatures where their structure changes from a ductile to a brittle nature.

Liquid hydrogen is more commonly used these days as a liquid fuel propellant in rocket engines, where it is often combined with an oxidiser, such as liquid oxygen, to produce the required energy thrust, thereby launching space craft and missiles to their intended elevations. In more recent times, the interest in using liquid hydrogen as a concentrated form of hydrogen storage has grown enormously. This has been driven by the search for alternative fuel sources, in particular for the fossil fuels used in automobiles, in a world where the realisation that its dependence on oil cannot be sustained. As its liquid density is low compared to the more common fuels, storing hydrogen in its liquid state has economic benefits. Likewise, the energy required to initiate combustion in hydrogen is much lower than for other common fuels, making it an idea candidate for replacing traditional petrol and diesel fuels. This ease of combustion however also carries with it the potential hazards of accidental ignition under certain conditions. In the cryogenics industries, it is therefore more of a specialist cryogen, with a very specific fuel storage objective which has to have many safety protocols in place wherever it is being used.

Hydrogen presents a number of potential risks to human safety. Its high combustibility can detonate and cause fires when it is mixed with the latent oxygen in the air. Hydrogen accumulations, for example from leaking systems, can quickly reach its extremely wide flammable concentration range of 4% to 75% in air. As the burning velocity of hydrogen is greater than most of the common fuels, the potential hazards of both explosion and detonation of the gas must always be considered as part of an overall risk assessment for its use. Cold hydrogen gas released from an enclosed liquid hydrogen cryogenic system can in turn form a concentrated cloud, which would eventually migrate horizontally and may, given the correct conditions, spontaneously ignite. Hydrogen has a spontaneous ignition temperature of around 450 °C, which means ignition can readily occur if the gas comes into contact with equipment likely to be operating in this temperature region. In addition, the spark ignition energy needed to bring about ignition is considerably low for hydrogen. As a comparison, the minimum ignition energy for methane is 0.28 mJ, whereas with hydrogen, this is as low as 0.02 mJ.

Hydrogen fires burn extremely hot, but unlike other burning gases, they are almost invisible, making them a potential cause of accidental burns to anybody in the vicinity who is unaware of their presence. In confined spaces, a typical hydrogen ignition can result in a rapid flame acceleration and the generation of a high-pressure blast-wave, capable of exploding buildings and ejecting shrapnel for some considerable distance. Hydrogen, in whatever form, should therefore always be treated with respect and any potential sources of ignition removed from its immediate vicinity. These ignition sources could be due to accidental sparks from electrical equipment, static electricity sparks, open flames or even extremely hot components. All electrical equipment used within the boundary of any liquid hydrogen installation must be classified accordingly, so that they cannot generate sparks during their use. Certain regulatory area classifications prevail, with the more stringent requirements being applicable to zones immediately next to any potential hydrogen gas source. Once ignited, hydrogen flames only require 5% of oxygen in the air to support continued combustion, making it difficult to extinguish unless the oxygen is almost completely removed from its environment. As such, a great deal of circumspection needs to be taken for any cryogenic application where liquid hydrogen is being utilised.

Where a hydrogen fire has come about due to a localised leak in a closed system, the resultant flame characteristics would typically appear as a torch or jet being emitted from the origin of the leak. As this flame will be virtually invisible, the traditional way of detecting it is to carefully move the bristles of a broom over the suspected source, watching diligently for any burning of the individual bristles. Detection is made easier in subdued lighting where the flame can be more readily distinguished. More sophisticated systems of detection are available using ultraviolet detectors, which are often then linked to alarm systems in order to warn of the presence of any such hydrogen flame. Once detected, the safest way to combat a hydrogen fire of this nature is to close off the supply of gas where possible. If this isn't feasible, a dry powder extinguisher should be used to dowse the flame. Caution should prevail however, just in case the flame is smothered without the flow of hydrogen gas also being halted, as this can continue to generate an explosive mixture in the surrounding work area. If this subsequent mixture comes into contact with a heat source or other form of ignition, the resultant explosion can be a more significant hazard than the original fire had been. The recommended firefighting procedure in these situations is to prevent the initial leaking gas flame from causing secondary fires and thereby spreading, and to simply allow this to burn until the source of hydrogen is completely consumed. It would be prudent during this operation to keep adjacent objects cooled with water, so that they themselves cannot be a further source of ignition. It is important however that water is not used on any pool of burning liquid nitrogen, as the water will evaporate in the process, creating addition gas in what is already a potentially volatile atmosphere. Suitable dry powder fire extinguishers should be made available wherever liquid hydrogen is being used, especially where it has the potential to convert to its gaseous form through leakage from any closed system. Significantly, liquid hydrogen leaks will typically produce a visible cloud of ice or fog from the surrounding air. The actual flammable mixture of hydrogen and air however is likely to extend beyond this region and will be completely invisible to the human eye. The potential for this mixture to therefore come into contact with any ignition source without the observer being aware of the fact is highly likely.

As with many of the other cryogens, liquid hydrogen has a relatively large liquid to gas expansion ratio of 865. This can lead to the unintended build-up of high pressures in closed systems where vaporisation of the liquid is allowed to happen. As ever, the importance of providing some form of self-operating pressure relief system is paramount in these situations. Where expanding hydrogen gas is allowed to escape from any closed system, it can easily cause suffocation by effectively diluting the oxygen concentration in the surrounding air atmosphere, especially where there is a significant amount of the liquid evaporated into its gaseous form. In many situations, this process of exclusion of the oxygen from the air can occur so rapidly that personnel may be unable to escape the affected work areas in sufficient time. On its own, hydrogen provides no support for life and is therefore considered to be an asphyxiant.

Another important safety factor relating to hydrogen is that its molecules are extremely small, which leaves containment systems prone to even the smallest leak path under normal operating condition. Added to this fact is the realisation that if they are being used under cryogenic situations, contraction leakages due to the sudden cooling of seal flanges for example, can further compound the creation of leak paths for the hydrogen. For many hydrogen arrangements, it is common place for a leak-test to be undertaken using helium gas before any hydrogen is introduced into the circuits. Helium has a similar sized molecule to that of hydrogen and in sharp comparison is an inert gas, making leak detection that much safer. Liquid hydrogen systems are also generally built as closed systems which can be initially evacuated and purged to eliminate the possibility of air entering any of the circuit which could have the potential of creating a dangerous flammable atmosphere mix of the hydrogen and air. These systems would then generally operate at above atmospheric pressure to prevent any further ingress of air into them if leaks did occur. As the liquid hydrogen is cold enough to solidify gases such as nitrogen from the air, any purging process is generally undertaken with helium gas instead. The solidification of gases due to the extremely low temperature at which liquid hydrogen exists can be an issue, as this can have the effect of causing the plugging of vital systems such as safety relief devices, which in turn can result in the rupture of pipework, storage vessels and transfer lines. Therefore, any venting system from liquid hydrogen storage vessels should be periodically examined to ensure they have not become plugged with frozen moisture from the air.

Liquid hydrogen transfer lines must always be vacuum insulated in order to prevent the formation of liquid air on the outside of the lines, which could subsequently create oxygen enrichment in small pockets along its length. This oxygen in itself poses an additional flammable risk. Likewise, liquid hydrogen should never be decanted by simply pouring it from one container to another in an atmosphere of air. This is due to the fundamental fact that the oxygen in the air may condense into the liquid hydrogen during the process, increasing the risk of an explosion occurring with the formation of this concentrated hydrogen and oxygen mixture. A pressurised withdrawal of the liquid hydrogen from one vessel to another through an insulated transfer line is by far the preferred option.

As hydrogen has the ability to migrate through even the smallest of openings, leakages from liquid hydrogen systems can be a real issue. As a result, the potential of forming and igniting any flammable mixture of the hydrogen with the surrounding air tends to be higher than for other flammable gas usage. As a consequence, it is common practice for all equipment associated with the use of liquid hydrogen to be electrically grounded in order to prevent the accidental creation of ignition sources, such as those created by static electrical sparks. This is especially so for activities such as the transfer of the liquid hydrogen using transfer lines where the chances of hydrogen leakage during the process is that much greater. Adequate ventilation in and immediately around the working area will help to reduce any possible formation of a flammable or explosive mixture of the hydrogen and the air. In addition, this ventilation will also serve to eliminate the likely occurrence of personnel suffering from asphyxiation due to the leakage or accidental spillage of liquid hydrogen during its use.

Electrostatic charges are one of the biggest concerns with the operation of liquid hydrogen systems. As these charges can be caused by as simple a process as pipe-flow friction, they can result in ignition of any hydrogen gas being vented to the atmosphere. It is important therefore that routine inspections are undertaken to ensure the integrity of any electrical earth and bonding systems employed on the associated liquid hydrogen cryogenic system. This is made manifest where road tankers are utilised for the transport of the cryogen. Before any transfer of the liquid hydrogen can be undertaken, it is critical that an earth-cable is connected to the Earth terminal of the receiving equipment in order to discharge any build-up of static electricity which the road tanker has generated during its journey. This earth-cable must remain connected throughout the process of transfer as further electrostatic charges can be generated by the flow of the cryogen itself as it is being transferred.

Due to the very low formation temperature of liquid hydrogen, the correct protective apparel needs to be worn to prevent exposure to both the extremely cold liquid, but also to any cold hydrogen vapours incidentally released. These would typically include cryogenic rated personal protective equipment (PPE), such as leather-backed gloves for hand protection, full-face visors and safety glasses for eye protection, and aprons to provide protection to the torso and the legs. Liquid hydrogen coming into contact with exposed skin will produce an immediate cold burn injury, where damage to the skin tissues due to the low temperatures will occur.

Another significant factor relating to the small molecular size of hydrogen is the fact that it can permeate into the crystal structure of certain metals, causing a phenomena known as hydrogen embrittlement [45]. It is also sometimes referred to as hydrogen-assisted stress corrosion cracking, as it is in reality a corrosion process which the hydrogen causes and propagates. The actual embrittlement mechanism is complex and depends on many influencing aspects, such as the stress regime the material is subject to and its basic chemical composition. In principle though, the hydrogen molecules have the ability to migrate into the metal matrix and locate on the metal grain boundaries, weakening them considerably and increasing the likelihood of crack propagation along these grain lines. This has the effect of reducing the material's ductility and tensile strength, but also increasing its brittleness at the same time. For hydrogen embrittlement to take place, there are some general mechanisms driving its formation. Clearly, hydrogen needs to be present in the first place, but the material must also be of such a composition, that it is defined as being susceptible to the diffusion of this hydrogen into its structure. Not all materials are, with some demonstrating a greater predisposition than others. High-strength steels in particular are known to be prone to this hydrogen embrittlement process. Another key factor needed in order to cause this damaging mechanism is the presence of stresses in the materials and whether they are static or dynamic. In its brittle state for example, a material is less likely to withstand impact stresses and the designs of equipment for hydrogen use needs to take this feature into consideration. It is also known that some materials become more prone to hydrogen embrittlement as their operating temperature is increased. This is due to the solubility of hydrogen increasing at higher temperatures, allowing it to migrate through the metal structure with more ease. In many cases, this rise in temperature need only be small for the enhanced diffusion effect to take place.

Specific materials which are susceptible to hydrogen embrittlement are varied, with the degree by which they absorb hydrogen equally diverse. Therefore determining whether a material can be used with hydrogen or not is generally a difficult process of selection. Many of the more vulnerable metals can often be protected from the embrittlement process by the addition of alloying elements at their production stage or through the provision of coatings which inhibit the diffusion of the hydrogen at source. It is important to remember, that whatever the material, if they are used under cryogenic conditions with hydrogen present, the natural toughening of the materials due to the low temperature regime will further compound any embrittlement effects. Some typical examples of materials susceptible to hydrogen embrittlement include.

#### *7.3.14.1 Hydrogen embrittlement in steels*

Certain grades of steel operating with hydrogen at high temperatures are susceptible to a process known as hydrogen attack where the migration of hydrogen into the metal structure leads to decarburisation, which ultimately reduces the strength and ductility of the steel. However, steels with an ultimate tensile strength of less than 1000 MPa are not generally susceptible to hydrogen embrittlement. Conversely, the higher-strength, precipitation hardened stainless steels such as grade 17–4PH undergo sever hydrogen embrittlement, suffering a dramatic loss in their elongation properties. Likewise, some of the 300 series stainless steels are moderately attacked by the hydrogen and are particularly susceptible to hydrogen embrittlement, especially where they are subject to high tensile conditions.

#### *7.3.14.2 Hydrogen embrittlement in copper*

Any copper alloy which contains oxygen as part of its chemical composition will suffer hydrogen embrittlement under elevated temperature conditions. Reacting directly with the oxygen, the hydrogen will diffuse into the copper, effectively forming bubbles of water on the grain boundaries. This formation of small particles of water has the undesired effect of reducing the ductility of the material in the process.

#### *7.3.14.3 Hydrogen embrittlement in vanadium alloys*

Most of the vanadium alloys in their solid form are relatively inert to hydrogen effects at room temperature conditions. However, those alloys containing quantities of either niobium or titanium as part of their chemical composition can suffer a loss of plasticity due to the hydrogen penetrating the metal matrix and causing damage to their crystal structure. This leads them to becoming ultimately brittle.

*7.3.14.4 Hydrogen embrittlement in nickel and nickel alloys*

Nickel will absorb significant amounts of hydrogen wherever it is present, resulting in damage to its crystal structure, with a corresponding loss of ductility and an increase in brittleness. This is especially so for the so-called nickel super alloys which suffer a marked loss in their fatigue cycle properties due to this take-up of hydrogen into their crystal structure.

As can be seen, the selection of suitable materials for their use with hydrogen needs careful consideration, especially where they are utilised for the containment of hydrogen in its liquid state. However, the storage of liquid hydrogen is further complicated by the fact that hydrogen molecules exist as two forms; para-hydrogen [46] and ortho-hydrogen [16]. At room temperature, the concentration of a quantity of hydrogen gas is typically 25% para-hydrogen and 75% ortho-hydrogen. Conversely, liquid hydrogen consists of 99.79% para-hydrogen and 0.21% ortho-hydrogen. The significance of this change in makeup is that the conversion process from the ortho-hydrogen to para-hydrogen as the gas is cooled releases a substantial amount of heat. If this release isn't suitably managed, the resultant boil-off can evaporate large quantities of the final liquid hydrogen being produced, causing an uncontrolled release of gas in the process. Any storage vessels used for liquid hydrogen must therefore be able to freely vent this gas in order to avoid mechanical failure of the containment structure. The standard way of avoiding this liquid loss is to catalyse the ortho to para conversion during the cooling and liquefaction process. Typical catalysts used would be iron oxide or activated carbon.

Risk prevention methods can be adopted for all applications of liquid hydrogen. In its gaseous form, any released hydrogen will rise upwards and collect under roof areas and overhangs. This can represent an explosive hazard and it is important that good ventilation is provided in order to immediately disperse any collected gas through this process. Ignition suppression systems for all electrical equipment in the workplace should also be employed in order to eliminate any potential source of detonation if this released gas mixes with the air. By providing hydrogen sensors in the workplace, both rapid detection of hydrogen leaks can be determined and a suitable warning given for the safe occupancy of the work area so long as adequate ventilation is provided. Due to its embrittlement properties, it is also possible for leaking hydrogen to permeate into any surrounding pipework transporting other services, damaging the crystal structure of the pipework materials as they are unlikely to have been selected for even temporary hydrogen use. To avoid this, hydrogen supply pipes should always be positioned above all other pipework, so that in the event of any leaks, the gas will safely rise without causing any structural damage to surrounding equipment.

Any liquid hydrogen storage system should be designed and installed in accordance with the relevant national and international safety codes and industry safe practices. This should include all aspects of structural design of the containment vessels, along with the flammable gas zoning and electrostatic discharge requirements associated with such a dangerously explosive gas. Anybody operating equipment containing hydrogen as either a cryogenic liquid or in its room temperature gaseous form, should be suitably trained and be instructed on how to deal with emergencies, such as leakages, spills and even fires. Any system containing hydrogen should be identified as such, ideally with its liquid or gaseous state also clearly defined. Any covered areas where hydrogen is stored or used should be provided with adequate ventilation to prevent the build-up of explosive mixtures of leaked hydrogen with the surrounding air. If possible, such facilities should be located outside of buildings, with suitable fixed and lockable cages provided to prevent unauthorised access. Above all, all spaces where hydrogen is being stored or used should be marked as a 'no smoking' and 'no naked flames' area.

As a growing technology of interest, the use of liquid or cold gas hydrogen as a land vehicle fuel is rapidly becoming an area of considerable research and investment. Fortunately, international standards have been developed to coincide with these studies [47], specifically aimed at the refuelling and dispensing systems which would have to be employed in order to integrate the technology into existing resupply infrastructures. Aimed at providing protection from loss of life and property, they cover the design and installation of any such liquid hydrogen fuelling and dispensing systems. This includes the safety protocols required for the incumbent pressures and temperatures applicable to the transfer of liquid hydrogen from storage vessels to the land vehicles. Safety is paramount in the formulation of these standards, with the fundamental aim of reducing the overall risks associated with the use of what can potentially be quite a dangerous commodity. As any future hydrogen economy offers significant challenges to both its storage and transport, these standards can be seen as the foundations of more specific legislation appearing as both hydrogen processes and applications evolve.

In concluding the narrative for hazards associated with using liquid hydrogen, it is important to recognise that the effects of the individual hazards already discusses can be more pronounced when taken collectively, especially where the low temperatures are also to be considered. The relevant safety information relating to the use of this relatively hazardous cryogen can be keenly curated by experts in the field, but practical experience often plays an important part in the implementation of safe systems where liquid hydrogen is to be employed.

### 7.3.15 Liquid oxygen [16]

Oxygen is the second most abundant component in the atmosphere, comprising 20.8% by volume of the air around us. By mass, it is the third most abundant element in the Universe after hydrogen and helium. As a gas, it is colourless, odourless, highly reactive, and is an oxidising agent which can readily form oxides with many other routinely used elements. A typical example being iron oxide, more commonly known as rust, which can readily form on ferrite alloys under prevailing conditions. Oxygen also provides the mechanism for energy release in combustion, demonstrated by the fact that materials which burn in air will burn more vigorously in pure oxygen alone. It is also vital for the aerobic cellular respiration in living organisms, providing the necessary and essential support for life. Oxygen is a highly reactive substance and must therefore always be segregated from combustible materials. It is often used in various aspects of industry as a means of providing a ready source of oxygen gas. Many medical applications for example use it for this specific purpose. For many years, it has also been used as part of the propulsion system for spacecraft rockets when they are in space. Both commercial and military aircraft also employ liquid oxygen as a transportable source of breathing oxygen. Other uses of oxygen include the production of steels and plastics, and as a source of combustion in welding and brazing processes.

Oxygen is highly flammable and can under certain conditions react explosively with organic materials such as oil and hydrocarbons. Very often, materials that are usually considered non-combustible may burn in contact with oxygen. As a result of this, in gas systems designed for pure oxygen applications, special care needs to be made in selecting the correct grade of thread-sealing tape. Normal PTFE thread-sealing tape contains volatile solvents, which can react with the oxygen in the system and could cause an explosion. A special oxygen compatible version known as 'Green' PTFE tape [48] should always be used instead. In addition, any equipment using oxygen must meet very stringent cleaning requirements to prevent the accidental inclusion of contaminants which could bring about accidental combustion. All systems must be manufactured from materials that have a very high ignition temperature and are non-reactive with oxygen under their normal operation conditions. All pipework and storage vessels must be designed and tested to applicable national and international standards, so that they can withstand both the process pressure and temperature and meet the requirements for use with oxygen.

For economic reasons, oxygen is very often stored and transported in its liquid state, although its main use will inevitably be as a gas. This is driven by the fact that storing oxygen as a liquid is less bulky and less costly than its alternative high-pressure gas storage form. This is realised by its large liquid to gas expansion ratio of 875, whereby enormous volumes of oxygen would have to be retained in it gaseous form compared to that in its liquid state. A typical liquid oxygen storage system would consist of an externally located cryogenic storage tank, a series of heat exchangers, along with the associated pipework to feed the produced gaseous oxygen to the required location. At the onset of its journey, liquid oxygen is passed through the heat exchangers from the storage vessel which converts the cryogenic liquid into its warmer gaseous form. Many hospitals for example routinely employ this method to provide a constant supply of pure oxygen to the various internal facilities and applications within its buildings.

Liquid oxygen has a pale blue colour and has a normal boiling point of 90 K, and as such, requires careful management of its storage containers in order to maintain this relatively low temperature. To do this, liquid oxygen storage tanks and containers are constructed as a two part assembly, with both an inner and outer vessel forming the complete structure. Between the two vessels is an annular space which normally contains some form of thermal insulating material, as well as a vacuum to further enhance its insulating properties. This separating layer prevents the external ambient heat from reaching the liquid oxygen which would otherwise heat up the liquid, thereby releasing gas in the process.

Liquid oxygen is considered to be another in the specialist cryogen category, which brings with it a tranche of specific hazards in addition to those relating to its flammable nature. Its boiling point is significantly below the safe temperature for human skin, and exposure to it in its liquid state, or as an extremely cold vapour, can cause cold contact burns to the tissues which make up the dermal layers of the skin. Similarly, this cold temperature can cause many of the common materials of construction to lose their ductility and strength and readily become brittle as a result. As this effectively reduces their impact toughness, many will shatter in this state if subjected to forceful contact, such as that likely to occur in the event of a collision. Careful selection of materials therefore needs to be considered for the construction of containment vessels as well as their associated pipework and equipment to ensure this embrittlement doesn't take place when they come into contact with any liquid oxygen.

The large liquid to gas expansion ratio of oxygen can cause over-pressurisation inside closed containment systems in the event of small amounts of liquid evaporating and expanding into the much larger equivalent volumes of gas. Unless the systems are adequately vented, this expansion ratio rapidly builds up the internal pressure to dangerous levels, eventually resulting in the explosive rupture of parts. It is critical therefore for any such arrangement to be protected with some form of pressure relief device which automatically vents this over-pressure as it is generated. Any such vent however must also consider the fact that this released gas will create an oxygen-enriched pocket of air to form in the region immediately around the relief mechanism. This in itself is a hazard, as the enriched atmosphere has a greater potential to ignite as the oxygen concentration is gradually increased above the normal value. Contact of this gas with any non-compatible materials could lead to a combustion reaction resulting in a fire or even an explosion. In fact, where the oxygen concentration in the air increases above 23.0%, materials which easily ignite in normal air will be more susceptible to ignition and will burn with a greater intensity if they are ignited. This would include materials such as clothing and human hair, both of which have structures which provide spaces between their fibres which can trap this oxygen-enriched air within. As the normal oxygen concentration in air is 20.8%, reaching a value of 23.0% can be achieved relatively easy around any venting points from a liquid oxygen cryogenic assembly. It is important that personnel working in this type of environment are therefore made aware of this fact and that suitable hazard controls are enacted to minimise the effects. Clothing for example, which has been accidentally splashed with liquid oxygen, or even exposed to oxygen-enriched gas, must be immediately removed and allowed to slowly release the excess oxygen trapped in its weave. In its oxygen saturated state, the clothing would pose a risk as it will readily ignite and consequently burn vigorously.

It is critical to recognise that as gaseous oxygen is odourless, colourless and tasteless, any atmosphere enriched by it cannot be detected by the human senses. Oxygen should therefore never be released into confined spaces unless they are provided with adequate ventilation and have oxygen level detectors installed. Both liquid oxygen and gaseous oxygen are heavier than air and any accidental releases will as a consequence preferentially accumulate in lower lying work areas, such as pits and deep trenches. In these locations, the gas will be slow to disperse, forming an established oxygen-enriched layer which would represent a significant fire hazard. Liquid oxygen will also quite easily seep through porous materials such as clothing. Some porous flammable materials will in fact absorb oxygen, making them highly dangerous in this enriched state. Generally, metals such as stainless steels, aluminium, copper, brass and bronze are normally safe for use with both liquid and gaseous oxygen.

Unlike many of the other liquid cryogens used with cryogenic technology, as liquid oxygen expands into its gaseous phase, there is no risk of causing a rarefied air local to the source of the gas. The liquid cryogen to gas expansions of inert substances will inevitably result in these particular gases diluting the oxygen concentration in the work place until dangerously low levels of oxygen remain. With liquid oxygen however, the expanding gas simply adds more oxygen to the local atmosphere, and should not under normal conditions represent a breathing problem other than increasing the likelihood of combustion occurring. Exposure to too much oxygen can however cause oxygen toxicity in the body which can affect the central nervous system and result in damage to certain cell membranes. Breathing in increased oxygen levels at normal pressure over a long period can also cause hyperoxia, where pulmonary and ocular toxicity can result in symptoms such as disorientation, breathing difficulty, and vision changes [49]. In extreme cases, seizures can occur. Oxygen toxicity is a particular concern for underwater divers, where they are routinely breathing in higher concentrations of oxygen at above normal partial pressures. For general use as a cryogen, oxygen toxicity can be managed by the correct installation of oxygen monitors which can be set to provide a warning for both low and high concentration levels. By reducing exposure to increased oxygen levels, avoidance of hyperoxia and oxygen toxicity can be achieved.

### 7.3.16 Liquefied Natural Gas (LNG) [50]

Natural gas is a hydrocarbon gas mixture which consists of mainly methane with a small percentage of other gases such as; carbon dioxide, ethane and nitrogen. The actual percentages will vary depending on the primary source of the gas. Natural gas is created deep under the surface of the Earth, as well as under the seafloor, over a timeframe of millions of years during which layers of decomposing plant and animal matter are subjected to both heat and pressure, causing them to release the gas in the process. As the main constituent of natural gas, methane is a commercially viable fuel which provides many homes throughout the World with a source of both heating and cooking as it is flammable between 5.4% to 17.0% concentration in air. It has a spontaneous ignition temperature of around 540 °C (813 K), although the exact value depends on the purity of the gas. Methane is a chemical compound made up of one atom of carbon and four atoms of hydrogen. Under room temperature and standard pressure conditions, it is colourless and odourless. As a safety measure therefore, natural gas supplied to the domestic market is blended with an odorant which makes the detection of leaks easier to undertake without the need for sophisticated equipment. It has a specific gravity of 0.56 measured with respect to air, meaning that under room temperature conditions, it is naturally lighter than the air surrounding it.

Liquefied natural gas (LNG) forms at the methane boiling temperature of 112 K and has a liquid to gas expansion ratio of 623. In its liquid form it can therefore be more readily stored or transported, as effectively it only requires 1/623 the volume of its equivalent gaseous state. In this form it remains odourless and colourless, but is also non-toxic and non-corrosive. However, any vaporised gas released from the liquid is flammable. Due to the low temperature at which the liquid is held, any initial release of gas will be heavier than the air around it and will tend to fall to the ground, spreading outwards until the gas temperature rises sufficiently in order to leave it in its natural buoyancy state. Typically, liquid releases will rapidly convert to gas, quickly reducing the local oxygen concentration value of the workplace in the process.

As with many of the other cryogens utilised for cryogenic technologies, any heat leaks into an LNG storage container can bring about a more accelerated boil-off of the liquid into gas. This in turn increases the internal pressure of the storage container, which can result in an overpressure occurring, eventually resulting in mechanical failure if the correct safety pressure relief systems have not been deployed. In addition, thermal expansion of the liquid itself can occur, especially in closed systems, which can likewise cause overpressure of the storage container. Ideally, LNG systems should be operated at above atmospheric pressure to prevent the accidental ingress of air into them, which can result in the formation of flammable mixtures. The boiling point of LNG is too high to cause atmospheric oxygen to condense but it can solidify carbon monoxide from any surrounding air. It is important therefore that any traces of carbon dioxide are purged from LNG systems before filling, in order to prevent the formation of carbon dioxide blockages.

Where gas is being released to atmosphere from an LNG system, it is possible for electrostatic charges to build up as a result of friction in the pipes. This can easily cause ignition of the gas if the 5.4% to 17.0% flammable concentration in the air is achieved. It is wise therefore that the integrity of all electrical earthing and bonding systems are routinely inspected to ensure they remain fully functioning. Likewise, where the transfer of LNG from road tankers is being undertaken, they must be equipped with the necessary earth-cable to discharge any potential electrostatic charges from their structure.

Where natural gas fires do occur, the best approach is to shut off the supply of the gas if possible, as extinguishing this type of fire is very often difficult to achieve. In many situations it is inadvisable to attempt to extinguish a natural gas flame, as the contents of the extinguisher can in itself result in the formation of potentially explosive mixtures with the already burning gas. The use of water for example, applied directly to a pool of burning LNG, will evaporate additional gas from the water into the local atmosphere. Ideally, foam type extinguishers should be employed, especially so-called medium expansion foams [51], which contain a detergent in the water-foam mixture. They are particularly useful in that they provide rapid blanket coverage over a large area during any firefighting activity. Any firefighting equipment should be regularly inspected and maintained, with records being kept to qualify this work has been carried out. In many facilities around the world, this is a legal responsibility of the owners of the LNG facility.

Some other typical hazards associated with the use of LNG relate to possible brittle fractures occurring in metals when they are used below their ductile, brittle-transition temperature. As the boiling temperature of LNG is relatively low, care needs to be made in the selection of suitable materials of construction which are compatible with this temperature regime. Likewise, the low temperatures can also cause the formation of ice, especially where water residue is present in the cryogenic equipment before being cooled down. This can result in mechanical fractures occurring, owing to the expansion of ice as it transforms from its liquid to solid state. Any sudden cooling of equipment can potentially lead to the differential contraction of flanges and pipework which can result in leakages occurring between sealing faces. By designing the parts specifically to overcome this issue, the problem will be mitigated, but a protocol of cooling down LNG systems at a uniform rate over a longer period should always be encouraged.

Common to similar cryogens, the main hazard associated with the use of LNG is due to the low temperature at which it is stored. If any of the liquid comes into contact with exposed skin or any other delicate surfaces of the body, tissue destruction in the form of cold burns, similar to that caused by an elevated temperature burn, will result. Uninsulated equipment containing LNG can also cause cold burns if the human body makes contact with the low temperature parts. In addition, releases of cold gas can cause significant damage to the lungs if they are accidentally inhaled for any prolonged period. Being non-toxic however, natural gas will not cause any chemical damage to both internal and external parts of the body. If a gas release is significantly large enough however, oxygen exclusion can result in the workplace which can lead to the more serious hazard of asphyxiation. If the released gas is due to a rapid boil off of the cryogenic LNG, this oxygen exclusion can take place with equal rapidity, often providing very little time for individuals to escape the immediate vicinity of the expanding gas.

Handling of LNG should only be undertaken with the appropriate personnel protective equipment (PPE), which must be rated accordingly for use with low temperature cryogens. In particular, special consideration should be made to the type of gloves, eye, and face protection used for any handling operations. In addition, suitable coveralls or aprons must also be considered to mitigate the effects of minor spillage of the LNG. Lace-up types of footwear should ideally be avoided, as the eyelets can become routes of entry into the footwear for any LNG cryogen spilled during handling.

As a flammable material, the use of LNG will require the zoning of areas immediately around any storage facilities. The higher the risk, the more specific the requirements are for these zones. In particular, spark arrest electrical equipment is mandatory in the zone closest to the source of the LNG. Where transport vehicle entry needs to be made into these areas, the vehicles must be fitted with additional safety systems in order to negate the likelihood of an ignition source becoming apparent. This would include features such as exhaust protection devices to avoid sparks and excessive temperatures coming into contact with any potentially released gases from the LNG storage.

The majority of LNG storage facilities are managed at an extensive commercial level, with typically large capacity containers holding huge quantities of the cryogen at any one time. Secondary spillage containment is often provided around these storage vessels in the form of a bund. These are normally high level concrete structures, designed to withstand any cold-shock should there be an unintended release of LNG from the storage vessel, and built sufficiently tall enough to capture any initial cold gas cloud which would form during the early stages of any such release. Less common, small volumes of LNG are occasionally stored under pressure within a vacuum insulated vessel. The positive pressure prevents any accidental ingress of air into the container during filling or decanting, with the vacuum providing a thermal insulation to limit the amount of external heat entering the LNG over time.

Where leakage from LNG systems is a possibility, some form of leak detection is required. This can be in a variety of forms and often depends on the specific layout of the facility and what has been considered as the greatest risk from such a leak occurring. For example, detectors are available which can determine the change in energy in pressurised liquids when a leak causes the liquid to convert to a gas. As LNG is stored at extremely low temperatures, any leak, even in its gaseous form, would also be at a low temperature. It is possible to detect this cold gas, by the use of thermocouples around storage vessels, which could trigger an alarm when a certain degree of change occurs with respect to the normal ambient temperature around the vessel. Once converted into a gas, the formation of a flammable mixture with the surround air is possible where a leak has occurred. This too can be detected, with the specific ratio of gas to air being constantly monitored, watching for any variance to occur which would indicate the presence of a leak. It is worth bearing in mind though, that very often, more than one cryogen or gas is often stored in the same area. Any gas detection system must therefore be calibrated to take this into account, normally set to detect the least sensitive gas being used in an area.

Other detectors which are useful where LNG is being employed would include oxygen depletion monitors, routinely checking for the oxygen concentration value in the surrounding workplace. These can be alarmed so that an audible warning is given when dangerously low levels of oxygen are found to be present, which could be either due to natural build-up of released gas from pressure relief systems, or an indication of a significant leak from a storage vessel. The other critical monitors to have in place are those used for the detection of fire. These can be for the detection of smoke particles created by a fire, or even devices which measure the radiant heat being generated during combustion. In practice, a range and combination of detectors are required for any LNG facility, many of which need to emit an alarm which warns any occupants of the need to immediately evacuate the facility.

### 7.3.17 Liquid argon [16]

Argon is the third-most abundant gas in the Earth's atmosphere and is chemically inert under most conditions, forming no known chemical bonds with other elements. It is colourless, odourless, tasteless, non-flammable and non-toxic in each of its solid, liquid and a gas states. It is produced industrially by the fractional distillation of liquid air in a cryogenic air separator. Commercially, it is frequently used as an inert shielding gas for welding operations, along with a variety of high-temperature industrial processes where, in the absence of an inert gas, materials can become reactive. An example of this is where an argon atmosphere surrounds a graphite electrical furnace in order to prevent the graphite element from burning at the extreme elevated temperatures of the process. Argon is in much demand elsewhere in industry due to several key properties it possesses. Being inert allows it to be used as a cheaper alternative to other gases such as nitrogen, as it is relatively inexpensive to obtain from the air around us where it is plentifully available at all times. It also has a low value of thermal conduction which makes it an ideal thermal insulator for critical systems.

Generally used in its gaseous state, argon is very often stored and transported as a liquid since this offers a more economic means by which a supply of the gas can be provided. Liquid argon has a boiling temperature of 87.2 K. Therefore, any materials used in the construction of storage vessels or pipework associated with the cryogen, need to be able to withstand the associated low temperatures. All such equipment should be designed to the relevant national standards, taking into account the operating conditions and the low temperatures involved. Storage vessels in particular need to be suitably insulated in order to prevent the surrounding heat transferring to the cryogen and thereby causing more rapid vaporisation of the liquid argon into gas.

Argon gas has a specific gravity of 1.40 measured with respect to air, which means any leakages from closed systems will preferentially fall to the ground and accumulate in lower lying work areas, such as pits and deep trenches. Argon also has a large liquid to gas expansion ratio of 860, whereby even relatively small amounts of liquid argon spillage can rapidly form into a significant quantity of gas. It is therefore considered as an asphyxiation hazard, especially in closed or confined spaces where this expanding gas can quickly reduce the breathable oxygen concentration value of the ambient atmosphere. Being odourless, colourless, tasteless, and a non-irritant, the human body is unable to detect the presence of argon. Therefore, any displacement of the oxygen from the surrounding air to levels below that which life can be supported, will typically cause dizziness, nausea, vomiting, loss of consciousness and ultimately death if a timely rescue is not put in place. Any area using liquid argon should always have some form of oxygen detection installed which should be consulted before anybody enters or approaches the liquid argon equipment.

Liquid argon has an extremely low boiling point and extensive tissue damage to the skin or other delicate surfaces of the body can occur when it comes into contact with them. The actual harm is similar in appearance to that of an elevated temperature burn, with the manifestation of blistering around the wound, only in this case the injury is as a result of a cold-burn. It is also important to remember that any released argon gases or vapours from a cryogenic system will also be at extremely cold temperatures and parts of the body, such as the eyes, are prone to similar, often irreversible damage. The correct cryogenic personnel protective equipment (PPE) should therefore always be worn when handling liquid argon, rated accordingly to suit the cold temperatures involved.

The storage and shipping of liquid argon is managed through the provision of containment vessels, which are available in a variety of types and sizes. Typically these would be cryogenic Dewars, cryogenic liquid cylinders and larger, cryogenic storage tanks. The capacity of these would normally be based on the end user requirements, ranging from small handheld Dewars, up to sizable containers holding thousands of litres of the cryogen. One common feature with all the types of containers is that heat leaks through the walls occur continuously, which in turns causes vaporisation of the liquid argon on a regular basis. A typical cryogenic storage vessel would be constructed as an insulated, vacuum-jacketed assembly in order to reduce this heat transfer. The actual rates of gas generated by this process depends on not only the design and construction of the containment vessels, but also the volume of the cryogen being stored. Where the containers are under pressure, these are designed and manufactured according to the applicable national safety codes and specifications to meet the essential requirements of the working cryogenic temperature and pressures involved. However, this continuous vaporisation must be controlled, otherwise pressure build-up over time can occur as the constant process of heat entering from outside the vessels causes the cryogen to liberate subsequently more and more gas. The fitting of protective devices, such as safety relief valves and rupture disc assemblies, are often stipulated as part of the encompassing safety codes for storage vessels. Where liquid argon is removed from any storage container, this is generally undertaken by using insulated withdrawal transfer lines which minimise the losses associated with any liquid cryogen converting to gas during the process. Where argon is withdrawn specifically as a gas from these containers, this is normally done by the use of an internal vaporiser.

### 7.3.18 Liquid neon [16]

Neon is a relatively common element in the Universe, although it is rare on Earth, representing only around 18 ppm of the air by volume in the atmosphere. The reason for this exceptionally low value is that neon is highly volatile and therefore forms no compounds which would allow it to fix itself to solids. As a gas it is colourless, odourless, tasteless and chemically inert, with a specific gravity of 0.697 measured with respect to air. In this form, it is frequently used in high-voltage illuminated signs where it produces a distinctive reddish-orange glow. It is commercially extracted by the fractional distillation of liquefied air. However, due to its rarity in the air, it is a relatively expensive commodity to extract and can be typically in the order of tens of times more expensive than helium to produce.

Liquid neon has a boiling point of 27.1 K, which makes it a useful cryogen to adopt where cryogenic cooling applications do not require the lower temperature range achievable with liquid helium. In addition, its latent heat of vaporisation is over 3 times that of hydrogen, and over 41 times that of helium, which means that it can provide more effective cooling power in comparison. Its high costs inevitably dictates the economics of its use and it tends to be limited to specialist cryogenic applications, such as those required as part of the space programme, where it is employed as a coolant for ultra-sensitive infrared imaging and detection equipment on space probes. Significantly, the triple point temperature of neon is 24.5561 K and is one of the defining fixed points used in the International Temperature Scale, ITS-90 established in 1990 [52].

On a safety level, neon has an exceptionally large liquid to gas expansion ratio of 1470, meaning that a small release of liquid neon will result in an enormous amount of gas being produced. Any storage vessels designed for the use of liquid neon must take this expansion factor into account and be fitted with an appropriate safety relief valve and bursting disc assembly to allow the automatic release of any internal build-up of pressure which this expanding gas will create. Through natural evaporation, small amounts of neon gas will be continually released from storage vessels via the safety relief valve, as heat from the surroundings steadily warms the contained liquid neon. Over time, this released gas can build up, with more significant releases occurring where there is a failure in the insulating jacket of the storage container. Neon gas will not sustain life and is therefore considered a potential asphyxiation hazard, especially where it is being used in enclosed work areas. In its gaseous form, it is lighter than air, but any initial release of gas from its cold liquid state will also be significantly cooled and will tend to collect around its source, falling to the ground until it has sufficiently warmed up and regained its natural buoyancy. This will have the effect of displacing the oxygen from the air in the immediate vicinity of any storage vessel, which can lower the oxygen concentration to a dangerously low level. As neon is colourless, odourless and tasteless, the human body cannot detect its increased presence in the air. Inhalation of excessive quantities of neon gas can result in a range of symptoms such as dizziness, nausea, vomiting, loss of consciousness, and ultimately, death. Where the oxygen concentration is excessively low in the working environment, unconsciousness and death can occur within seconds, generally without any indicative warning.

The guidance for the safe handling of liquid neon must follow the same principles adopted for the use of all other cryogens. Due to its extremely low boiling point, tissue damage to the skin or other delicate surfaces of the body can occur should liquid neon come into contact with them. Similar in appearance to that of an elevated temperature burn, injuries will appear as cold-burn blisters around the wound area. It is also important to remember that any released neon gases or vapours from a cryogenic system will also be at extremely cold temperatures and parts of the body such as the eyes are prone to similar, often irreversible damage. The correct personnel protective equipment (PPE) should therefore always be worn when handling liquid neon, rated accordingly to suit the cold temperatures involved.

### 7.3.19 Liquid fluorine [16]

Fluorine exists under normal conditions as a highly toxic, pale-yellow gas. It is extremely reactive with most other elements, except argon, neon and helium. Materials such as Teflon, Mylar and Kapton are also found to be highly resistant to fluorine, along with certain grades of steel and Monel metals [53]. Purity of these metals however is important, as under certain conditions, low level contaminants can act as a catalyst with the fluorine to initiate combustion. Fluorine reacts violently with most hydrogen-containing compounds, such as water, ammonia and organic chemical substances. Flourine is the 13th most common element in the Earth's crust, found in minerals such as fluorite, fluorspar and cryolite.

Its uses are specialised, mainly as a refrigerant and as a rocket and missile propellant. Derivatives of methane and ethane, in which the hydrogen has been replaced by fluorine and chlorine, have been found to exhibit ideal refrigerant properties. In this state they are relatively non-toxic and non-flammable and are typically used in air conditioning systems and as aerosol propellants. However, in more recent times the use of these refrigerants have been internationally prohibited due to their association with the Earth's ozone depletion. Liquid fluorine and oxygen–fluorine mixtures can be effectively used as oxidisers for various fuels, providing propellants with enormous impulse forces which can be used in space rockets and missiles. Fluorine is also used in parts of the production process of steel. Its largest application however, is as a gas for uranium enrichment in the nuclear industry, where it is used to make uranium hexafluoride ($UF_6$). It is also used in its gaseous form to make sulphur hexafluoride ($SF_6$), which is a very effective electrical insulating gas employed in high-powered electricity transformers. Refining pure fluorine is a relatively expensive process, so much so that most commercial applications only over use fluorine compounds.

The economic storing and transport of fluorine is undertaken in its liquid state through the provision of containment vessels, which are available in a variety of types and sizes. As fluorine reacts with most materials, these containment vessel have to be manufactured from specific grades of steel or Monel metals. They must not be used for the containment of other cryogens as they have been precisely selected for fluorine use only and are safe to operate at the low boiling temperature of the liquid. The boiling point of liquid fluorine is 85 K, which is in the similar region to that of nitrogen and oxygen. It also has a specific gravity of 1.108 measured with respect to water in this liquid state and a specific gravity of 1.336 measured with respect to air in its gaseous form. It also has a liquid to gas expansion ratio of 888, which means that any small release of liquid fluorine can manifest itself as an equivalent large volume of gas on evaporation. Likewise, any heating of it in its liquid state can have a significant effect on the quantity of gas being evaporated as a result. Any storage containers must therefore be provided with some form of safety relief system to protect against the potential for an internal over pressurisation occurring due to this heating of the liquid. It is important to remember however, that fluorine is a reactive, corrosive and highly toxic gas, and any released gas under these circumstances must be controlled in such a way to prevent exposure to any employees or vulnerable equipment in the immediate vicinity.

As the use of liquid fluorine tends to be considerably less frequent than that of the other major cryogens, the potential exposure to its associated hazards is equally rare and would only occur under normal circumstances through accidents. This being said however, accidental releases of liquid fluorine can occur and can either be as a direct liquid spill, or a concentrated cloud of cold gas evaporating from the liquid. This vapour would be heavier than air and tend to collect at ground level, migrating into any low level facility such as pits and trenches. Any large spills of liquid fluorine would also cause the formation of hydrogen fluoride gas through a natural reaction it would have with moisture in the atmosphere. Both the fluorine gas and the hydrogen fluoride gas can have the effect of displacing the oxygen in the air and thereby reducing the oxygen concentration value to a dangerously low amount. Fluorine and its derivatives have no known metabolic role for the human body and would therefore pose an extreme hazard if found in significant quantities due to the accident release of gas under these circumstances.

Industries using fluorine on a regular basis, such as steel, aluminium and fertiliser manufactures, consider the chronic effects caused by long-term exposure to high concentrations of airborne fluorine. Under these conditions, fluorine can be retained in the human body, typically in localised regions of the skeleton. Symptoms such as the calcification of ligaments and the thickening of the bone ends can result. Both liver and kidney damage can occur at relatively low concentrations, with permanent eye and nose damage occurring at higher values. The inhalation of 1000 ppm of fluorine in the air will cause death within a matter of minutes. Short-term exposure to a sudden release of fluorine gas from a cryogen storage container, would on the other hand cause choking and suffocation if the concentration is sufficiently high. This is due to the fact that the fluorine causes a swelling of the mucous membranes in the two bronchial tubes feeding the lungs which in turn causes an obstruction, making it hard for any sufferer to breathe.

Due to the extremely low temperature at which liquid fluorine exists, dermal cold-burns can occur if the skin is exposed to either the liquid or cold evaporated gas. In addition, chemical burning of the skin will take place as the fluorine reacts directly with the surface flesh. The fluorine will also react with the moisture of the skin to create hydrofluoric acid, which will result in profound tissue poisoning. In fact, low level exposures may not demonstrate any damage for several hours until this poisoning evolves, after which the injured person will become aware of pain around the wound. It is important therefore, that every suspected exposure is managed as though it were a chemical burn, with immediate medical treatment being administered. The eyes in particular are sensitive to irritation caused by fluorine gas and would generally be the first indicator that the gas is present in the air. Wearing of the correct personnel protective equipment (PPE) is essential to protect both the skin and the eyes, from not only the extremes of temperature involved, but the corrosive nature of the gas and its derivatives.

As fluorine is reactive to most materials, liquid fluorine spillages onto things like structural steel and concrete can be damaging to them to differing degrees. In fact, the so-called alkali metals; lithium, sodium, potassium, rubidium, caesium and francium, will react particularly explosively to the gas, and wood will burn in the gas without the need for any ignition source. The main consideration however for materials subject to a liquid fluorine spill is the sudden, dramatic change in temperature as the cryogen causes the rapid contraction of the material's structure. Brittle failures are not uncommon where any cryogen spill encounters a particularly incompatible material, and this needs to be considered at the design stage of any cryogenic assembly. In addition, the boil off of fluorine gas from the liquid phase will initially be high for any such spillage, as the cryogen hits the significantly warmer materials around it. This can cause an immediate asphyxiation risk, where a dense cloud of cold fluorine gas fills the surrounding air, excluding the breathable oxygen from the atmosphere [54].

### 7.3.20 Liquid carbon dioxide [16]

Carbon dioxide is a colourless gas, which at low concentrations is considered odourless. At high concentrations however, it has a distinctive sharp, acidic odour.

In its gaseous form, it has a specific gravity of 1.53 measured with respect to air, implying that it is heavier than air and will fall to the ground when released. Carbon dioxide occurs naturally in the Earth's atmosphere from sources such as active volcanoes and hot springs. It is soluble in water and therefore occurs naturally in rivers, lakes and the oceans. It is also produced as part of the respiratory process of both humans and all other air-breathing land animals. It has many industrial uses, such as being provided as an inert gas for welding applications and as the extinguishing gas in certain types of fire extinguishers. It is also typically added to drinking water and carbonated beverages such as beer and lemonade to provide their fizziness.

Carbon dioxide gas is classified as non-toxic or harmful, but is considered to be an asphyxiant wherever it is released in sufficient quantities to displace the oxygen in the surrounding air. Concentrations of up to 1% in the air can cause drowsiness and the lungs will start to feel laboured whenever inhalation is made. Concentrations of 7% to 10% can cause suffocation, even where there remains sufficient oxygen in the air. Symptoms under these circumstances would be dizziness, headaches, visual and hearing difficulty, and eventually unconsciousness. As the gas is heavier than air, any releases will tend to collect on the ground, forming relatively high concentrations local to its source. Unless the gas is dispersed through ventilation, this concentration can gradually increase, until dangerous levels are achieved.

Generally, the absence of adequate ventilation is the primary cause of excessive carbon dioxide concentrations forming in confined spaces. This is especially demonstrated in situations such as those found in poorly ventilated coal mines, which tend to be susceptible to the formation of a particular mixture of carbon dioxide and nitrogen gas. The miners traditionally called this 'blackdamp' or 'choke damp' [55] and it comes about when the coal is exposed to the air in the mine, thereby absorbing oxygen, and at the same time releasing carbon dioxide and water vapour. Many years ago, miners would monitor for the presence of blackdamp by taking caged canaries into the mines with them as they worked. Seemingly, the dangerous gases in the mines would affect the birds before levels became sufficiently dangerous for the miners. Any signs of distress from the canaries would be an indication that the prevailing gas conditions were no longer safe. Often the canaries would stop singing as they became sensitive to the blackdamp, eventually falling off their perches if the concentrations became too high. Similarly, miners would also employ a flame protected device known as a Davy lamp [56] in order to detect the blackdamp, as the flame of the lamp would burn less brightly wherever the gas was found. Likewise, the Davy lamp would burn brighter with the presence of methane in the air.

Carbon dioxide is used in both its liquid and solid states as refrigerants, especially for the food industry. The boiling temperature of liquid carbon dioxide is 217 K, which is relatively high in comparison with other cryogenic technology and is not strictly a cryogenic temperature by the true definition. However, it offers the similar cooling properties as most commercially used cryogens, but specifically, at a temperature more suited for the transportation, storage and preservation of products such as; ice cream and other frozen foods. Liquid carbon dioxide has a specific gravity of 1.18 measured with respect to water and is colourless and odourless. Being heavier than the surrounding air, it is often employed by the film and theatre industries where it is used for the production of haze and fog effects in studios and on the stage.

Liquid carbon dioxide is generally stored and transported at a constant temperature and pressure, normally at around 20 bar. As such, the storage vessels have to be compliant to various national regulations and standards, ensuring that suitable materials of construction are selected for both strength to withstand the pressures, and compatibility to the cold temperatures at which the liquid exists. Storage tanks are typically insulated to reduce the amount of ambient heat warming up the liquid. Most static storage tanks have a small refrigeration plant attached which offsets any of the heat gains, keeping the liquid carbon dioxide at its boiling temperature at all times. As such, these systems tend to be closed circuits, albeit they should still be fitted with pressure relief devices should the refrigeration system fail at any given time. Liquid carbon dioxide cannot exist at atmospheric pressure, and so any release from these safety devices will therefore appear as a mixture of cold gas and small flakes of solid carbon dioxide. Any prolonged exposure to liquid carbon dioxide, cold gas, or solid carbon dioxide can result in severe skin damage through the effects of cold burning. The evaporating gases resulting from a liquid spill are also likely to reduce the oxygen concentration levels immediately around the spill.

Liquid carbon dioxide has a particular property as a dry cleaning agent as it is able to dissolve stains without producing hazardous waste products or harmful emissions. It is now one of the leading methods by which the dry cleaning of delicate fabrics is achieved.

Solid carbon dioxide is often referred to as dry ice, and again, in the purist sense, is not truly a cryogenic material as its sublimation temperature is relatively high at 194.6 K. At this temperature, handling the dry ice without the correct personnel protection equipment (PPE) carries the high risk of individuals suffering cold contact burns. Under atmospheric conditions, the dry ice will convert into large quantities of carbon dioxide gas. It should only ever be used therefore in well-ventilated environments or in the outdoors.

As with its liquid form, dry ice is generally used as a means of keeping items cold or frozen, such as foods and biological samples, without the use of mechanical cooling systems, such as refrigeration. It especially provides a useful means of cooling in remote locations where power for mechanical cooling is generally not available. It also has the advantage over water ice due to its lower temperature and therefore preserves items at the below ambient temperatures for longer. It is also used as a means of cooling gases in order to trap water vapour and oil vapour out of suspension, and is another relatively easy and cheap commodity used for contracting metal parts for assembly work.

Solid carbon dioxide is not considered to be toxic. However, the gradual evaporation from it as it warms up can release significant quantities of carbon dioxide gas, which if inhaled, can cause hypercapnia [17], which is where there are abnormally high carbon dioxide levels in the blood. This can especially occur progressively over time if the dry ice is being used in a confined location. For this reason, many applications will involve a sealed system where the solid carbon dioxide is normally maintained under vacuum conditions.

### 7.3.21 Superfluidity [16]

Superfluidity is a specific property of a fluid when it has zero viscosity. Effectively, such a fluid will, under these conditions, be able to flow without any loss of kinetic energy. When stirred for example, a superfluid would form a vortex, which in essence would continue to rotate ad infinitum. Likewise, if the superfluid's container is rotated, the fluid inside will not move, as its zero viscosity has no frictional reaction against the container's walls.

However, superfluidity only occurs at extremely low temperatures, typically in the region below 2 K. It can therefore occur in both liquid helium-3 and helium-4, at which point the liquids behave unexpectedly, 'creeping' along surfaces in order to find their own level. In fact, if the liquid helium is held in an open container, it will be seen to rise up the sides of the container and flow over the top, seemingly defying gravity in this state. In addition, if a capillary tube is placed into a bath of superfluid helium and subsequently heated, the helium will flow up through the tube, emerging from the top as a fountain of extremely cold liquid helium.

As the temperatures at which superfluidity occurs are in themselves a challenge to achieve, the associated safety precautions which need to be applied are specific to the technology being used. The main area of concern however is where liquid helium mixes are being used as a means of achieving below 1 K temperatures. Clearly, at this temperature, some of the helium will possibly begin to exhibit superfluidity, and heat exchange chambers filled with the cryogens, may suddenly appear empty as the fluids 'creep' from them. Also, with zero viscosity at this stage, the helium is able to flow through the smallest of apertures, through which normal liquids would be prevented by the nature of their surface interactions and their relative viscosity. Any containment vessels therefore considered as sealed against helium under normal conditions, may suddenly appear to have developed leaks where superfluids are present. It is important to also recognise that the thermal conductivity of superfluid helium is millions of times the value of that in its normal liquid phase and is in fact greater than the best metallic conductors. Clearly, this can give rise to unwanted thermal conductions into the cryogenic system and must be studiously considered as part of any extremely low temperature cryogenic application.

### 7.3.22 Ice formation [57]

The use of cryogenic technology, especially those operating in the very low temperature range of the liquid helium boiling point of around 4 K, brings with it the potential to solidify many of the gases which making up the air surrounding us. Accidental air leakage into a liquid cryogen storage vessel will for example result in the introduction of moisture from the water vapour in the air being condensed by virtue of the extremely low temperatures of the cryogen. This water may under certain circumstances gradually solidify, forming ice which can in turn eventually form an ice plug in the neck of open Dewars. Evaporation of the cryogen beneath this ice plug, would over time, result in a gradual and significant build-up of pressure which will have no means of release as the storage vessel is now effectively sealed. Over time, as this pressure further increases as more and more of the cryogen warms, thereby releasing more gas into the volume above the liquid surface, there becomes an acute danger of the ice plug suddenly being expelled at high velocity, not unlike the effect of a cork being released from a bottle of pressurised sparkling wine. It is advisable therefore that operators of cryogenic equipment keep both the hands and face away from the neck of the storage vessels at all times if an ice plug is discovered. In some extreme cases, the ice plug remains intact and the pressure is able to increase sufficiently to eventually rupture the storage vessel.

Because of the extreme risks involved, clearing ice blockages of this nature is a specialist task. Under normal circumstances, this should only ever be carried out by someone with the appropriate knowledge and training. As a precaution against over-pressurisation due to the formation of an ice plug, most cryogenic storage Dewars are fitted with a redundant relief passage so that if the main neck of the Dewar is blocked, an escape route is still provided for any excess pressure being created by boil-off of the remaining stored cryogenic liquid. Under these circumstances, it should be possible to safely vent this excess pressure whilst slowly defrosting the ice plug. This can be done by applying a flow of room temperature, or heated helium gas onto the ice plug, preferably through a small tube so that a small localised hole can be melted in the ice plug. Once this small opening has been formed, the rest of the blockage can then be carefully chipped away, or melted by applying further gas over the affected area. It is worth remembering that the neck of storage Dewars are normally made from thin metal and can be very easily damaged or punctures during this clearing process. The wearing of the appropriate personal protective equipment (PPE) throughout this procedure is critical.

After any ice plug has been removed, the remaining cryogen must also be emptied from the storage Dewar as it is likely that both ice and water would have fallen into the cryogen during the clearing process. Importantly, any residual ice and water found at the bottom of the storage reservoir must be removed before the Dewar is refilled again. Where helium Dewars are concerned, once cleared of all traces of ice and water, they must be sealed and evacuated to remove all the air from them before a fresh charge of liquid helium is introduced.

### 7.3.23 Manual handling and ergonomics [58]

The manual movement of cryogenic equipment, such as storage Dewars, tanks, gas systems and vacuum pumps, is a common feature of most facilities where cryogenic technology is employed. Undertaken incorrectly, or even repetitively, this manipulation of equipment can give rise to manual handling injuries, such as those to the back, the neck, hands and fingers. Whether these manual operations use specialist equipment or are undertaken individually, it is essential that they are enacted with health and safety in mind at all times. By taking an ergonomic approach to the movements involved, individuals can make sure that manual operations involving cryogenic equipment are done in a comfortable and efficient manner.

Proper lifting and moving procedures should be adopted and lifting aids, such as trollies, carriers and specific lifting equipment like cranes and hoists, used where possible in order to avoid physical injuries. As a result, by adopting ergonomic principles, the correct pushing, pulling, lifting, lowering and manoeuvring practices will safeguard the wellbeing of those undertaking the tasks. As a percentage of workplace days lost due to injuries, manual handling is representative of typically a third of the overall total [59], making it a particularly hazardous area of what should nominally be a routine task. Of the so-called high-risk manual handling activities undertaken as part of the use of cryogenic technology, rather than specifics, certain categories prevail. These would include forceful exertion, awkward posture and repetitive motions. Likewise, the level of exposure must also consider factors such as the peak forces involved, the sustained force, the weight of the objects being handled, the cumulative weight being handled over a period of time, the duration of the exposure, the frequency, and the detailed working height. Each of these can have an effect on the body position during the handling operation and will produce differing potential injury characteristics depending on the individual's current health or previous history of manual handling injuries.

Whatever the task, the preferred option it to always push, rather than pull, equipment where possible, with the arms bent and as close to the body as possible. This will absorb any impact forces if there are any sudden changes in the movement of the equipment being pushed. This method also allows the legs to be used as a driving force and will tend to keep the back straight and the body well balanced over the legs as movement is made. Adopting this method also means that should the equipment being pushed happens to tip over, the person pushing it will be able to automatically let go and back away from any immediate danger.

Pulling the equipment on the other hand, increases the risk of injury as individuals are more likely to be leaning away from the equipment, using their weight rather than your legs as the driving force. Their arms will not be bent, which will put additional stress on the shoulders and the back should there be any sudden change in movement of the equipment. Pulling will also mean that the body will be adopting an off-centre balance which makes it more likely for individuals to fall back and for the equipment to roll onto them. There is also the risk that their feet will be close to the equipment as they pull, increasing the likelihood of them becoming trapped under the equipment, especially if the terrain suddenly presents a downward slope.

The movement of fully charged cryogenic storage vessels in passenger lifts should be avoided at all times as the potential gas released from the Dewar within such a small volume could quickly create an asphyxiation risk. The preferred option instead, is for an open caged type lift to be used where the frequent movement of cryogens is required. Any spillage, or accidental release of liquid cryogen which then converts into gas in this type of lift will be inherently safer as there is no enclosed walls in which the evaporating gas can accumulate. Critically, as an added precaution, no personnel should accompany the storage vessel in any type of lift as the risks of accidents occurring are greater during any movement of cryogen storage vessels as any disturbance of their contents causes internal mixing which significantly increases the normal evaporation rates significantly. It is also worth noting that the rolling of Dewars should never be undertaken as this will likewise rapidly increase the evaporation of the contained cryogen and also result in likely spillage of cryogens from the Dewar.

By definition, ergonomics is the study of how equipment and workplaces can be arranged in such a way that staff can undertake work activities efficiently, but more importantly, in a comfortable manner, without the risk of injury occurring. The layout of any cryogenic work area should therefore be designed with these ergonomic doctrines in mind, ensuring that a safe system of work is employed and that manual handling activities are minimised or even eliminated. An example of this would be where static cryogen storage tanks, and piped liquid/gas supplies are used in preference to mobile storage tanks and equipment which would need to be regularly transferred from one station to another. Where costs preclude such a system, simple changes to layouts and relocation of delivery points can greatly reduce the associated risks, especially where the movement of mobile storage tanks is undertaken. Where these types of Dewars are employed, it is essential that any fitted handles and wheels are securely attached and maintained so that the wheels in particular always turn freely. Wheels and axles employed in this way should be subject to routine maintenance and inspection to ensure their continued integrity.

Many cryogen Dewars these days are appropriately supplied with mobility handles which can be used to manoeuvre the storage vessels from place to place. Unless otherwise stated however, these handles must never be used as lifting points for lifting the Dewars by way of cranes. The welds on these mobility handles are unlikely to be sufficiently strong enough to take the weight of a fully charged Dewar, and failure during lifting could occur as a result. If fitted, lifting points will be appropriately identified and will be subject to statutory inspections, where the continued safe use of the lifting points is evaluated on a routine basis.

Whatever the manual handling task, the commitment to workplace safety should be part of the overall health and safety policy of any cryogenic related establishment. This includes all tiers of management, where well-defined roles and responsibilities should be clearly outlined. It is not for example sufficient to state that manual handling aids need to be routinely checked and inspected if nobody is given the task to do this important inspection role. Tied in with this therefore must be the correct education of all parties involved in the manual handling of cryogenic equipment. This can be in the form of informal instructions or more detailed training. But whatever way, it must inform them of the correct methods of safe handling, as well as enlightening them of their responsibilities and providing an outline of what they are specifically accountable for. All employees should be given the basic understanding of body mechanics so that they are able to recognise high-risk manual handling activities and what the potential consequences would be to the human frame. Any symptoms of manual handling injuries, near misses or accidents, should be reported in a formal way so that any necessary actions to remedy the problem, or prevent a repeat occurrence, can be taken at the earliest opportunity.

Where the careful consideration of tasks is routinely made, manual handling can be undertaken safely, without the risk of injuries occurring. This is a proven canon, demonstrated by the vast number of workers in the cryogenic industry who do not suffer manual handling related injuries whilst undertaking their day to day duties. That is not to say however that such injuries never happen, as they can often still prevail, especially where infrequent or one-off high-risk activities are being carried out. By carefully engineering out the remaining problems related to these high-risk manual handling activities, it is possible to further reduce the likelihood of somebody suffering injuries from these tasks. Simple adaptations, such as implementing work and rest schedules, adopting job rotations, and reducing the weight and size of the cryogenic equipment being used can have a direct impact on the risk factors. It may be that several of these options have to be jointly applied in order to significantly reduce the risks of injuries, which can incidentally incur costs as a result. These expenses however should always be balanced by the hope that both productivity and quality of output can be increased if people are not getting injured and have to therefore spend time off work convalescing.

### 7.3.24 Low viscosity [60]

The viscosity of a fluid is its ability to resist deformation for a given rate and would be seen in practical terms as the speed with which the fluid flows as it is poured from a container. For liquids, this is also often referred to as its thickness, where something like clear honey would have a higher thickness, or viscosity, than that of liquid water. In some extreme cases, fluids with particularly high viscosity, such as tar, may actually appear to be a solid as their flow rates are indeterminately slow, taking long periods to manifest themselves as definable movements.

All fluids have a positive viscosity value, measured as the shear force per unit area required to move the fluid with unit velocity. Zero viscosity can however be observed at extremely low cryogenic temperatures in superfluids, where flow will occur without the loss of kinetic energy in the fluid. Superfluidity occurs in liquid helium at a temperature of 2.17 K. This is called the Lambda point [61] and it is where the density of the liquid helium drops and elements making up the cryogen attain this zero viscosity state. As a result, under these conditions, the liquid helium will be able to rapidly move through any small openings within its containment system. Typically, in this zero viscosity state, vacuum containers which initially are considered to be helium leak tight will suddenly appear to leak.

All cryogenic liquids have low viscosity as demonstrated in figure 7.8 [62]. Water for example at room temperature has a viscosity of $1.0016 \times 10^{-3}$ Pa.s, whereas liquid nitrogen only has a viscosity of $0.1579 \times 10^{-3}$ Pa.s in comparison. For general applications using liquid cryogens, their relatively low viscosity compared to water means that they will penetrate woven or other porous materials much faster than expected. Any spillages on clothing would quickly run through the textile fabric and make rapid contact with the unprotected skin beneath. This is contra-intuitive, as

| **Liquid** | Nitrogen | Helium | Neon | Air | Fluorine | Argon | Oxygen | Methane |
|---|---|---|---|---|---|---|---|---|
| **Viscosity** Pa.s x $10^{-3}$ | 0.1579 | 0.00357 | 0.124 | 0.168 | 0.2447 | 0.2521 | 0.188 | 0.1186 |

**Figure 7.8.** Showing the viscosity values of typical cryogenic liquids.

the cryogens have a general appearance of water and the human mind will tend to automatically assume any spillages will react as water would, absorbing the liquid before it fully permeates the clothing. For this very reason, personnel protective equipment (PPE) employed for working with cryogens will normally have a dense material, such as leather, as its primary protection which will prevent the cold liquids from reaching the insulating fabrics beneath. Where spillage of cryogens has fallen onto clothing, it is important that this clothing is removed as quickly as possible before the low viscosity of the cryogen allows it to migrate through the fabric onto the skin beneath. This is an equally important consideration for any footwear worn in the cryogenic environment. Any shoes or boots should not have the means by which the liquid cryogens can gain ingress into the footwear, as their low viscosity will allow them to flow easily through the smallest of gaps made available. Eyelets and the open tops of footwear are particularly prone to this problem.

### 7.3.25 Oxygen enrichment [16]

The low boiling temperature of the two specific cryogens; liquid nitrogen and helium, both have a lower boiling point compared to that of oxygen. As a result, and given the correct prevailing conditions, oxygen can be condensed out of the surrounding atmospheric air wherever these cryogens are being used. This is especially so around cold pipework, valves and even in the neck of open storage Dewars. This condensed gas has the potential of gradually collecting as an enriched amount of pure oxygen and may result in increased local flammability and pose a potential explosion risk. Under no circumstances must this oxygen enriched liquid come into contact with oils, grease or flammable materials, as spontaneous combustion can readily occur.

As a routine safety measure, any potential collection points or exposed contact areas should be cleaned to specific oxygen clean standards. Where possible, insulating materials should be used on exposed cold surfaces to reduce the potential for the condensed oxygen forming. Care should however be made that the insulation itself does not create trapped volumes of air, which themselves could be the catalyst for the formation of the oxygen enriched liquid.

In addition, where the very use of liquid oxygen can result in situations whereby oxygen-enrichment of the workplace environment can occur, this will bring about an increased fire hazard. Normal atmospheres contain about 21% concentration of oxygen. Enrichment of the atmosphere to as little as 25% can give rise to an increase in the rate of combustion of any materials which are already burning. In fact, many materials, including some of the more commonly used metals, which are considered to be non-flammable in normal air, may burn in oxygen-enriched atmospheres if they are ignited. As the human senses are unable to detect oxygen-enriched environments, it is important that if there is a possibility of an oxygen gas release occurring, that the workplace is well ventilated in order to allow the quick dispersal of the excess oxygen. Both liquid oxygen and cold gaseous oxygen are heavier than air and will tend to collect in low-lying areas, such as pits and trenches. From here, they will tend to be slow to diffuse and additional care needs to be made so that no sources of ignition, such as electric motors, are located in these potentially high-risk locations.

### 7.3.26 Physiological effects [17]

Particular exposure to the cold vapours released through the evaporation of liquid cryogens, can have both short-term and long-term effects on the human lungs. Short exposure, whilst not causing any permanent damage, may produce an immediate discomfort in breathing, but will generally be quickly reversed when the person working in the cryogenic environment is removed from the source of the vapour. Prolonged exposure however, and in particular where the regular inhalation of the extremely cold vapour or gas occurs, can produce serious lung effects such as pulmonary edema or pneumonia. Pulmonary edema is a condition where the lungs fill with fluid and become congested [16]. Sufferers will typically struggle to gain enough oxygen and would demonstrate a general shortness of breath whilst undertaking the simplest of tasks. This in turn can result in an increase in their heart rate, as their rapid breathing will attempt to supply the much needed oxygen into the body. Likewise, pneumonia is where the small air sacs of the lungs become inflamed, with typical symptoms demonstrated as a dry cough, possible chest pains and a general difficulty of breathing [16].

Other negative consequences to the body's respiratory tract can occur in these cryogenic environments, especially where existing medical conditions prevail, such as asthma and chronic obstructive pulmonary disease (COPD) [63]. These are normally exacerbated by sudden air temperature drops, which effectively prevent the lungs from sufficiently adapting to these thermal changes. Typically, a drop of only 5 °C or more over a short period is enough to affect the respiratory system in this way. Setting this against a sudden release of extremely cold gas from a cryogen storage Dewar for example, where the immediate temperature of the air can abruptly drop by upwards of 20 °C or more, these adverse conditions can quickly be achieved. In addition, the airways in the body are lined by a thin layer of protective liquid which would be susceptible to exposure to these cold gases and vapours. Any cryogenic vapours inhaled will have the effect of both cooling this surface fluid, but also drying the airway as well. Over a short period, the body can sufficiently replace this surface fluid, but longer exposures can cause permanent damage to the lining of the airways as a result of this dramatic drying process.

Ironically, in more recent times, there has been a rapid growth in the commercial use of cryogenics as a therapy for treating conditions such as muscular tension, pain relief and the recovery from exercise-induced muscle damage to name but a few. Here, the patient is specifically placed in a chamber, surrounded by cold vapours generated by liquid nitrogen or refrigerated air. Clearly, such treatment runs the risks of patients inhaling extremely cold vapours and gases in such devices and users are generally protected by the head remaining above the vapours at all times to prevent this occurring.

### 7.3.27 Explosion due to pressure

Liquefied gases are generally stored at or near their boiling point, and hence, there is always some gas present in the storage container directly above the liquid surface. Due to the large expansion ratio of these cryogenic liquids when they vaporise, a build-up of pressure can occur as the liquid warms up and evaporates more and more of its volume. The majority of cryogen storage vessels are not designed to operate at pressures significantly above ambient pressures as their normal operating range would not require this. If any internal pressure from the evaporating liquid is significantly high enough however to exceed the strength of the containment vessel, an explosion or rupture of the container can occur, especially if safety assemblies, such as pressure relief devices are not installed or fail to operate correctly.

The evaporation rate of the liquefied gas will depend on the actual cryogen, the storage container design, the quality of any insulating material applied to the vessel, as well as the environmental conditions in which the container is stored or used. It is important therefore that the container's capacity must always include an allowance for the evaporation of its liquid contents into the gaseous state. Any pressure relief device should be maintained and checked on a regular basis for any signs of leaks or damage which could impair its correct function.

All containment vessels must be designed and manufactured according to their respective national regulations and standards. This provides a level of assurance that the correct materials of construction are selected, that fabrication and manufacture is undertaken in a processed manner and that under normal operating conditions, failures and ruptures of the vessels are not possible. Albeit rare, pressure equipment explosions do happen however, and particularly with gases, due to the potential energies involved, they can be violent and noisy. Very often they are likened to the detonation of a bomb and are often rated in terms of equivalent explosive potentials. Where the rupture of a storage vessel occurs, the rapid increase in volume of the contained gas releases an enormous amount of energy in an extreme manner. This is likely to eject fragments of the storage vessel at high velocity from its source, which can typically travel hundreds of metres in all directions. These can seriously injure or kill personnel in the vicinity, along with causing damage to the immediate structures and buildings. In addition, a high-energy blast pressure wave will be generated as a result of the explosion, often travelling likewise away from the source at near supersonic speeds. Exposure to these pressure waves can result in serious injuries to the auditory, respiratory and gastrointestinal systems of the body. A pressure wave of as little as 1 psi (0.07 bar) is sufficient to knock a person over, whilst one at 5 psi (0.35 bar) is very likely to cause eardrum damage. In comparison, the pressure wave threshold where fatalities occur sits at around 35 psi (2.4 bar) [64].

### 7.3.28 Explosion due to chemical reactions

Any explosion is defined as a sudden, violent change of potential energy into kinetic energy, which is transferred to the immediate surroundings in the form of a rapidly moving pressure increase, more often referred to as a blast wave. Chemical explosions are specifically brought about through a rapid and violent oxidation reaction within a compound or mixture that suddenly produces large amounts of hot gas in the process. The speed of this oxidation reaction is often instantaneous, with violent results. In the workplace, chemical explosions are generally initiated by an electrical spark or flame, especially where the presence of oxygen is sufficient to cause combustion. They can also occur through the application of heat or a catalyst, and in some extreme cases, due to the influence of mechanical shocks. During the explosion, the reaction volume increases enormously over a short period of time, converting solid or liquid chemicals into a rapidly expanding gas. This conversion and the resultant large solid/liquid to gas expansion ratio also increases the pressure of the gas at the reaction zone.

Some chemical explosions are brought about through the combination reactions of two or more components which respond together to produce the hot gases. Classically, gunpowder and fireworks work in this way, along with the types of explosives used in industries such as mining. The level of damage caused by any chemical explosion depends on how fast the explosive reaction takes place and this will tend to depend on the mixture of the components and their propensity to oxidise. Accidental chemical explosions are more common than other physical explosions, such as those due to the over pressurisation of pressure equipment. More often, workplace chemical explosions are due to vapour cloud explosions, where a combustible fuel in gaseous form is directly mixed with the surrounding air. If ignited, the burning rate of this cloud may be sufficient to form a pressure shock wave which can be significant enough to cause damage to or destroy structures. Clearly, in the cryogenics technology forum, the use of flammable and combustible gases of this nature is often a prerequisite to the cryogenic processes being conducted. The associated risks of chemical explosions occurring are therefore high and suitable control measures should always be implemented under these conditions.

More specifically, cryogenic fluids with a boiling point below that of liquid oxygen are able to readily condense oxygen from the atmosphere. Repeated replenishment of any given system can thereby cause this oxygen to accumulate as an unwanted contaminant, especially in recesses and corner positions of assembly structures. Violent reactions caused by the rapid combustion or explosion of this oxygen may occur, especially if any materials which come into contact with the oxygen are themselves combustible. This is especially so where a source of ignition, such as electrical supplies or even heated surfaces is also present in the local environment.

In the strictest sense, an atmosphere that contains more than the normal 21% of oxygen by volume creates a dangerous fire hazard. Higher concentrations of oxygen will increase both the chance of a fire and its intensity and can involve materials normally regarded as being non-flammable. It is of the utmost importance therefore that there is no source of ignition in areas where oxygen enrichment is likely to occur. Similarly, other flammable and combustible gases, such as methane and hydrogen, can condense in the same manner as the oxygen, which in turn can lead to explosions and the consequential fires occurring.

### 7.3.29 Explosion due to ionising radiation

Ionising radiation, in fundamental terms, is a type of energy released by specific atoms in the form of electromagnetic waves of particles [65]. We are all exposed to low-levels of natural sources of ionising radiation, often manifest in the soil or water around us, as well as occasionally manmade sources, such as those generated by medical x-rays. Acute health effects, such as skin burns and radiation induced cancers can occur when doses of ionising radiation exceed certain levels. The type and severity of any resulting damage to human tissue and internal organs depends on the amount, or dose, of radiation received.

The effects of ionising radiation on materials and chemicals can likewise be damaging in the extreme. The ionisation of the molecules in these materials and chemicals can lead to a breaking of their chemical bonds, along with the formation of highly reactive free radicals. These highly chemically reactive free radicals can then react chemically with neighbouring materials, even after the original source of radiation has been removed. Ionising radiation can also have the effect of accelerating existing chemical reactions, such as those associated with corrosion. Similarly, optical materials will demonstrate a deterioration of their optical properties under the effects of ionising radiation [66].

The cryogenics world is therefore, by default, prone to the effects of ionising radiation in these essential ways. The implicit use of ionising radiation is often found as a primary research tool in facilities where irradiation conditions dominate the landscape of engagement. Of greater concern however, is that under these radiation conditions, such as those produced in a proton beam, there have been cases recorded where liquid nitrogen can explode when the cryogen is subsequently warmed. The actual mechanism causing this is not thoroughly understood, but it is generally accepted to be due to reactions with ozone formed from trace amounts of oxygen in the cryogen. The explosions, it is believed, are caused by the very rapid decomposition of this ozone which has been formed by the action of the ionising radiation on the oxygen within the body of the liquid nitrogen [67].

To some presumed level, the belief has also been held that chemical reactivity of any nature ceases at temperatures below 100 K. This is not the case however, as many chemical fusions continue to take place where cryogenic temperatures are a critical factor in the process. As well as the ozone formed through the irradiation of liquid nitrogen for example, other nitrogen oxides can also be produced, which in turn collect as a sludge at the bottom of Dewars and other storage vessels. These may not be the ultimate cause of any irradiated explosions, but can be a catalyst to enhancing the yield of the ozone produced, which in turn promotes such explosions to take place. Likewise, it is also believed that the materials used in the construction of cryostats, Dewars and other storage containers can play an equally important role in the formation of the ozone when the liquid nitrogen has been irradiated. It is concluded therefore, that the formation of the ozone has a direct correlation to surface reactions with these materials and their alloying components when they are in contact with the liquid cryogen. The ozone it is believed, will concentrate at the interface between the liquid nitrogen surface and the containing walls of the storage vessels due to the evaporation of the cryogen at this point. Ideally, materials of construction should have a low wetting property in order to limit this taking place.

### 7.3.30 Fire

By description, fire is the rapid oxidation of a material due to the chemical process of combustion, which in turn releases both heat and light, along with other reaction products [68]. The process releases large amounts of energy, which is the principal cause of fire being recognisably hot. Fires also produce flames which consist primarily of carbon dioxide, water vapour, oxygen and nitrogen. If sufficiently large enough, these flames can cause physical damage through the process of burning. Fire stands as a potential major hazard in all situations of domestic, leisure and industrial environments, posing a risk to life and property, and causing potential pollution to the atmosphere, as well as being a source of water contamination.

Fire has many uses and applications, none more so, than to provide warmth when the surrounding natural temperature drops to dangerously low levels. It is also used for cooking, for propulsion, for smelting and forging, along with many more specialist practices. The risks to life when accidental fires occur are mitigated by national regulations and safety standards, which define the safety devices and protocols needed to make an escape from a burning building as easy as possible. The provision of the most appropriate means of extinguishing a fire forms part of this requirement. For many countries throughout the world, firefighting services form part of national provisions, their role being to extinguish or at least control fires in order to save lives. These firefighters are trained specifically to undertake their role, and will adopt a range of fire appliances, including water and foam extinguishers to combat all nature of fires.

In the working environment, fire prevention should form the basis of every workplace risk management strategy. This would include reducing where possible the sources of ignition which could instigate the fire in the first place. Likewise, limiting the amounts of combustible materials near to any ignition sources, reduces the potential of these becoming fuel to any latent fires accidentally created in those areas. Fire prevention should also include the education of all employees on how to avoid causing fires, along with how to safely evacuate a building in the event of a fire breaking out at any point in time. Often established as national law, many establishments must practice these evacuation procedures on a regular basis to prepare the occupants of buildings on how to react to a real building fire.

Further prevention is provided by the various national and regional building standards and regulations, and construction codes. These typically stipulate requirements, for things like the provisioning of passive fire protection, such as smoke alarms, as part of a building's infrastructure. Some situations also merit the installation of active fire protection, an example being sprinkler systems, which would automatically operate in the event of a fire occurring. Likewise, all building materials and any internal furnishings will be required to comply with fire-resistant standards for their combustibility and flammability.

Each of these methods addresses the ethos of preventing fires occurring and this is the best way of managing the risks associated with fire hazards. Where fires do occur however, strategies should be employed which makes the evacuation of buildings and the extinguishing of fires both safe and easy to adopt. These are general ideologies, whereas the technique of extinguishing fires in relation to the use of cryogenic technology needs to be considered in the context of the actual cryogenic materials being used in the workplace. $CO_2$ extinguishers for example can cause static discharge of sufficient magnitude to ignite some flammable gas mixtures. This fact should be considered in the understanding that several of the cryogens when in gaseous form are defined as flammable, such as oxygen, hydrogen and methane. Add to this, it is not uncommon for other flammable and combustible materials to be used in conjunction with cryogenic technology, such as flammable oils used in compressors or even kerosene, which is commonly used as the pressurising medium in equipment such as high-pressure intensifiers.

It is also important to realise that in the event of a fire occurring where cryogens are being stored, there will be an increase in the evaporation rates of the cryogens from their storage containers due to the inevitable rise in the external ambient temperature. Under these conditions, pressure relief systems will operate more frequently in ordered to accommodate this heating of the cryogen and the resultant increase in the expanding gas being released from the liquid phase. If these gases are themselves flammable, they could add potency to any established fire, resulting in conflagrations of enormous magnitude.

It is worth remembering as well, that the very nature of some of the cold environments created by cryogenic technology can bring into existence enriched liquid oxygen from the surrounding air. This tends to be small in quantity, but is in such a concentrated form that self-ignition can occur under certain ideal conditions. If ignited, this liquid oxygen will burn intensely and there is an increased risk of explosions occurring if sufficient oxygen is present.

### 7.3.31 Combustibles storage

All combustible materials should ideally be stored externally to the cryogenic work area where they are to be used. This is especially so if they are volatile and flammable, such as solvents and liquefied petroleum gas (LPG). If they need to be stored inside buildings, suitable storage cabinets, in well-defined areas, should be provided so that they pose no threat to staff working in the vicinity. Any such internal storage should be arranged in a way to prevent the spread of fire if it ever did occur. This could involve the separation of storage areas from the rest of the cryogenic workplace by the provisioning of partitions that are designed to provide at least 30 min of fire resistance. Any doors to these storage cabinets and to the areas themselves must be self-closing, as well as fire resistant. The single objective of all these measure, is that in the event of a fire incident, people will be able to safely escape from the working area without coming into contact with the fire. In this context, by storing flammable and combustible substances in specifically designed cupboards in dedicated storage areas of appropriate construction, they provide a physical fire resistant barrier which would delay any associated fire for sufficient time to allow the safe evacuation of the workplace of all personnel.

Where highly flammable and volatile liquids are being stored, extra precautions need to be taken to mitigate the risks of any build-up of dangerous vapours and gases from the products. The easiest methods of managing these risks is through the provision of good ventilation to the storage area, with the vents opening to the outside, where any fumes can be readily dispersed into the natural atmosphere. Where pressurised gases are being stored inside of buildings, ideally they should only be kept in small volume containers and in minimal overall quantities. They should be kept in lockable wire cages, as far away from the cryogenic working area as possible. Whatever system of storage is employed, it is important to have separate stores for flammable solids, liquids and gases, with each being clearly marked as such. No other materials, other than flammable materials, should be kept in these flammable materials stores. Typical recommendations stipulate that the maximum quantities of flammable materials stored inside of buildings should be limited to 50 litres for those products which have a flashpoint around the ambient temperature of the working area, and no more than 250 litres for other flammable products with a higher flashpoint temperature [69].

In the cryogenic technology workplace, because of the potentially hazardous consequences in the event of fire occurring, flammable substances and combustible materials should not be stored or allowed to accumulate in the vicinity of cryogenic liquid installations. This is particularly important where cryogenic oxygen is being stored due to issues of spontaneous combustion that can occur if the oxygen concentration in the air reaches a sufficient point. Likewise, if flammable gases or flammable cryogens are being utilised, additional fire prevention or mitigation should be securely in place before they are used. Very often, primary risk management will already exist for the usage of cryogenic materials in the working environment, in particular those applicable for the use of liquid cryogens. These would include the deployment of oxygen monitors, forced ventilation and fire and smoke detectors. These will in turn provide some measure of control in their association with preventing fires occurring in the first place. As a general policy, good housekeeping of the workplace will ensure that the accumulation of combustible waste in particular does not reach dangerous levels.

### 7.3.32 Cryo-pumping

Cryo-pumping is the method by which the deliberate use of cryogenic temperatures produces a vacuum in an enclosed chamber. In the broadest sense, it is the removal of gases in any such chamber through the condensing of the gases as they stick to an extremely cold surface generated by some cryogenic means. This capturing of the gases has the effect of lowing the measurable pressure inside the chamber, as more and more of the gas is progressively removed from circulation. This capturing of the gases can be through either; cryosorption, in which the gas particles attach to a physical material, such as activated charcoal; cryocondensation, in which the gas particles attach to other gas particles that have already attached to the cold surface; or cryotrapping, in which a non-condensable gas is deliberately mixed with a condensable gas, which in turn can then be condensed onto the cold surface.

Commercial cryo-pumps will operate at a variety of temperatures ranging from around 80 K all the way down to sub-Kelvin levels. Often, the cryogenic cooling is provided by a reservoir of liquid cryogens, such as liquid nitrogen, or my mechanical coolers, such as closed-cycle-refrigerators. Cryo-pumping systems have advantages over other vacuum pumping systems in that they often have very high pumping speeds and they create a clean vacuum without the possibility of contamination by hydrocarbons from the types of mechanical pumps employed to generate the low pressures. Extremely clean vacuum systems are of utmost importance to industries such as those involved in the manufacturing processes of semiconductors.

As a deliberate cryo-pumping process, the associated hazards are identical to those found with all other cryogenic systems employed as part of the cryogenics industry. However, where the accidental creation of vacuums do occur, or the general lowering of ambient pressures in a chamber due to the cryo-pumping effects of cold surfaces, these effects need to be considered as part of an overall risk assessment of the operating system. As it is, most cryogenic systems will need to operate in a vacuum environment for the simple reason that the condensing of gaseous contaminants from the air is an unacceptable consequence of the cooling process. The formation of ice from water vapour contained within the air as it is condensed at the cryogenic temperatures is a typical example of how this would be a problem without the deployment of some level of vacuum. Where systems employ an exchange gas as part of the cryogenic process, a genuine hazard comes about should this exchange gas be able to condense on any cold surfaces as cryogenic temperatures are achieved. This will have the consequential effect of lowering the pressure of the containing chamber. If this chamber has only been designed to take positive internal pressures due to the presence of the exchange gas, there is a risk that under vacuum conditions, this effective negative pressure can cause the collapse of the chamber. Thin walled constructions of pressure equipment can withstand significant internal positive pressures. The same equipment, subject to internal vacuum conditions however, is more than likely to collapse due to the combined axial and circumferential loading imposed by the external atmospheric pressure. This is not only catastrophic for the equipment, but could result in the failure of subsequent elements of the cryogenic system, for example, the insulating jacket of a cryostat which could then cause a rapid boil-off of any contained cryogen. This in turn poses a more significant hazard due to the potential oxygen depleted atmosphere this bulk release of cryogen can cause.

### 7.3.33 Decanting

Many applications within the field of cryogenics will use liquid cryogens as the means of achieving the low temperatures required. As such, very cold cryogenic liquids are regularly transferred from larger storage containers and dispensing systems to smaller containers, or directly to equipment, such as cryostats. These transfers are of a higher-risk nature and can often pose a serious hazard. Therefore, essential precautions need to be taken for all transferrals from one container to another, irrespective of the size and type of container, and the cryogenic liquids being conveyed. Cryogen storage containers are commercially available in a range of sizes, with some having a capacity of thousands of litres. Smaller Dewars and flasks are often utilised for the temporary storage of liquid cryogens and tend to have capacities limited only to the few tens of litres range.

For significantly larger quantities, static storage tanks are the preferred method of providing a ready supply of cryogens. These tend to be located outside of buildings and will typically provide both supplies of cryogenic liquids and gas from their capacity. The filling of these vessels is undertaken on a regular basis by supply tankers by the suppliers' own drivers. Albeit the tanks are installed at customer locations, they are normally owned and operated by industrial gas supply companies.

Where the transfer or decanting of liquid cryogens from storage containers is undertaken, some basic safety rules should be employed. Transfer lines for example, are specifically designed for this purpose, so that any risks with their use are minimised. It is important however that these transfer lines are regularly checked to ensure any vacuum insulation is still intact. Where transfers are being made into smaller Dewars, they must be suitable for the cryogen they are receiving and should be clearly labelled so that the eventual contents are easily reconciled. Dewars tend to be designed and manufactured these days for very specific cryogens and care should always be made to ensure that only the correct cryogenic liquids are being dispensed into them.

As the risk of coming into direct contact with the extremely cold cryogens is proportionally higher during decanting, here, more than ever, personnel decanting cryogens must wear the correct personal protective equipment (PPE) in order to protect themselves against accidental exposure. Decanting and transfers of liquid cryogens is a particularly risky operation, so it is essential that cryogenic rated gloves, glasses, goggles or face masks, the correct footwear and a laboratory coat or apron are worn throughout the transfer. All receiving vessels should be checked to ensure they are clean and dry before any filling process is started, and ideally they should be pre-cooled beforehand by first adding a very small quantity of cryogenic liquid and allowing this to safely boil-off. Whatever the method of transfer, it is important that cryogenic liquids do not have to fall through a significant distance to reach the receiving vessel. This runs the risk of spillages and splashes occurring during the transfer. Where manual pouring of liquid cryogens is undertaken, the receiving vessel must be secured in such a way that it cannot tip over during filling, and any assisting personnel encouraged to stand some distance away, just in case spillage does take place. Any dispensing should be done slowly, again to reduce the chances of splashing or spilling of the cryogens. Certain storage systems will incorporate a withdrawal device which facilitates easier transfers without the need for tipping or pouring processes to be adopted.

All transfers should be undertaken in a well-ventilated area as the accidental release of cryogens is more likely during the process, which can in turn create localised vapour clouds as the evaporating liquids convert to gas. If the transferred quantities are sufficiently small enough, these vapour clouds pose only a small risk. For larger transfers however, spillages and leaks during the process could have a profound effect on the overall oxygen concentration in the workplace, increasing the likelihood of an oxygen depleted environment being created.

Cryogenic storage vessels must only ever be filled inside a suitably ventilated area, preferably in an external location in order to prevent the creation of an oxygen depleted environment. Likewise, decanting from these fully loaded storage vessels should be done in a ventilated environment and should only be undertaken in accordance with predetermined safe operating procedures. Essentially, the decanting should only be carried out by personnel who have received practical training on how to undertake the task safely.

Before starting the decanting process, a check should be made to ensure that the supply vessel is currently within its statutory inspection date. If this cannot be determined and the last inspection date cannot be identified, the vessel should not be used. The inspection is undertaken for a reason, as this qualifies that the storage vessel is safe to use and that no concerns about its construction have been observed. Continued use of the storage vessel beyond its inspection period carries the greater risk of failure of the vessel. Even if the storage vessel is still within its inspection date, it is prudent to also check any pipework, and that the pressure relief valve and bursting disc are free from damage and leaks.

Any decanting process must also ensure that the storage vessel supplying the cryogen is properly labelled so that the contents decanted are the cryogen to be expected. The relief valve and bursting disc should also have a valid date ring attached, as these may differ from the dates recorded for the overall inspection of the vessel. These safety valves should also be free from ice which would imply a slight, but constant leak from the vessel. The vessel itself should also be checked for obvious damage or any ice build-up which would indicate a breakdown of the vessel's internal insulation. The storage vessel must be in a good working condition, with specifically no damage or twisting to the neck, which is where the highest structural stresses are found in its construction.

If there appears to be any modifications to the vessel, it would be wise to investigate further before using it to decant from. It is not uncommon for example for additional shut-off valves to be installed ahead of leaking relief valves by untrained personnel. By doing so, this prevents the automatic release of over-pressurised gas from the vessels, leaving it in a dangerous state.

The storage vessel must not be used if it is incorrectly labelled for its actual contents. It must also not be used if it has been previously filled with a different cryogen to that being decanted, especially for those used for liquid oxygen, where there is a flammable risk of any remaining cryogen being condensed out of the contents. If there is evidence of water or ice inside, this must be removed before filling. If there is excessive frosting around the neck, this could imply a breakdown in its insulation or internal damage to the Dewar itself.

Only purpose designed transfer hoses should be used which have the correct construction and are made with cryogenic suitable materials. Once again, these must not be damaged or kinked as this would increase the likelihood of heat getting into the cryogen during transfer, causing a more rapid boil-off than expected. All fittings and connections should be in a good working order. Operators should try any threaded connections before use to check that the thread forms are in good condition.

If decanting liquefied oxygen, there must be no sign of normal PTFE tape around any connections as the oxygen can react violently to the hydrocarbons in the tape. Personnel making hose connections should be correctly trained and must use the correct tools to undertake the task. Any transfer hoses must fit easily into the neck of the Dewar. Once again, it is suggested that the hose is tested in the Dewar prior to the decanting taking place. Some Dewars have a liquid withdrawal device and it is important that the Dewar is vented to atmosphere before removing the device.

Hoses should be initially purged to clear any moisture or dust. To do this, the hose should be held away from the body whilst wearing cryogenic gloves and the decanting valve opened slightly to introduce a low flow of cryogens into the hose. The valve must be closed as soon as frosting appears on the hose. It is now safe to insert the hose into the Dewar. Following this operation, the fill valve can now be partially opened. If the receiving Dewar is warm, there will be some initial boiling of the liquid cryogen as it turns to gas. This is normal, but can initially appear as though the Dewar is full. Gradually, as the Dewar begins to cool, the fill valve can be fully opened to provide a steady flow of liquid cryogen from the storage vessel.

If the decanting is to be made into Dewars without a neck-tube construction, they should only ever be filled to a level below that which the corresponding insulating bung will reach. These insulating bungs can be quite long in length, so a trial fitting and even measuring the length is advisable to limit the fill level. Likewise, any decanting into Dewars with a neck-tube construction, the filling should stop when the liquid cryogen reaches the bottom of the neck. As there is likely to be a lot of condensed gases surrounding the Dewar, it may actually be difficult to see this position. The simplest way to judge this fill height is to listen for a change in sound as the Dewar is gradually filled with the cryogen. The tone will change in such a way that it will become obvious as the liquid level reaches the bottom of the neck.

When the Dewars have been filled, the cap or insulating bung should be immediately replaced. This not only reduces the heat effects from the surrounding environment getting to the cryogen, but also to protect against accidental spills from the now full vessels. If on closing, the cap is found to rattle, this means that the Dewar is overfilled and the cryogen is boiling off at a higher than expected rate. In this case, the cap should be removed in order to allow this boil-off to continue for some time until the cap can then be refitted securely.

To fill small vessels with liquid cryogens, a stainless steel funnel should ideally be used. Under no circumstances should a plastic funnel be used as this is likely to crack and cause spillage. Better still, use of a purpose built tipping trolley is recommended when pouring liquid cryogens from a Dewar into a working flask. Always position the receiving vessel at the correct height to avoid splashing of the cryogens during filling.

It is important that domestic screw-capped thermos flasks are never used to store or transport cryogenic liquids. These containers have no means of negating the effects of pressure build-up from the warming cryogen and using them runs the risk of an eventual explosion. These containers are not designed as pressure vessels and will not be supplied with a valid statutory pressure test certificate. Small Dewars will either have a non-sealing cap to release the inevitable build-up of pressure as the contained cryogens evaporates, or an insulating bung which has longitudinal grooves in it, allowing the expanded gas to escape.

### 7.3.34 Spillage during decanting

Inevitably, some spillage during the decanting process is likely, even if control measures have been employed to mitigate the likelihood. If liquid cryogens of a significant quantity are spilt, all personnel must be evacuated immediately from the area affected by both the liquid and the resulting evaporating gas. Particular attention should be made to any basement rooms and cellars where the cold gas is more likely to accumulate. If the correct risk assessment has already been conducted, these locations should have alarmed oxygen monitors fitted and the rooms should be immediately evacuated if a spillage occurs and the alarms are sounded.

If forced ventilation systems have been fitted in preparation for the likelihood of liquid cryogen spills, it is critical that this ventilation does not simply spread the cryogen gases into another work area. Ideally, the system should extract from one level and introduce fresh air at another, allowing some mixing of the new air with the accumulating cryogen gas. Even if a cryogen vapour cloud is low-lying and appears innocuous, personnel should not enter the cloud as it is likely to remain extremely cold for some significant time. There also remains the danger of an oxygen depleted atmosphere existing slightly above the level of the cloud without personnel being aware of the fact.

### 7.3.35 General handling of cryogenic materials

In any operation assumed as part of using cryogenic technology, it is important that personnel are suitably trained and understand the risks of what they are being asked to do. This edict should be given even more prominence in tasks where the movement of cryogenic liquids are taking place. Individuals undertaking the work need to be trained in not only the correct cryogenic related procedures, but also those applicable to the manual handling of the storage vessels and any associated equipment. Accordingly, they must also be physically capable of carrying out the task in a safe manner, otherwise injuries are more likely to be incurred. It would be imprudent for example, to expect a person with an existing back injury to move heavy storage vessels from one location to another.

Many of the responsibilities relating to the general handling of cryogenic materials belong to the remit of the employer, who must ensure the safety and wellbeing of their staff in whatever task they are being asked to enact. This does not mean that employee responsibilities can be overlooked, as they too have a duty to protect their own health and wellbeing, without causing harm to those working around them in the process. Individuals moving liquid cryogens for example must ensure that the route they intend to use is clear of obstructions and that all floor surfaces are suitable for the movements of the cryogens to be undertaken safely. This would for example, imply that there should be no uneven surfaces which could cause issues with things like the wheels on the storage vessels or any trolleys being employed. More specifically, personnel are required to use any system or equipment that is provided in the workplace which assists with the movement of cryogenic storage vessels, such as purpose designed trolleys or transport vehicles.

Where possible, moving cryogenic storage vessels through doorways, up and down stairs or slopes, should be avoided, as these present hazards which could lead to a higher risk of personnel slipping or tripping during the process. There is also a greater risk of dropping or tipping the cryogenic storage vessels due to these same hazards. In fact, for every change of direction, or manipulation of the storage vessels the chance of an incident occurring increases accordingly. It is prudent therefore that any such movement is thoroughly planned and the route checked before actually moving the cryogens. Any plan of this nature should be easy to modify if local hazards have changed in such a way to hinder the proposed movement.

Where possible, cryogenic storage vessels should only be moved on an appropriate trolley where the vessel is firmly attached, so that there is no risk of it detaching itself from the trolley in transit. Importantly, the process of loading the storage vessels onto the trolley should not in itself create a secondary hazard. It is vital for example, that cryogenic storage vessels are never rolled on their base in order to load them onto these trolleys as this can effectively churn the contents, which will have the effect of increasing the amount of evaporated gas being released from the safety relief valve. The trolleys should be specifically designed for transporting cryogenic storage vessels and should not be used for any other equipment.

Emphatically, the movement of cryogenic storage vessels should only ever be carried out by trained staff. It would not be wise for example, to expect a new recruit to understand all the safety requirements of this task until they have been trained and gained useful experience. All staff moving cryogens must make use of the Material Safety Data Sheets (MSDS) [70] pertaining to the cryogenic liquids involved in order to establish the relevant safety controls needed for the substance being handled. This document will provide a detailed account of any inherent risks associated with the particular cryogens and will recommend the appropriate personal protective equipment (PPE) that should be worn at all times to prevent injuries. This would automatically include gloves and eye protection, but it may also be worth considering individuals being equipped with personnel oxygen monitors as well if the delivery point of the storage vessel is in an area where the likelihood of an oxygen depleted atmosphere being created is significantly high.

### 7.3.36 Cryogenic liquid transport

Moving any pressurised cryogen storage vessels will have the effect of accelerating the rate of evaporation as the liquid inside the vessels will be effectively stirred and mixed in response to the movements. This will result in an increase in the internal pressure as above normal amounts of gas will be generated from the liquid as it is disturbed in this fashion. Agitation of the liquid cryogen provides it with kinetic energy, especially at the surface where splashing and wave formation can naturally occur. This kinetic energy is in turn converted into heat energy within the liquid. This has the effect of raising the temperature of the cryogen, which results in gas being produced from its mass. This needs to be seriously considered for all planned transporting of cryogens, as the resultant cryogen gas released over time during movement will be higher than the manufactures stated evaporation rate for the respective storage vessel.

If several storage vessels are being transported at the same time, the anticipated gas released needs to be calculated and the internal volume of the transport vehicle assessed for the likelihood of an oxygen depleted atmosphere being created over time due to this total release. As a result, the quantity of cryogens being transported may well have to be reduced to safe levels, or additional ventilation provided to the transport vehicle. A commonly adopted policy is that the transport of any filled cryogen storage vessels is only ever undertaken using open-sided transport vehicles which automatically provide the required ventilation throughout the process. Likewise, when transporting open-topped Dewars and cryogen storage vessels, it is advisable that they should only be filled to a maximum of 90% of their capacity to prevent spillage during movement. Any spillage would add additional quantities of gas being released as the liquid cryogen evaporates instantly into its gaseous phase.

When transporting cryogenic storage vessels in equipment lifts, additional consideration needs to be made to the resultant effects of gas releases, especially where these lifts are small in capacity and are not constructed as open-sided units. It is recommended that the internal head-pressure in the inner vessel should be vented to at least 50% of the relief valve setting. This is to allow for subsequent pressure build-up inside the vessel as the liquid is agitated during transit. This venting must be done in the open air or within a well-ventilated area, and specifically not inside the lift. Any equipment lift used should be a key-controlled type to prevent the lift being called without the cryogen storage vessels having been first secured in place.

Wherever possible, it is advisable not to use passenger lifts for transporting liquid cryogens unless a robust procedure is in place and under no circumstances should staff travel with the cryogens in the lift. Some facilities, such as those in older University buildings, may not for example have equipment lifts installed and therefore they would need to use passenger lifts for the movement of liquid cryogens between levels. In these circumstances, the sender should remain outside the lift during transportation with a receiver likewise waiting at the destination floor. Where possible, a portable alarmed oxygen monitor should be temporarily attached to the cryogen vessel so that the receiver can determine immediately if it is safe to enter the lift on its arrival. Any procedure would also require individual guards stationed at each intermediate floor to prevent accidental ingress of personnel into the lift with the cryogen. The lift should therefore not be available to other users throughout the transport.

Whatever method of movement being employed, the correct personal protective equipment (PPE) must be worn during the transportation of liquid cryogens. Especial care should be made to protecting hands and eyes from accidental splashes of cryogens as this is more likely to occur during the transport operation.

Moving liquid cryogens increases the chance of accidents occurring. It is important therefore to check if any intended route takes the cryogens through populated work areas where the likelihood of accidental spills would have a greater consequence and the risk of being accidentally impeded during the process, or even knocked into, would be higher. The specific things to consider would be any slip or trip hazards along the route which by consequence, could increase the risk of spillage. Stairs in particular increase the likelihood of spillage, as the combined linear and vertical movements involved can cause the cryogen surface to oscillate within any storage vessels and containers, increasing the likelihood of spills, but also increasing the gases being released. Before any movements are started, it is critical that a check is made to ensure that the destination is ready to receive the cryogen, otherwise temporary safe storage of the liquid cryogens would have to be provided when they arrive.

By far the safest way to transport liquid cryogens is to use a purpose built trolley which will reduce the likelihood of spills and would also securely hold any cryogen storage vessel in one place during transit. However, here again, a check needs to be made to ensure that the planned route remains passable throughout, as restricted or narrow pathways would hinder the trolley's movement. Even if the route is one which is normally undertaken and is generally clear of any impedances, a final check on the day of transport should be made, as circumstances may have changed and the route may no longer be suitable for the trolley being used. If possible, transport liquid cryogens during quieter periods, such as the evenings, when the general population along the route is more likely to be lower. Clearly, try to avoid using routes which have steps or kerbs which have to be negotiated, as these increase the risk of the trolley and the storage vessel being toppled over during movement.

### 7.3.37 Vehicular transportation of cryogenic materials

As already seen in chapters 3, 4, 5 and 6 which detail the national and international regulations put in place for the use of cryogenic technology, the transportation of cryogenic substances on public highways is covered by a range of Regulatory standards and national laws. These define the specific volumes and masses of dangerous goods that may be transported and places duties of responsibility on the sender, the transporter and the recipient for specific transport operations. One important aspect is the correct packaging and labelling of the cryogenic goods being transported, so that everyone involved in the process knows exactly what is being moved. The type of vehicle used is also critical, along with the training the driver has undertaken. It would be unfair for example to simply instruct a driver to transport liquid cryogens without their understanding of the risks involved and what they should do in the event of an incident, such as a spillage, or a rupture of a storage vessel.

These regulations and laws generally come into force if cryogenic substances are transported on public roads, for however brief a period. This has to be a serious consideration, if for example, where a facility using cryogens is located on two neighbouring sites, divided by a public road. Even though the transport of the cryogens would be between the two sites, the presiding regulations have to be adhered to.

#### *7.3.37.1 Company vehicles transporting liquid cryogens*

Undertaking the transport of liquid cryogens represents a significant hazard and the appropriate control measures need to be in place in order to mitigate the risks of incidents occurring during the transit. Specifically, a written risk assessment should be in place for any activity being undertaken which involves the use of a vehicle to transport any liquid cryogen. Appropriate labels should be on the transport containers describing the contents so that suitable risk control measures can be enacted in the event of an accident occurring. Only suitable vehicles designed for the purpose of transporting liquid cryogens should be used. These would normally be ones with an open cargo-bay, and where the driver is separated from the load by a fully sealed bulkhead. This prevents any asphyxiation risks if there should be an accidental spillage during the transport as the driver is protected by the bulkhead from any oxygen depletion effects encroaching into the driver's cab. The driver must also be made fully aware of the nature of the load they are transporting, all the associated hazards and the important emergency procedures should there be an incident involving the substances.

Appropriate personnel protective equipment (PPE) must be provided for the driver and any passengers travelling with them. This would include any protective clothing and gloves suitable for using with the liquid cryogens being transported. The PPE must be applicable to any processes which the driver may be expected to undertake, such as the loading and unloading of the cryogen storage vessels to and from the vehicle. They may also be expected to refill the storage vessels at the destination and therefore need to have the applicable PPE for this task, as well as the necessary training to undertake this hazardous role.

An information sheet should be carried within the vehicle to inform the emergency services of the foreseeable hazards of the goods being transported. It would be naive to believe that firefighters and ambulance crews attending an incident involving cryogen spillages would fully understand the inherent dangers of such an event. By providing them with as much information as possible about the severity of the hazards imposed by cryogenic technology, they will be able to implement appropriate strategies to combat the risks.

*7.3.37.2 Third party vehicles transporting liquid cryogens*

It is not always possible for organisations to provide liquid cryogen transport facilities within their existing infrastructure. More commonly, third party companies are employed to undertake this hazardous task. Any transport company contracted to deliver liquid cryogens should be supplied with very specific information about the product they are being expected to handle:

- The product designation, which should describe what the cryogen is and in what state it is to be transported, i.e. nitrogen, refrigerated liquid.
- The product classification code, which relates to the nature of the hazard which the cryogen poses, i.e. Class 2 covers gases, with class 2.2 being specific for non-flammable, non-toxic and non-poisonous gases [71].
- The product UN number, which is a unique identifier relating directly to the type of material being transported, i.e. UN 1977, which is for nitrogen, and UN 1963, which is for helium [72].
- The volume of each type of cryogen storage vessel and the total number of vessels being transported.
- The consignor's (the sender's) name and address, required in the event of an incident involving the products being transported, so that they can be contacted to provide additional information about the hazards involved.
- The consignee's (the recipient's) name and address, required in order to be contacted if there is a delay in delivering the products, as this may have a consequence on a process at the destination which depends on the delivery being made on time. An example of this could be where temperature dependant samples have been cooled to cryogenic temperatures inside a helium cryostat, which would require periodically refilling with liquid helium. A failure to do so could result in the sample overheating and therefore being irretrievably spoilt.

- A statement signed, or at least authenticated by, or on behalf of, the consignor, confirming that in accordance with the relevant provisions of the respective national, Carriage of Dangerous Goods by Road Regulations and the Carriage of Dangerous Goods (Classification, Packaging, Labelling and Provision of Information) Regulations or equivalent, that the dangerous goods as presented may be transported and that the storage vessels are properly labelled, providing all necessary safety information about the product.

*7.3.37.3 Transport of large volumes of liquid cryogens, typically greater than 500 litres*

The transport of cryogenic liquids in large amounts can represent a significant hazard due to the nature of the products and the quantities involved. By association, the larger the quantity of a hazardous material being moved, there is a resultant increase in the risks involved. For this reason, the transport and distribution of large volumes of liquid cryogens makes use of purpose-designed equipment for each corresponding cryogen. Transporting quantities above the 500 litre capacity would normally involve the use of road tankers, ISO containers [73], ships, or even pipelines if the movement is to be undertaken over relatively short distances.

For the use of road tankers, specifically designed vehicles are employed which have been made to exacting standards, taking into account the extremely low temperatures of the product being transported. These vehicles would be used to undertake the bulk transport of liquid cryogens and will typically be owned by specialist companies who have well-established protocols for undertaking this type of work. In order to reduce the risks associated with the movement of large quantities of liquid cryogens, both the drivers of the road tankers and the operators will have to undertake formal training in the correct operating procedures. These would include instructions on how to prevent product contamination, along with the possible loading of the wrong liquid cryogen into the respective storage container. This is particularly important where flammable or oxidising cryogens are concerned, as they require more stringent safety protocols to be enacted. All road tankers must be clearly identified with the type of cryogen being transported and must also carry the appropriate safety and emergency instructions to suit that product.

ISO (International Standards Organisation) containers are box-shaped devices used for the shipping or transport of a whole range of goods. They are manufactured to very detailed size specifications in terms of their exterior height, exterior width, and exterior length, to ensure that any such container can be accommodated by equipment handlers, chassis, rail cars, and storage holds of ships wherever in the World they are being used. They are made to be resilient and reliable for repeated use and for long-term storage and come in a variety of designs to suit a range of products. For the movement of liquid cryogens in large quantities, the ISO containers comprise an internal storage tank, with a supporting structure around this which matches the container size requirements. The containers are of a rugged, weatherproof construction, which allows them to be transported worldwide by rail, sea or road. Designed and manufactured to ISO standards and compliant with the necessary national and international regulations for storage tanks, they provide a ready means of moving liquid cryogens in bulk. As part of the accreditation requirements, the storage tank will undergo initial testing and will then require periodic inspections to confirm continued compliance.

For the distribution of cryogenic liquids over relatively short distances, the safest option is through the use of fixed pipeline installations. These automatically eliminate the hazards associated with the movement of quantities of cryogens through more conventional means, by providing a direct supply of cryogenic liquids straight to the receiving destination. Many facilities employ this method by having fixed pipelines installed as part of their building infrastructure. Insulated pipelines can convey the cold liquids over distances ranging from a few metres to several hundred metres. The limiting factors to the length of pipeline will be dependent on the product being transferred and the quality of the insulation on the pipe runs. The longer the length, the more heat gains will occur in the cryogens, which in turn will result in losses of liquid as it warms up and converts to gas. For short distances, solid insulation tends to be the more practical means of isolating the liquid cryogens from the ambient temperatures. Over longer distances, this becomes impractical and vacuum insulated pipelines are more commonly employed, although there is additional costs incurred to provide the vacuum to this type of insulation.

Albeit the use of fixed pipeline installations removes many of the associated hazards with transferring cryogenic liquids, significant hazards still prevail. The general safety considerations relating to working around extremely cold surfaces are clearly applicable, especially where there may be a risk of insulation breakdown due to the nature of the transfer process. The wearing of all appropriate personnel protective equipment (PPE) must be mandatory wherever pipelines are being employed for the distribution of the cold liquids. Also, at both the distribution point and the receiving outlets, spillages and leakages of liquid cryogens can be manifest. Any operation and maintenance of these services points, along with the pipelines themselves, must therefore strictly follow designated procedures. Only trained personnel should be permitted to undertake the transfer of cryogens using this method. Likewise, the pipeline systems will require routine inspections and testing in accordance with the respective national Regulations.

### 7.3.38 Disposal of liquid cryogens

As with any industrial and commercial commodity, there will be inevitable costs incurred in the initial production of liquid cryogens into their usable state. Expenditures can be quite significant depending on the liquefied gases being employed. It is therefore difficult to imagine a situation where cryogenic liquids would be squandered in any way, but there could realistically be an occasion where the disposal of these products is a relevant outcome and possibly a preferred option under prevailing conditions. Credibly, this could be under circumstances where the recovery of surplus cryogens would in itself incur exorbitant additional costs, or even where any retention of the cryogens would pose a greater hazard to people and infrastructure, than simply discarding the excess. Disposal of any liquid cryogen has many potential risks and should be done only in a controlled and approved manner.

The most common misconception made in respect to the discharge of cryogenic liquids is that it can be undertaken using the normal means of disposal in the workplace. This is a mistaken belief and under no circumstances should cryogenic liquids be poured down normal domestic drains. The reason should be obvious, in that the materials of constructions of these drains are unlikely to be able to withstand the extremely cold cryogenic temperatures without some form of failure occurring. Domestic, and even laboratory drainage plumbing typical use plastic fittings of grades which would only be rated for use down to around −10 ° C in the extreme. Even external drainage systems, traditionally made from vitrified clay or even cast iron, would be prone to the low temperatures involved and could easily fracture due to the effects of rapid contraction on their structure. Modern versions of these drains will inevitably be manufactured from plastics, such as polyvinyl chloride (PVC), which itself would become brittle at cryogenic temperatures, and very often feed directly into the main concrete sewage systems.

The other associated hazard of disposing of cryogenic liquids into any drainage system is the rapid boil-off of the cryogen into what would chiefly be constricted volumes. This could cause a pressure build-up in any such drainage system, which in turn could possibly vent at the same entry point where the cryogen had been discharged into. In addition, if the cryogens involved are flammable, it is quite plausible for the build-up of gases in the drains to form the catalyst to an explosion which may very well vent through manhole covers and open drainage grills.

The safest, and more practical means of disposing of waste cryogens is to allow them to evaporate naturally in a fume hood if possible, where an element of control over the gases being released can be managed. If this method isn't possible, then pouring the cold liquid slowly onto gravel or bare earth in the open air, where is can evaporate without causing damage, is an acceptable alternative. Clearly, precautions need to be applied to make sure people are excluded from this area of ground until the evaporation has been completed. Notably, it is not advisable to pour any liquid cryogens onto grass or any other vegetation, as this will irreparably freeze the living cells making up the plant life. On thawing, the flora will be found to have died from the experience. Likewise, do not pour any liquid cryogens onto pavements, concrete surfaces or Tarmac surfaces. The extreme and sudden change in temperature is likely to cause damage in the form of fracture cracking of the material surfaces.

### 7.3.39 The human factor

Many of the hazards associated with the use of cryogenic technology relate to substantive physical properties which, if presented to the human body, pose a risk of injury or death occurring. There is one particular type of hazard however, pertinent to any application in the workplace, which is often inadvertently neglected and yet it is probable the more likely to be present in any working environment. Simply applied, there is a potential for hazards to be created directly by the people themselves as they perform any particular task. This is often known as the 'human factor' and the potential hazards associated with human interaction with cryogenic systems can dramatically increase the risk of accidents occurring. These interfaces can often be difficult to anticipate, but are borne from the very fallibilities which our species possess. Put rudimentarily, we are all capable of making errors in what we undertake, which can have a consequential effect on those associated processes and our surroundings. Likewise, humans are also capable of deliberately deviating from the defined rules of a procedure, which can in turn give rise to even greater risks where blatant abuse of systems can effectively negate well-intended safety protocols. These unsafe acts, whether considered to be deliberate or accidental can be conveniently grouped into five defining categories;

- Lapses of attention
- Mistaken actions
- Mistakes of perception
- Getting priorities wrong
- Breaking the rules

*7.3.39.1 Lapses of attention*

Would occur when a person has the correct objectives and intentions, and has decided on the correct course of action, but is distracted during the performance of the action. An example of this would be where an operator forgets to close an outlet valve on a decanting system as they have been distracted by an incoming phone call during the process. Lapses of attention can happen anytime and for a variety of reasons. In a growing technological world, our sight and our hearing are forever being drawn towards the output on our computer screens, handheld devices and our mobile phones, and it is easy to reconcile how these types of accidents, due to a lapse of attention, would result from the distractions placed before us on a regular basis.

*7.3.39.2 Mistaken actions*

Would occur when a person does the wrong thing, believing it to be the correct thing to do. In these cases, individuals know exactly what needs to be done but, for whatever reason, ultimately chose the wrong method. An example of this would be where a technician handling a small Dewar of liquid nitrogen correctly decides to wear gloves to protect their hands but does not chose the right cryogenic rated gloves for the task. Once again, mistaken actions can be attributed as a result of a person being distracted or preoccupied, which places them in a situation where they eventually inadvertently chose the wrong thing to do. This can more often happen where individuals are expected to operate systems by undertaking a complex series of sequential tasks, such as pressing a set of control buttons in the correct order. In these situations, it is easy to imagine confusion occurring, by which the operator could simply forget which control button has already been pressed. It is important therefore that depending on memory alone may not be sufficient to guarantee a successful result and the use of checklists or detailed instructions may be a more certain way of achieving the objective.

*7.3.39.3 Mistakes of perception*

Would occur when a person misreads the situation of the task they are undertaking, believing that one thing is happening when in fact something totally different is occurring. An example of this would be where a person fixes the lid to a small transportable cryogen Dewar, which on inspection appears to be loose and they therefore assume that the lid is somehow flexing due to its contents evaporating. However, it is in fact unsecured and the liquid cryogen contents could quite easily be spilt during subsequent transporting of the Dewar. Very often, mistakes of perception are borne from the repeat undertaking of a routine task where a single expectation is assumed, based on the number of times the task has already been undertaken. It can also happen when a person has a preconceived idea of what the outcome is likely to be and they can gain so-called tunnel vision, naively ignoring the contraindications telling them otherwise when they believe something to have occurred.

*7.3.39.4 Getting priorities wrong*

Would occur when the relative importance of an organisation's different objectives and goals are not clearly communicated or understood. This can include both the principal business aims of an organisation, but also its health and safety policy as well. An example of this would be where a delivery of cryogenic equipment needs to be urgently made to a certain location and the person managing the task forgoes the wearing of personal protective equipment (PPE) in order to speed up the process. Such errors of judgement can often be caused by the conflict between general and specific safety and other management functions within organisations. Time pressures, along with limiting resources and the lack of sufficient personnel can often exacerbate the problem, as the greater the emphasis on one priority becomes, there is a higher risk of safety being compromised. Even the simple action of failing to take a break from display screen equipment activities due to the pressures of workloads can often have a consequential effect on safety as tiredness and fatigue can bring about a consequential lack of concentration.

*7.3.39.5 Breaking the rules*

Would occur when a person deliberately disregards procedures or blatantly forgets the rules in order to achieve an output. An example of this would be where a technician fails to wear cryogenic gloves and a face guard in a mandatory area because it is too much trouble and would slow their processes, or that the PPE is uncomfortable to wear, so they chose not to. Once again, workloads and time dependent work pressures can often lead to shortcuts being taken which not only contravene policies but also compromise the safety protections provided. Unlike the other unsafe acts which result from careless mistakes, breaking the rules is a conscious decision made by the person undertaking the transgression. Whether guided by their managers or undertaken for individual gains, indiscretions of this nature not only cause risks to the individual, but can result in the manifestation of a greater risks to equipment and personnel in their immediate vicinity.

It is worthwhile noting that these collective unsafe acts do not just occur because people are lazy. Often, unsafe acts are caused by the safety culture of an organisation, where staff are routinely encouraged to cut corners in order to get things done by a critical deadline. Likewise, if the safety culture is poor within an organisation, it is often the case that employees are not provided with enough information and training to understand what they are doing is unsafe. It is vitally important therefore that any new person starting work in cryogenic technology are guided through the processes by the provision of the correct training and instruction in order to highlight the associated risks of what they are being asked to do.

Unsafe acts are often considered as irrational responses which are not related directly to equipment hazards, but they are none the less vital to consider as part of the overall working environment. A really good example which demonstrates this doctrine and puts into context how weaknesses in systems and processes can often be dominated by the human factor, is reflected in a list which defines the five common causes of United Kingdom traffic accidents [74]. These are listed as:

- Drink driving
- Losing control of the vehicle due to speeding or poor weather conditions
- Driver fatigue
- Carelessness
- Failure to look properly

Significantly, of these five, none are related to the failure of any component parts or systems within the vehicle being used. In fact, the data shows that 90% of motor vehicle crashes are actually caused by human error, highlighting how the 'human factor' dominates the risks associated with driving a motor vehicle on the United Kingdom's roads. Similar statistics probably prevails throughout the world, not only for motor vehicle activities, but for other types of undertakings where the human factor has a key role in the accident data. Some factors, such as the weather, can exacerbate the risks, but the prevailing cause of these particular road accidents still remains the driver's inability to respond to changing road conditions due to the change in the weather. Each of the five unsafe act categories are represented in these common causes of traffic accidents, emphasising that they can all readily occur and are of genuine concern to all other road users.

In the context of the working environment and in cryogenic technology in particular, these unsafe acts must be reconciled in every task undertaken. Many of the processes involved in the cryogenics world revolve around procedures and certain actions taking place at given periods in time. Often these actions depend on a person to instigate the action through the opening of valves, or the pressing of buttons, actions which are easy prey to the propensity of the listed unsafe acts. Clearly, the role played by any individual undertaking a task, and their fallibility has to be considered as a major potential to cause any accident in the workplace.

## Cryogenic hazards—summary

*Hazards*—By definition, a hazard is something that has the potential to cause harm in the form of injuries, diseases, property damage and damage caused to the environment. Hazards exist in everything we undertake, from the simplest of tasks, to the extremes associated with the handling of dangerous substances. Hazards can be either natural in characteristic, such as a violent storm, or those created by humans, such as spilled oil creating a slip hazard.

*Identifying hazards*—It is often difficult to identify all the particular cryogenic technology hazards for very specific situations. A useful method therefore is to use a four point checklist to ensure that all relevant hazards in the workplace have been deliberated. The key points are:

*Premises*—A check of the physical environment in which cryogenic technology is being used, such as the size, the condition of the work area and the quality of lighting.

*Plant and substances*—This is to check the type of cryogenic equipment being used in the work area, ensuring that they are fit for purpose, as well as looking at any chemicals and substances which may form part of the overall process being undertaken.

*Procedures*—A check of any relevant safe systems documentation, instructions on how to use cryogenic equipment and the method undertaken for any work being proposed for a particular task.

*People*—A check on training levels, competencies and general health of individuals undertaking specific cryogenic work.

**Specific hazards associated with cryogenic technology**

*Pressure build-up*—Related specifically to cryogen storage vessels, where a pressure is built up continuously above the liquid cryogen surface due to the gradual warming over time of the cryogenic liquids within the vessels. This hazard can be easily mitigated by the installation of pressure relief devices.

*Expansion ratio*—The ratio of the volume a cryogen will occupy in its liquid form compared to the volume it will occupy as a gas. Each cryogen has a different expansion ratio with values ranging from 1:600 to 1:1400.

*Hot burns*—Injuries which can arise due to the direct contact with flames or as a result of the radiated heat from a hot source. A hot burn will cause immediate damage to any exposed skin in the form of blistering and swelling.

*Cold contact burns*—Injuries which can arise due to the direct exposure of the skin to the extremely low temperatures of liquid cryogens and other equipment operating in the cryogenic region. The effects to the skin are similar to those of hot burns, with the addition of decreased circulation and the loss of feeling around the injury.

*Frostbite*—A cold contact injury which occurs when the skin is exposed to low temperatures for a prolonged period of time, causing it to freeze. More commonly affects the fingers and toes and can in extreme situations cause permanent damage requiring amputation of the affected limb.

*Hypothermia*—A condition caused by long exposure to extremely cold environments, where the human body core temperature drops below 35 °C for any extended period of time. It can be prevented by controlling the exposure period and providing suitable PPE and warm clothing.

*Toxicity*—The level by which particular substances can cause damage to an organism. Most of the gases and liquefied cryogens used in the context of cryogenic technology have relatively low toxicity, but some may cause dizziness or nausea if they are accidentally inhaled in sufficient quantities.

*Embrittlement of materials*—The inherent Young's modulus value of most materials will increase by approximately 20% as they are taken to the liquid helium (4 K) temperature. This will also increase the material's strength and stiffness, reducing ductility and increasing its brittleness.

*Material stresses*—The cooling of materials down to cryogenic temperatures will cause physical changes which in turn can create high internal stresses within the materials. As these materials will also have lost ductility at these temperatures, under extreme conditions these stresses can cause them to shatter. In order to prevent this loss of ductility, metal alloys should be selected so that they do not undergo a phase transformation on cooling.

*Contraction of materials*—Most materials undergo a negative expansion when they are cooled which is defined as a thermal contraction. Typically, contractions of 0.3% of the original length can take place between room temperature (300 K) and the liquid nitrogen temperature (77 K).

*Liquid nitrogen*—An inert, non-toxic, non-corrosive and non-flammable cryogen used in many cryogenic applications, with a relatively low boiling temperature of 77 K. If allowed to expand in sufficient quantities to its gaseous phase it can act as an asphyxiant through the displacement of the oxygen in the breathable air.

*Liquid helium*—An inert, non-corrosive and non-flammable cryogen used as a cooling mechanism in cryostats and high-field superconducting magnets. It can be sourced as the more commonly used helium-4 isotope, as well as the specialist helium-3 isotope. It has an extremely low boiling temperature of 4.2 K, and if allowed to expand in sufficient quantities to its gaseous phase it can act as an asphyxiant through the displacement of the oxygen in the breathable air.

*Liquid hydrogen*—A highly flammable, colourless and odourless cryogen with a boiling temperature of 20.28 K. As a gas, it is relatively easy to ignite and requires only 5% of oxygen in the air to support continued combustion. If allowed to expand in sufficient quantities to its gaseous phase it can act as an asphyxiant through the displacement of the oxygen in the breathable air. Hydrogen can also permeate into the crystal structure of certain materials causing hydrogen embrittlement.

*Liquid oxygen*—A highly flammable, colourless, odourless and highly reactive cryogen with a boiling temperature of 90 K. Oxygen is highly flammable and can under certain conditions react explosively with organic materials such as oil and hydrocarbons. If allowed to expand in sufficient quantities to its gaseous phase it will increase the oxygen concentration of the workplace to dangerously high levels where upon materials will more easily ignite and burn with a greater intensity once ignited.

*Liquefied natural gas (LNG)*—A highly flammable, colourless and odourless cryogen, mainly consisting of methane, with a boiling temperature of 112 K. As a gas, it is extremely easy to ignite and has a relatively low ignition temperature. If allowed to expand in sufficient quantities to its gaseous phase it can act as an asphyxiant through the displacement of the oxygen in the breathable air.

*Liquid argon*—An inert, colourless, odourless and tasteless cryogen with a boiling temperature of 87.2 K. As a gas it has a very low value of thermal conduction, making it an ideal thermal insulator for critical systems. If allowed to expand in sufficient quantities to its gaseous phase it can act as an asphyxiant through the displacement of the oxygen in the breathable air.

*Liquid neon*—A highly volatile, colourless, odourless, tasteless and chemically inert cryogen with a boiling temperature of 27.1 K. Its latent heat of vaporisation is over 41 times that of helium, which means that it can provide a more effective cooling power. It has a very large liquid to gas expansion ration which means if allowed to expand in sufficient quantities to its gaseous phase it can act as an asphyxiant through the displacement of the oxygen in the breathable air. The triple point temperature of neon is used as one of the defining fixed points in the ITS-90 temperature scale.

*Liquid fluorine*—A highly toxic, reactive cryogen with a boiling temperature of 85 K which will react violently with compounds such as water, ammonia and organic chemical substances. If allowed to expand in sufficient quantities to its gaseous phase it can act as an asphyxiant through the displacement of the oxygen in the breathable air. Any direct exposure to airborne fluorine will result in chemical burning of the skin, where it will react with the moisture in the skin to produce hydrofluoric acid, resulting in profound tissue poisoning.

*Liquid carbon dioxide*—A non-toxic, colourless cryogen with a boiling temperature of 217 K. Any high concentration of carbon dioxide gas has a distinctive sharp, acidic odour. If allowed to expand in sufficient quantities to its gaseous phase, liquid carbon dioxide can act as an asphyxiant through the displacement of the oxygen in the breathable air. Concentrations of up to 1% can cause drowsiness and breathing will feel laboured, whilst concentration of 7% to 10% can cause suffocation, even where there remains sufficient oxygen in the air.

*Superfluidity*—A specific property of a fluid when it has zero viscosity. It only occurs at extremely low temperatures in the region below 2 K and only with liquid helium-3 and helium-4. In its superfluid state, the liquid will appear to creep along surfaces in order to find its own level.

*Ice formation*—A condition where the very low temperatures used with some cryogens will solidify many of the gases from the surrounding air. This can cause ice formations in the venting port of safety relief valves and ice plugs in the necks of open Dewars. Clearing ice blockages of this nature is a specialist task.

*Manual handling and ergonomics*—If undertaken incorrectly, the manual handling of cryogenic equipment can give rise to back, neck, hands and finger injuries. Ergonomics is the study of how equipment and workplaces can be arranged in such a way that work activities can be undertaken in an effective and comfortable manner without the risk of these manual handling injuries occurring.

*Low viscosity*—All cryogenic liquids have a low viscosity, which means they will penetrate woven or other porous materials much faster than expected. Any spillages onto clothing would quickly run through the textile fabric and make rapid contact with the skin beneath.

*Oxygen enrichment*—Both liquid nitrogen and liquid helium have a boiling point lower than that of oxygen. As a result, where they are used, oxygen can be condensed out of the surrounding air, gradually collecting as an enriched quantity of pure oxygen. This can result in an increased danger of combustion occurring and pose a potential explosion risk.

*Physiological effect*—Exposure to the cold vapour released through the evaporation of liquid cryogens, can have both short-term and long-term effects on the human lungs. Short exposure may produce discomfort in breathing, which will generally reverse as the source of the vapour is removed. Prolonged exposure can produce serious, permanent lung effects where they become congested with fluid, leaving the sufferer constantly short of breath.

*Explosion due to pressure*—There is always gas present in cryogen storage vessels above the liquid surface, which over time can build up in pressure if it is not vented. If this pressure is high enough to exceed the strength of the containment vessel, an explosion or rupture of the container can occur.

*Explosion due to chemical reaction*—Chemical explosions are brought about through a rapid oxidation reaction within a compound or mixture that produces large amounts of hot gas. In the cryogenic workplace, chemical explosions are generally initiated by an electrical spark or flames, especially where the presence of oxygen is sufficient to cause combustion.

*Explosion due to ionising radiation*—There have been cases recorded where under ionising radiation conditions, liquid nitrogen can explode when it is subsequently warmed up. This is thought to be due to reactions with ozone forming from trace amounts of oxygen in the cryogen.

*Fire*—Fire is the rapid oxidation of materials due to the chemical process of combustion and can come about due to combustible materials coming into contact with an ignition source. Fire prevention in the working environment should therefore form the basis of every workplace fire management strategy which is often part of national regulatory requirements.

*Combustibles storage*—All combustible materials should ideally be stored externally to any cryogenic work area, or provided with suitable storage cupboards inside building. Any internal storage area should be arrange in such a way to prevent the spread of fire if it ever occurred.

*Cryo-pumping*—Under certain circumstances, unintended cryogenic-pumping can occur in cryogenic systems where the removal of gases is made as they are condensed and then become stuck to extremely cold surface. This can have the effect of lowering any system pressure to a point where a vacuum remains, causing thin-walled sections to collapse under the effect of the unopposed external atmospheric pressure.

*Decanting*—Cryogen liquids are regularly transferred from large storage vessels to smaller containers. These transfers pose a serious hazards and therefore essential precautions need to be taken for all such transferrals, such as the wearing of suitable personal protective equipment.

*Spillage during decanting*—Some spillage during any decanting process is inevitable, even if control measures have been employed to mitigate the likelihood. If significant quantities of liquid cryogens are spilt, all personnel must be immediately evacuated from the affected area. Suitable oxygen monitors should be fitted to any decanting area.

*General handling of cryogenic materials*—It is important that personnel are suitably trained and understand the risks associated with the handling of cryogenic materials. It is the responsibility of employers to ensure the safety and wellbeing of their staff in whatever cryogenic related tasks they are being asked to undertake. Any movement of cryogenic equipment should therefore be fully assessed before being carried out.

*Cryogenic liquid transport*—Moving any pressurised cryogen storage vessels will have the effect of accelerating the rate of evaporation from the liquid. This will result in an increase in the amount of gas being released from the vessels due to pressure build-up above the liquid surface. Any planned transporting of cryogens therefore needs to take this into consideration, as the resultant gas released over time during movement will be higher than the manufactures stated evaporation rate.

*Vehicular transportation of cryogenic materials*—The transport of cryogenic substances on public highways is covered by a range of regulatory and national laws. These define the specific quantities that may be transported and places duties of responsibility on the sender, the transporter and the recipient for specific transport operations. These regulations and laws come into forced if cryogenic substances are transported on public roads, for however brief a period.

*Company vehicles transporting liquid cryogens*—The transport of liquid cryogens represents a significant hazard and the appropriate control measures need to be in place in order to mitigate the risks of incidents occurring during transit. A written risk assessment needs to be in place for any activity being undertaken which involves the use of a vehicle to transport any liquid cryogens.

*Third party vehicles transporting liquid cryogens*—It is not always possible for organisations to provide transport facilities for cryogenic liquids within their existing infrastructure. More commonly, third party companies are employed to undertake this hazardous task. Any transport company contracted to deliver liquid cryogens need to be supplied with the product designation and classification code, the product UN number and all information relating to the consignee and the consignor.

*Transport of large volumes of liquid cryogens, typically greater than 500 litres*—The transport and distribution of large volumes of liquid cryogens makes use of purpose-designed equipment. Transporting quantities above 500 litres would normally involve the use of road tankers, ISO containers, ships or even pipelines if the movement is over relatively short distances.

*Disposal of liquid cryogens*—Under no circumstances should cryogenic liquids be poured down domestic drains as their material of construction are unlikely to be able to withstand the extremely cold temperatures. Instead, they should be allowed to evaporate naturally in a fume hood, so that the gases being release as the cryogen warms up can be controlled.

*The human factor*—There is the potential for hazards to be created directly by the people performing any particular cryogenic task. Humans are capable of making errors in what they are doing, which can have a consequential effect on the processes involved. Likewise, humans are also capable of deliberately deviating from the defined rule of a procedure, which can in turn give rise to even greater risks.

## References

[1] Institution for Occupational Safety and Health (IOSH) *Managing Safely—Module 3—Hazard Definition* (Coalville: Clwyd Associates)

[2] International Labour Organisation (ILO) *Chemical Exposure Limits* (www.ilo.org/safework/info/WCMS_151534/lang–en/index.htm) (Accessed 2020)

[3] Wikipedia *Acceptable Daily Intake (ADI)* (https://en.wikipedia.org/wiki/Acceptable_daily_intake) (Accessed 2020)
[4] United Kingdom Health and Safety Executive (HSE) *Reporting Accidents, Incidents and Diseases* (www.hse.gov.uk/toolbox/managing/reporting.htm) (Accessed 2020)
[5] European Chemical Agency (ECHA) *Material Safety Data Sheets (MSDS)* (https://echa.europa.eu/safety-data-sheets) (Accessed 2020)
[6] Safeopedia *Safe System Of Work* (www.safeopedia.com/definition/485/safe-system-of-work) (Accessed 2020)
[7] hsedocs *Method Statements* (www.hsedocs.com/documents/method-statements/) (Accessed 2020)
[8] United Kingdom Health and Safety Executive (HSE) *Permit to Work System* (www.hse.gov.uk/comah/sragtech/techmeaspermit.htm) (Accessed 2020)
[9] Physics.stackexchange *Energy of Liquid and Gas at the Same Pressure* (https://physics.stackexchange.com/questions/81156/energy-of-liquid-and-gas-at-same-pressure-is-different) (Accessed 2020)
[10] Instrumentationtools.com *Pressure Relief Valve (PRV)* (https://instrumentationtools.com/pressure-relief-valve-principle/) (Accessed 2020)
[11] United Kingdom Health and Safety Executive (HSE) *Written Schemes of Examination Pressure Systems Safety Regulations 2000* (www.hse.gov.uk/pubns/indg178.pdf) (Accessed 2020)
[12] Princetoncryo *Static Evaporation Rates for Liquid Nitrogen Storage in Cryogenic Freezers* (https://princetoncryo.blogspot.com/2009/01/static-evaporation-rates-for-liquid.html) (Accessed 2020)
[13] Wikipedia *Pressure-Release Rupture Discs* (https://en.wikipedia.org/wiki/Rupture_disc) (Accessed 2020)
[14] The National Board of Boiler and Pressure Vessel Inspectors (NBBI) *Specification of Rupture Disc Burst Pressure* (www.nationalboard.org/index.aspx?pageID=164&ID=381) (Accessed 2020)
[15] Monroe Brothers Limited 2005 *Liquid Methane Expansion Ration MB/05801—Module 2* (Moreton-in-Marsh: Monroe Brothers Ltd)
[16] Thipse S S 2013 *Cryogenics: A Text Book* (Oxford: Alpha Science International Ltd)
[17] Martini F H 2004 *Fundamentals of Anatomy and Physiology* 6th edn (New York: Benjamin Cummings)
[18] Nuclear-power.net *The Leidenfrost Effect* (www.nuclear-power.net/nuclear-engineering/heat-transfer/boiling-and-condensation/leidenfrost-effect-leidenfrost-point/) (Accessed 2020)
[19] United Kingdom National Health Service (NHS) *Frostbite* (www.nhs.uk/conditions/frostbite/) (Accessed 2020)
[20] Wikipedia *Toxicity* (https://en.wikipedia.org/wiki/Toxicity) (Accessed 2020)
[21] Wikipedia *Hyperoxia* (https://en.wikipedia.org/wiki/Hyperoxia) (Accessed 2020)
[22] European Chemical Agency (ECHA) *Material Safety Data Sheets (MSDS)* (https://echa.europa.eu/safety-data-sheets) (Accessed 2020)
[23] Roark R J and Young W C 1984 *Formula for Stress and Strain* 6th edn (New York: McGrawhill)
[24] ScienceDirect *Brittle-transition Temperatures* (www.sciencedirect.com/topics/engineering/brittle-transition-temperature) (Accessed 2020)

[25] Britannica *Face-centred Cubic (FCC) Crystal Structures* (www.britannica.com/science/face-centred-cubic-structure) (Accessed 2020)

[26] Physics in a nutshell *Body-centred Cubic (BCC)* (www.physics-in-a-nutshell.com/article/12/body-centered-cubic-bcc) (Accessed 2020)

[27] British Stainless Steel Association *Selection of Stainless Steels for Cryogenic Applications* (www.bssa.org.uk/topics.php?article=41) (Accessed 2020)

[28] Acerinox *Stainless Steel Grade 430 (1.4016)* (www.acerinox.com/en/productos/stainless-steel-grade/EN-1.4016—AISI-430-00001/) (Accessed 2020)

[29] The World Material *Stainless Steel Grade 410 (1.4408)* (www.theworldmaterial.com/aisi-410-stainless-steel/) (Accessed 2020)

[30] Nationwide Stainless *Stainless Steel Grade 304 (1.4301)* (www.nationwidestainless.co.uk/materials/304_304l_stainless_steel/) (Accessed 2020)

[31] Nationwide Stainless *Stainless Steel Grade 316 (1.4401)* (www.nationwidestainless.co.uk/materials/316_316l_stainless_steel/) (Accessed 2020)

[32] Science Direct *Martensitic STRUCTURE* (www.sciencedirect.com/topics/chemistry/martensitic-structure) (Accessed 2020)

[33] Science Direct *Body-centred Tetragonal (BCT)* (www.sciencedirect.com/topics/engineering/body-centered-tetragonal) (Accessed 2020)

[34] Nondestructive Testing (NDT) Resource Centre *Coefficient of Linear Thermal Expansion* (www.nde-ed.org/EducationResources/CommunityCollege/Materials/Physical_Chemical/ThermalExpansion.htm) (Accessed 2020)

[35] Engineering Toolbox *The Volumetric Thermal Expansion Coefficient* (www.engineeringtoolbox.com/volumetric-temperature-expansion-d_315.html) (Accessed 2020)

[36] Roark R J and Young W C 1984 *Formula for Stress and Strain* 5th edn (New York: McGrawhill)

[37] Science struck *Nitrogen Gas in the Atmosphere* (https://sciencestruck.com/nitrogen-in-atmosphere) (Accessed 2020)

[38] Wikipedia *Atmosphere of the Earth* (https://en.m.wikipedia.org/wiki/Atmosphere_of_Earth) (Accessed 2020)

[39] Trusted Informer *Liquid Nitrogen Costs* (https://trustedinformer.com/how-much-does-liquid-nitrogen-cost/) (Accessed 2020)

[40] Theilacker J C and White M J 2005 *Diffusion of Gases in Air and Its Affect (sic) on Oxygen Deficiency Hazard Abatement* (Fermi National Accelerator Laboratory USA—Fermilab Conference)

[41] The Physics Fact Book *The Price of Liquid Helium* (https://hypertextbook.com/facts/2007) (Accessed 2020)

[42] Kramer D Helium shortage has ended, at least for now *Physics Today* (https://physicstoday.scitation.org) (Accessed 2020)

[43] Airships.Net *The first Zeppelins* (https://www.airships.net/zeppelins/) (Accessed 2020)

[44] Airships.Net *Hindenburg airship* (https://www.airships.net/hindenburg/) (Accessed 2020)

[45] Raymond L 1988 *Hydrogen Embrittlement: Prevention and Control* (Philadelphia, PA: American Soc for Testing and Materials (ASTM))

[46] Difference Between.com *Difference between Ortho and Para-Hydrogen* (www.difference-between.com/difference-between-ortho-and-para-hydrogen) (Accessed 2020)

[47] International Organisation for Standardisation (ISO) *ISO 13984:1999 (en)—Liquid Hydrogen—Land Vehicle Fuelling System Interface* (www.iso.org/obp/ui/#iso:std:iso:13984:ed-1:v1:en) (Accessed 2020)

[48] Wikipedia *Thread Seal Tape—'Green'—Oil-Free PTFE Tape* (https://en.wikipedia.org/wiki/Thread_seal_tape) (Accessed 2020)

[49] Wikipedia *Hyperoxia* (https://en.wikipedia.org/wiki/Hyperoxia) (Accessed 2020)

[50] British Cryogenics Council (BCC) 2017 *Cryogenics Safety Manual* 5th edn (Oxford: Butterworth-Heinmann)

[51] Wikipedia *Firefighting Foam* (https://en.wikipedia.org/wiki/Firefighting_foam) (Accessed 2020)

[52] Wikipedia *International Temperature Scale, ITS-90* (https://en.wikipedia.org/wiki/International_Temperature_Scale_of_1990) (Accessed 2020)

[53] Wikipedia *Monel Metals* (https://en.wikipedia.org/wiki/Monel) (Accessed 2020)

[54] Bond D T, Guenther M E, Stimpson L O, Toth L R and Young O L 1979 *Experience with Fluorine and Its Safe Use as a Propellant* (Pasadena: CA: JPL Publication—NASA Jet Propulsion Lab, California Inst of Tech)

[55] ChemEurope.com *'Blackdamp' or 'Choke Damp'* (www.chemeurope.com/en/encyclopedia/Blackdamp.html) (Accessed 2020)

[56] The Royal Institution *Humphry Davy's Miners' Safety Lamp* (www.rigb.org/our-history/iconic-objects/iconic-objects-list/davy-lamp) (Accessed 2020)

[57] Cryomagnetics Inc. *Ice Formation* (www.cryomagnetics.com/support/ice-formation-safety/) (Accessed 2020)

[58] Wikipedia *Ergonomics for Manual Material Handling* (https://en.wikipedia.org/wiki/Ergonomics_for_manual_material_handling) (Accessed 2020)

[59] Health and safety Executive (HSE) *Working Days Lost in Great Britain* 2019/20 (www.hse.gov.uk/statistics/dayslost.htm) (Accessed 2020)

[60] Thermopedia.com *Cryogenic Fluids—Relative Viscosity Values* (http://thermopedia.com/content/676/) (Accessed 2020)

[61] Wikipedia *Lambda Point* (https://en.wikipedia.org/wiki/Lambda_point) (Accessed 2020)

[62] Thermopedia.com *Cryogenic Fluids—Relative Viscosity Values* (http://thermopedia.com/content/676/) (Accessed 2020)

[63] American Academy of Allergy Asthma and Immunology (AAAAI) *Chronic Obstructive Pulmonary Disease (COPD)* (www.aaaai.org/conditions-and-treatments/related-conditions/chronic-obstructive-pulmonary-disease) (Accessed 2020)

[64] Karl Zipf Jr R, Kenneth P E, and Cashdollar L *Effects of Blast Pressure on Structures and the Human Body* (www.cdc.gov/niosh/docket/archive/pdfs/NIOSH-125/125-ExplosionsandRefugeChambers.pdf) (Accessed 2020)

[65] National Radiological Protection Board (NRPD) 1998 *Living with Radiation* (www.gov.uk/government/collections/national-radiological-protection-board-nrpb-report-series) (Accessed 2021)

[66] Wei Q, Liu H, Wang D and Liu S-X Degradation in optical reflectance of Al film mirror induced by proton irradiation (www.sciencedirect.com/science/article/abs/pii/S0040609011000083) (Accessed 2020)

[67] Gregory C R and Nuttall C W European Organization for Nuclear Research *Report on the Explosive Risks in Cryogenic Liquids Exposed to Ionising Radiation—CERN TLS-CFM/95-06* (http://kirkmcd.princeton.edu/mumu/target/CERN/cern_at9506.pdf) (Accessed 2020)

[68] Wikipedia *Fire* (https://en.wikipedia.org/wiki/Fire) (Accessed 2020)

[69] Health and Safety Executive (HSE) *Storage of Flammable Liquids in Process Areas, Workrooms, Laboratories and Similar Working Areas* (www.hse.gov.uk/fireandexplosion/storageflammliquids.htm) (Accessed 2020)

[70] European Chemical Agency (ECHA) *Material Safety Data Sheets (MSDS)* (https://echa.europa.eu/safety-data-sheets) (Accessed 2020)
[71] United Nations (UN) *Central Product Classification (CPC)* (https://unstats.un.org/unsd/classifications/unsdclassifications/cpcv21.pdf) (Accessed 2020)
[72] Wikipedia *UN Number* (https://en.wikipedia.org/wiki/UN_number) (Accessed 2020)
[73] Wikipedia *Intermodal Container* (https://en.wikipedia.org/wiki/Intermodal_container) (Accessed 2020)
[74] Onecall Accident Management *Common Causes of Car Accidents in the United Kingdom* (http://onecallam.co.uk/common-causes-of-car-accidents-uk-and-how-to-avoid-them/) (Accessed 2020)

# Chapter 8

# Controlling cryogenic hazards

## 8.1 General

In all aspects of life, unmitigated hazards pose the greatest risks of causing physical injuries or even damage to property. Accidents don't just happen though, there always has to be a primary cause, no matter how divorced this is from the eventual accident. Very often this is because hazards are not being sufficiently controlled in order to reduce both their consequences and the likelihood their effects can have on people, property and the surrounding environment. Typically, we take for granted the many things in our daily lives which in reality are hazards to our wellbeing and yet they are so well controlled that they no longer pose a threat to us. Electricity for example, in its exposed state can cause instant death on contact. Yet it is controlled by not only physical barriers and mechanisms to prevent accidental contacts, but is defined by legislation which establishes the correct parameters by which equipment has to be appropriately wired in the first place to ensure absolute safety.

In the workplace, these same ideals are abound in the shape of the simplest form of safety devices, all the way through to the ideology of a company's health and safety policy. As already explained, accidents don't simply materialise out of thin air, they are linked directly to hazards and are specifically caused through unsafe acts or unsafe conditions. Unsafe acts are invariably due to someone doing something unsafely, or not taking the necessary action to prevent unsafe conditions occurring. Unsafe conditions are generally caused by poor design of equipment or processes, or even due to wear and tear of systems which have not been routinely maintained or inspected. Yet ironically, both of these foremost causes can be readily mitigated by addressing the specific hazards they can generate and controlling them in such a way that they reduce the associated overall risks of any undertaking.

Therefore, in order to mitigate the hazards explicitly associated with the use of cryogenic technology, it is essential that any establishment implements particular control measures when dealing with given cryogenic applications. The aim of doing this is so that activities where the risk of harm is unacceptable are identified and the

doi:10.1088/978-0-7503-3123-4ch8  

appropriate actions to reduce the risk applied. The overall aim of this procedure is to reduce the general risks as well as the specific risks and adopting precautions at different levels forms part of the process. This is often known as the hierarchy of controls and would involve several notable levels of implementation [1]:

- Eliminate the hazard
- Substitute the hazard
- Combat the risks at source
- Reducing the risk by applying safe systems of work, training and supervision
- Personal protective equipment

### 8.1.1 Eliminate the hazard

Eliminating the hazard sits at the apex of the hierarchy of controls as its implementation can immediately negate all the effects which the hazard posed to the working environment. By completely removing the hazard from the process being applied, any potential incidents associated with that hazard cannot therefore occur. Put fundamentally, if there is no hazard, an incident cannot occur. In any risk management process, this elimination process must always be the first option to be considered as it offers the greatest reduction in overall risk. An example of how this could be applied for the use of cryogenic technology might include contracting out certain low frequency specialist activities to organisations who are more competent to do the task, especially if they routinely undertake this type of high-risk work and have gained valuable experience from this. This could include the maintenance of cryogenic systems and the removal of ice-plugs in cryogen storage vessels. Another example could be where the filling of cryogen storage vessels is automated, thereby eliminating the need for personnel to undertake the majority of tasks. This will reduce the overall risks associated with the filling process, especially where there is the potential of creating an oxygen depleted environment in areas where people are routinely working.

### 8.1.2 Substituting the hazard

Substituting the hazard is the obvious alternative to eliminating the hazard if this course of action isn't deemed possible. Substituting in this case involves replacing the specific hazard with something which is known to be less hazardous in nature. This is very often the type of solution employed where chemical hazards have been identified and alternative substances can be used which have a lower risk factor, providing they function in the same manner to the original chemical. Adopting this particular method does not directly eliminate the particular hazards associated with the use of chemicals, but by substituting it for something less hazardous, the overall risk is hopefully reduced. It is important that overall risks are considered when substituting, as alternative solutions may not be as effective and may therefore need a longer process. The substituted product may be less hazardous, but when duration of use is factored in, they may in fact be a greater risk than originally thought. Any comparison with the original product therefore needs to evaluate the total risks with their use.

Other workplace substitutions typically adopted are where higher-voltage electrical equipment have been replaced by equipment using significantly lower voltages. This is seen more and more as battery technology has advanced to the stage where almost all hand-tools traditionally requiring mains electricity to power them, can now be replaced by low voltage, battery powered tools instead. Likewise, noisy industrial plant equipment, such as compressors, made several years ago, can now be substituted with more modern ones which are specifically designed to generate less noise. Even the processes involving the movement of heavy items around the workplace can now be substituted with the provision of a wealth of handling equipment such as sack trucks and motorised forklift trucks.

A simple example where substituting the hazard can be applied for the use of cryogenic technology would be where the capacity of cryogen storage vessels are reduced, replacing large capacity Dewars with small sizes, so that the potential risks are proportionally reduced. This reduces several risk factors, such as those associated with the manual handling of the much larger Dewars. It also reduces the potential quantities of gas produced should any spillage of cryogens occur from these smaller Dewar sizes, thereby lowering the chance of an oxygen depleted environment being created in the local workplace. Again however, if this substitution in turn results in more frequent decanting processes and an increase in the movement of these smaller Dewars through the workplace, overall risk may not necessarily be reduced.

### 8.1.3 Combating the risks at source

Combating risks at their source is the process also known as the provision of engineering controls and can be done by either reducing the hazard severity or by directly reducing the likelihood of an incident ever occurring through any interaction with the hazard. This could be as simple as providing guards on equipment, so that the hazard severity is reduced to an absolute level, or providing warning signs where a hazardous activity is being undertaken, thereby lowering the likelihood of ingress towards that particular activity. Providing separate pedestrian and vehicle routes is another example where the risk is reduced by the existence of some form of physical regulation mechanism. Engineering controls are obviously less effective than the elimination or substitution processes, as the basic hazard remains and therefore retains the potential to cause harm if these engineering controls are compromised in any way. This can easily occur, as engineering or workplace controls can very often fail due to aspects such as lack of maintenance or general wear and tear of things like guards, or they can even be deliberately circumvented, thus exposing the hazard in its unmitigated state and its original high-risk potential. Whether through a mechanical failure or considered human intervention, removal of the engineering controls can render hazards back to their maximum prospective, thereby putting all immediate personnel in an unsafe environment. In their simplest form, engineering controls serve to separate employees from hazards and are generally applied through a combination of legal requirements, as in the case of guards being fitted to rotating equipment, or more usually, due to a local risk assessment, which highlights the needs for these controls to be implemented.

Examples of where the risk is combated at source for cryogenic technology would be where signs and barriers are installed around a temporary cryogenic installation, thereby warning casual bystanders of the potential risks in the immediate area. This effectively reduces the likelihood of personnel being in the proximity of the potential hazards. Likewise, the fitting of pressure relief valves on cryogenic systems which have the potential to produce excessive pressures are also an example of an engineering control which would automatically operate in the event of an over pressure situation occurring. A pressure relief valve is also an example of an engineering control which has to be compliant with national pressure equipment regulations or directives, and in many circumstances would also have to be periodically maintained and checked in order for this compliance to continue.

Hazard warning signs are a useful method of alerting people to the potential hazards of cryogenic technology in the workplace. Having these in place is an important method of alerting individuals in advance of their entry into hazardous environments, so that they in turn can apply additional and appropriate control measures before they enter. Seeing the signs may even have the effect of deterring casual ingress into the work area, more so if the cryogenic technology being used requires a level of competency and training from the operators working in those areas. Unfortunately, there is as yet no approved regulatory hazard warning sign for cryogenic equipment. Many facilities across the world have however adopted the snowflake symbol as a depiction of the associated cryogenic features and this has by default become the nearest thing to a recognised common standard. There are many variations of the image, all of which conveys the principle that the snowflake represents something extremely cold in temperature.

A cautionary note should be recognised in relation to the use of warning signs, as they are only useful if they can immediately convey information about the hazard being notified. As in the case of the snowflake symbol, seen in isolation, this can sometimes be confusing, as it doesn't explicitly define a cryogenic hazard. It could be used simply to warn of something being held at a cold, but not necessarily under cryogenic conditions. This confusion can be more pronounced for individuals who do not routinely work in a cryogenic environment. Many warning signs therefore use script along with the symbols to add credence to the message being conveyed. In fact, some warning signs use only script and dispense with any potentially misinterpreted symbol completely. The only drawback with utilising words to convey the warning is that there is an assumption that the person viewing the sign is literate in the language portrayed on the sign. There are many facilities across the world for example where teams are comprised of multi-national individuals. In order for the comprehensive understanding of warning signs to be made therefore would necessitate several translations of the warning script on the one sign.

One of the biggest hazards associated with the use of cryogenic technology is the accidental creation of oxygen depleted environments when handling liquid cryogens. The resultant risk of an individual suffering asphyxiation is therefore extremely high. Here again however, there is no approved hazard warning sign for this particular risk. The European Union and other European countries have generally accepted one specific warning sign depicting a figure seated on the floor with their head slumped forwards, representing someone who is assumed to be unconscious due to the asphyxiation hazards they have been working in. Once again, variants of the sign exist, with the principle characteristics common throughout.

### 8.1.4 Reducing the risk by applying safe systems of work, training and supervision

The higher level hazard controls have up to now considered only physical methods of limiting contact with hazards, such as providing guards on equipment or replacing the hazard with something less hazardous. Sitting beneath these approaches are the procedural means by which risks can be reduced, such as; safe systems of work, training and supervision. Clearly, implementation of these is not a guarantee that hazard management is in place at all times. Technical measures of this nature can often be forgotten, especially if training isn't routinely updated, or even deliberately ignored if constraints and priorities are changed during a process activity. Even the concept of supervision can introduce a composite variable, as it is a profound fact that nobody can be exclusively overseen all of the time. The main aims however are to develop procedures for the safe undertaking of any task, and then training people to follow these procedures, along with the monitoring of their activities to ensure deviation from the procedure does not take place.

A common way of preventing incidents from occurring in any facility is to have so-called safe systems of work in place for all applications which are deemed hazardous. These can take the form of various processes, such as; method statements [2], standard operating procedures [3], or even permit-to-work systems [4]. In addition, and generally associated directly with these safe systems, are the training and supervision needs of employees undertaking the actual work. It is fundamental that the people carrying out the tasks should be suitably trained to follow the procedure. Having safe systems of work can reduce specific risks, but they rely on the fact that each trained individual must remember and religiously follow the procedures. It is a known reality however that people have lapses of attention, and on some occasion, even decide to break the rules. This procedural method is consequently a less effective way of controlling risks than through eliminating, substituting or by using engineering controls. Therefore, in order to ensure that safe systems of work remain effective after training, people need to be supervised so that their activities can be monitored and compared with what has been defined in the supporting documentation.

Safe systems of work are often the catalyst for creating a good health and safety culture in any organisation, which in turn, will lead to lower overall risks for the processes being undertaken. These safe systems should list out all associated hazards for the work taking place, the risks involved with the work, and what needs to be implemented in order to keep these risks as low as possible. Most are common sense principles, but by deliberately stopping and thinking about them in more detail, risks can be effectively reduced without the need of employing other risk control methods, such as complex mechanical safety systems. By developing the essential procedures that control any hazardous activities, standards can be established, which when applied, should guarantee some level of risk reduction. This could be as simple as adopting standard operating procedures, which should be a proven safe method, and can therefore be implemented each time a particular task is carried out. Typical examples would include work instructions, which define how a task should be undertaken safely and the preparation of induction procedures for new starters, who are unlikely to be aware of the work environment and its associated hazard. These procedures may also involve the application of a permit-to-work system, where critical safety protocols would have to be in place before a particular hazardous activity can begin. Examples of permit-to-work systems are those employed as part of the electrical isolation requirements when working on high-voltage equipment, where the electrical currents are physically locked-off before any work can begin. Likewise, the routine retraining of forklift-truck drivers can be a prerequisite to them operating the vehicles, which ensures that standard operating procedures for forklift use would always be followed.

Once any safe system of work has been established, the people carrying out the work need to be suitably trained so that they can correctly follow the prescribed procedure. Having such a safe system of work can reduce risks, but it does rely on the trained individuals ultimately remembering and then adhering to the defined procedure. As already stated, it is therefore a less effective method of controlling risks than the other hierarchy of controls, as it is easy to see how even the slightest of contraventions by an operator could result in a transgression away from the set procedure. In order to ensure that any safe system of work remains effective after training has taken place, it is necessary to supervise people doing the work so that their activities can be checked and monitored for compliance.

An example of how this would be applied with the use of cryogenic technology would be where a closed cycle refrigerator needed to be prepared in order to cool a material sample to cryogenic temperatures. The safe system of work could be very specific, outlining how the preparatory task should be undertaken, but also defining the requirements where the operator would be expected to have some level of training in order to run the device. This implies a degree of competency in how the closed cycle refrigerator is therefore prepared, which is linked to the training and experience required of the operator. The use of a checklist, which has a programme of sequential sub-tasks, would also aid compliance with the cool-down procedure, ensuring that each of the individual task are undertaken in a logical and successive manner. If in addition, this checklist has to be countersigned by a supervisor on successful completion, an extra level of validation is factored in.

In a similar manner, in the event of a cryogenic injury occurring, it is useful to have a first aid checklist on hand which addresses the essential treatments required and more importantly, what should not be done. As with all first aid protocols, whatever the incident being responded to, it is important that individuals ensure their own safety before attempting to assist any injured person. The hazard causing the incident may still be present and it would be unwise to attempt a rescue or to provide first aid without first considering the likely consequence of that intended action. This is never more apparent than where the release of cryogens over time have resulted in a potentially oxygen depleted atmosphere in a work area which increases the risk of asphyxiation to anyone entering that atmosphere. It is not uncommon for a situation to arise where an individual falls unconscious due to a reduction in the oxygen levels in their work area and for a second person to then attempt a rescue, only to fall victim to the same insufficiency of oxygen. Only when it is considered safe to do so should the injured person be removed from the site of the incident to a safer location where treatment can be better administered. By doing this, the risk of a secondary incident occurring by remaining close to the hazard has been eliminated.

Adopting these safe working practices should be constantly reviewed to match the current workplace dynamics and where applicable, they should be incorporated into a suitable safe system of work for use with cryogenic technology. Examples of this would be where any vessel containing a cryogenic material must be clearly marked with the name of the contents and should only ever be used for that particular material. Storage vessels are specifically designed for the cryogen they will contain and so will not necessarily be compatible with the storage of alternative cryogens. If these containment vessels are then to be transported to another area, the route should be assessed in detail prior to the movement as part of this safe system of work. Wherever possible, the transporting of liquid cryogen containers should be avoided where there are slopes, steps or stairways as these types of terrain increase the hazardous conditions for the storage vessels and thereby raise the risk of tipping the container, or spilling the contents. Notably, any movement of a cryogen storage vessel will automatically increase the evaporation rate of its contents and moving over uneven or sloping ground will significantly magnify this increased rate.

Another example of a safe system of work would be where a method statement is employed for the movement of any cryogens through a workplace environment. As the contents of any cryogenic liquid vessel is in itself a hazard, they should not be moved through densely populated areas where the risk of causing harm could therefore be significantly amplified. A safer method would be to move the vessels during the evening when work areas are less likely to have staff occupying them. Be aware however, moving the storage vessels during darker hours where there is low levels of lighting brings its own danger as the risk of slips, trips or falls where visibility is poor would be pointedly increased. The proposed route should also be checked for any restrictive access which may limit the size of the vessel which can be transported. These factors can be written into the method statement, which would detail a checklist of the main points of the movement and when these would need to be implemented.

As has already been stated in earlier chapters, certain types of containment vessels must undergo regular inspections in order to continue their operational function. In addition, if the storage vessel is fitted with wheels to allow easy movement from its delivery point to its user position, these wheel would also need to be inspected and maintained on a routine basis to ensure they remain in good condition. The focus should be on the wheels, bearings and axles, as these could cause the vessel to topple if poorly maintained. This requirement could be instigated through a permit-to-work system, whereby proof of any inspection would need to be presented before any movement of the storage vessels can be undertaken.

All operators of cryogenic equipment should be familiar with the systems and procedures, in particular, those which are applicable to the handling of cryogenic liquids as these have the greatest potential to cause harm. This should normally be provided through specific equipment training and regular supervision of their work. This can also be attained through general cryogenic technology experience, where individuals would assist a more competent member of staff until they are themselves capable of undertaking the specific cryogenic tasks safely. They should also have an understanding of how to use any cryogen storage vessels, in particular how they work, what the properties of any released gases are and what relevant and mandatory personal protective equipment (PPE) they should wear. These are especially important for processes such as decanting procedures where the risk of direct contact with the cryogens is significantly heightened. The operators must also be knowledgeable of the emergency procedures for their respective work areas, periodically practicing these measures so that they are never put in the situation where they feel unfamiliar with the correct emergency techniques.

As part of this training process, an often neglected factor is that of making sure the operators of cryogenic technology are thoroughly familiar with the equipment supplier's operating manual. It is all too common for situations where people are trained in how to use this type of equipment by a previous member of staff who may unintentionally pass on their own lack of knowledge of a system. It makes sense therefore to question every detail of a cryogenic system, ensuring that there is an understanding of how and why each element is included and how it should be operated correctly.

Some key requirements forming part of any safe system of work is having a process in place whereby appropriate clothing and PPE must be provided and worn by anyone undertaking work with cryogenic technology. Risk assessments must consider the actions required in the event of a sudden release of cryogenic liquids and what happens as this warms up and converts to gas. In order to reduce the likelihood of asphyxiation, adequate ventilation must be in place for all areas containing cryogenic liquids. Likewise, facilities must ensure that there are suitable control measures in place if there is a potential for any lone working with cryogenic technology. This may simply be the setting up of routine contact with the individuals who are lone working, or providing some means of communication, such as a walkie-talkie or a mobile phone, so that in the event of an accident, they are more likely to be able to contact the emergency services.

Risk assessments for the use of any cryogenic materials, which should form part of the safe system of work, should include oxygen depletion calculations and must be undertaken prior to any work starting which involves cryogenic technology. If identified by the risk assessment, adequate oxygen depletion monitoring systems, such as alarmed oxygen level monitor systems [5], must be in place before any work is initially undertaken in the work area. Thereafter, these systems must be regularly maintained and serviced to ensure they are continually working. Safe methods of working with the various elements of cryogenic technology and suitable training must be provided for the actual cryogenic materials used. For example, the various cryogens available differ in many ways, so it is important that the training focuses specifically on the materials which are likely to be used, even if it is only very occasionally that they are handled.

### 8.1.5 Reducing the risk by using personal protective equipment (PPE)

The use of PPE to protect against particular hazards is the final level of risk control in the hierarchy of controls. It is particular to the individual undertaking any task, where the risks are directly reduced by providing equipment that protects the individual from a very specific hazard. As a result, it is the least effective control mechanism, as it does not mitigate the hazard but offers direct protection for the person carrying out those tasks considered to be hazardous. It therefore follows that it does not provide protection for other workers in the same workplace, unless they themselves are wearing similar PPE. Importantly, the wearing of PPE should not be the sole risk management strategy adopted, but should be seen as part of an overall risk control approach employed in a work situation, with other controls supporting this concluding layer of safety defence. In fact, a variety of controls should always be implemented for every identified hazard, thereby providing different levels of safeguarding which provides supporting backup should one aspect ever fail. The protection offered by PPE may be for the whole body, but is more commonly targeted towards the detailed areas such as; the hands, the face and the eyes. Typical examples of PPE worn for general workplace use are; safety gloves, safety glasses, high-visibility vests and hearing protection. In addition, specialist PPE for very detailed applications are also available. These would include items such as; respiratory protective equipment, which are designed to protect individual wearers from hazardous substances in the workplace; harnesses and fall-arrestors type equipment, used for work at heights; and electrostatic discharge (ESD) shoes, which provide a safe and a controlled method of discharging the body's static electricity.

The correct size and fitting is critical for any PPE being worn, as a poorly worn item will effectively negate the protection being offered. Many standard PPE items therefore are available in a variety of sizes to accommodate the range of heights and sizes to suit employee requirements. As such, the wrong size of PPE should never be worn, even if it is a larger size than required, as ill-fitting PPE can itself become a hazard in many applications. Where items are only generally available in one-size, they will normally have some form of adjustment, such as adaptable straps and interchangeable clips, so that the PPE can be tailored to suit the wearer.

For cryogenic technology, the type of PPE used will have to be explicitly designed and rated to withstand the extreme low temperatures involved. Protective gloves would for example be of a design where adequate thermal insulation is provided throughout their use, whilst still offering dexterity of movement so that the unhindered operation of valves and other small devices can be successfully made. Non-absorbent leather-backed gloves with elasticated wrists tend to be the preferred option when handling cryogenic liquids and other extremely cold equipment, protecting the hands in particular from accidental splashes. Alternatively, the gauntlet type gloves, whilst offering comparable protection, are not as ideal, as during their use cryogens may splash into the wide cuff and become trapped. Under these circumstances, it is possible for the cryogens to eventually penetrate the fabric of the gauntlet and cause cold burning to the exposed hands beneath. Whichever type chosen, it must be remembered that gloves rated as suitable for cryogenic application are for use with vapour only and under no circumstances can they be immersed into a cryogenic liquid for any period of time. Their insulation in particular is only designed for short duration exposure to the extremely cold temperatures. In addition, clothing should be selected with long sleeves which can then be pulled down over the gloves, again preventing the possible ingress of accidentally splashed cryogens.

It is also essential that the face and eyes are protected from accidental splashing of cryogens. A face mask, goggles or cryogenic rated safety glasses should be worn when undertaking any operation involving the handling of cryogenic liquids, thereby preventing injury to the skin of the face and especially the delicate membranes of the eyes. Likewise, any spilled cryogen is more than likely to migrate downwards towards the floor, so the correct footwear should be used to prevent injury to the skin of the feet due to any exposure to these spills. Waterproof shoes or boots with no lace-holes are the preferred option, as lace-holes can provide a potential leak-path for the cryogens to the feet. Special footwear with the tongue sewn into the boot are readily available for these specific applications.

Cryogenic aprons are also readily available and are usually made from multi-layer insulating materials which provide protection to the torso and the upper legs from any contact with cryogenic surfaces. They are useful for protecting against possible splashes and spills of cryogenic liquids and would typically be rated to suit the cryogens being used. They would normally have a waterproof outer layer which further protects against any splashes of the cold liquids. Likewise, they are fitted with adjustable straps and quick-release buckles at the neck and the waist to make emergency removal a simple and effective process.

Protecting the face is normally made by the use of face shields which generally incorporate a small supporting flange worn on the head, which in itself offers some protection to the top-front of the head. They normally consist of a moulded, clear splash shield, wrapped around the face to provide coverage to both the neck area and the sides of the face. They are generally available in a variety of sizes, but have adjustable headbands to suit the wearer and adjustable pivots for the shield so that it can be lifted out of the way when the hazards are not present in order to provide unimpeded visibility. Typically, the shield is replaceable should it ever become damaged or scratched in any way. They are lightweight and comfortable to wear, providing an unobstructed view of the work area. They are normally made from materials which would not withstand being immersed into cryogenic liquids, but they are capable of providing protection from accidental splashes of the extremely cold liquids.

## 8.2 Maintaining and monitoring the risk controls

Irrespective of where any specific control measure sits in the hierarchy of risk controls, many have weaknesses and can be accidentally or even deliberately bypassed. The consequence of this would directly increase the risks associated with the undertaking of tasks, especially where this initially appears to be a protected process. It is important therefore that some form of monitoring risk control system is employed to ensure that any of the risk controls adopted are actually effective. This not only includes checking their effectiveness at the point when they are initially implemented, but also over their lifetime period, during which they need to continue to function as a safety control.

An example of this would be the fitting of a pressure relief valve on a cryogen storage vessel to prevent over pressurisation as the cryogen warms up and gas evaporates from the liquid. Any new relief valve will be supplied with a certificate of conformity which states that it is fit for purpose and that it will function as described by its specification. Over time however, due to the aging of parts, accidental damage and general wear and tear, the relief valve may no longer operate as it should do so. This would increase the risks associated with the use of the storage vessel. A programme therefore needs to be implemented whereby the function of the relief valve is periodically checked to reduce the risk of it ever failing during use. It may even be prudent after a given length of time that the relief valve needs to be completely replaced in order to ensure the correct function of this required control measure.

Bearing this in mind, risk control systems are typically used to ensure that the implemented workplace precautions are being employed and used correctly. For them to be effective however, employees need to fully understand the purpose of the risk control systems, how they work and who is responsible for undertaking the checks. Equally so, it is also important that the actual risk control systems themselves are audited from time to time, in order to make sure that they are working correctly. Examples of risk control system audits would include:

- Workplace or housekeeping inspections
- Behavioural or activity audits
- Equipment inspections
- Health surveillance

### 8.2.1 Workplace or housekeeping inspections

The general health and safety philosophies within an establishment are considered to be a good indicator of the types of workplace hazards that have been identified and how they are being addressed in order to reduce the risk of injury occurring to staff. The monitoring and evaluation of these standards is commonly done through defined procedures, such as safety tours, where spot-checks of current housekeeping can be judged and any immediate actions relating to remaining risks progressed or prioritised. Safety tours are typically viewed as general inspections of the workplace, undertaken to obtain an appraisal of the working environment and the safety practices being adopted. As such, they can be done informally by internal teams, or formally through independent inspectors. Both have different merits, as the internal team would hopefully provide a greater knowledge of the establishments' processes and therefore would be able to look for specific high-risk activities as part of the safety tour inspection, capturing any flaws which become apparent. Independent inspectors on the other hand, by their very objectivity, would be able to offer an impartial judgement of what their safety examination uncovers, with no bias towards the process or the establishment they are looking at.

More formal inspections can also be undertaken which would include detailed aspects, such as safety sampling. Here the systematic sampling of any particularly dangerous activity, process or work area, is made as a means of inspecting the effectiveness of any control measures being implements. These inspections are more limiting in scope as they only capture a particular sample which can give both false positive findings and false negative results. A more general inspection of these same dangerous activities, processes and work areas, can be made through safety surveys, which capture a more holistic view of what an organisation is undertaking as part of their work output. These look at larger profiles of the high-risk processes, looking for trends of both good and bad safety practices. Where incidents have occurred in the workplace, incident inspections should be carried out following any accident that has caused a fatality, injury or significant near miss which could have resulted in an injury or ill health. The aim of this type of inspection is to discover the cause of the incident and learn from this, followed by implementation of further strategies to prevent such an event occurring again in the future.

The frequency of workplace or housekeeping inspections will depend on the nature of the work being assessed. For example, inspections tend to be performed less often, especially if the working environment is deemed low-risk, such as in a predominantly administrative office. If an organisation however has certain high-risk areas in their workplace, or undertakes specific activities that are considered hazardous or are frequently changed or modified, more recurrent inspections may be justified. Construction type projects would be a good example where regular inspections are required, as the dynamics of the work being undertaken would be changing with each progressive phase of the project. Likewise, cryogenic working environments are more likely to be areas where equipment is regularly being moved into and out of the area, and therefore they too would require more frequent inspections, especially of the control measures implemented to reduce risks. Safety tours within cryogenic work areas should not only address the obvious cryogenic related hazards and control systems, such as personal protective equipment (PPE) and oxygen monitor installations, but should look at general housekeeping risks as well. A cluttered cryogenic laboratory for example can be a hazardous environment, where slips and trips may dominate the level of accidents occurring.

### 8.2.2 Behavioural or activity audits

A similar review system is that of behavioural or activity audits, where the actual process being undertaken is periodically assessed in order to check compliance with the original specification and to ensure established risk controls remain relevant. This is important if risks are to remain low, as over time the way a particular process is carried out could be changed due to the inclusion of new technology, or even the person undertaking the work may be replaced. As a result of these variables, any risk controls originally put in place to address specific hazards may no longer be effective and a review of their function would be urgently required. The change of operator in particular carries with it higher risks, as it may be that a new member of staff does not have the experience and competency which the original operator had. This is where individual training and the establishment of safe systems needs to be reviewed.

It must be emphasised that behavioural or activity audits are not a test of the operator or their skills and should not assess non-safety related factors, such as output and performance. They are there to review both the hazards and the mitigating risk controls which may have an impact on the operator's safety as they undertake specific tasks. That being said however, where these risk controls relate to safe systems, the established policies, rules and procedures can be directly tested for their effectiveness in reducing risks. But likewise at the same time, an operator's understanding and implementation of these safe systems should also be appraised. These audits are normally achieved through surveys and interviews, along with location specific observations of how a particular process is carried out. Ideally, these audits should be undertaken at least once a year, but at more frequent intervals if any part of the overall process is modified. This would include any staff changes made during the period in between the respective audits.

A typical example of a behavioural or activity audit undertaken in a cryogenics environment, may be to check something like the way a cryostat is routinely filled from a storage Dewar. Not only would the audit look at the method employed, but would scrutinise the method statement, the training of the staff undertaking the task and any other safe systems employed, such as access controls to the work area. This audit could be undertaken by a manager of the work area, but ideally should not be carried out by an operator who would normally do the task, as they would tend to bring preconceived ideas of the process. The best person to undertake the audit would be an independent observer, who had detailed knowledge of the cryogenic hazards associated with the task, but impartial enough to know how the task should be safely performed.

### 8.2.3 Equipment inspections

Any equipment used as part of a process must be intrinsically safe and suitable for the intended purpose. This ensures that any associated risks created from using the equipment are mitigated to a suitably safe level. It is also a fact that equipment of all varieties can be used over long periods of time, during which, the expectation is that it will hopefully retain its original safety credentials, so that operators can use it without the danger of injury occurring. Deterioration and failures of equipment do take place however, as over time, component parts are likely to be subject to wear and tear which can eventually result in breakages and other forms of damage occurring.

The purpose of equipment inspections therefore is to make an assessment and identify whether work equipment can continue to be operated and maintained safely. The inspections are looking specifically for evidence of deterioration to the equipment, so that these can be remedied before they result in a health and safety risk. It is worth noting that not all equipment will need a formal inspection to ensure its safety. In many cases, an informal quick, visual check before use will be sufficient. However, for equipment where significant risks to health and safety may arise from the incorrect installation or deterioration of parts, a formal inspection is required. The need for inspections and their frequency are determined through the risk assessments pertaining to the equipment. Some items of equipment are covered under the requirements of statutory periodic inspections as part of their associated regulations and directives, such as pressure and lifting equipment.

Where the equipment being employed is an integral part of the engineering controls for a particular process, inspections of these safety devices are important. These inspections need to be instigated at both a local level, as well as those associated with statutory requirements, as this is the opportunity where component safety parts are checked periodically to ensure they continue to function correctly. Safety devices over time may develop faults which may manifest themselves as a failure of the component to operate as originally installed, thereby negating the safety control function they were originally selected for. It may also be that the process being undertaken has evolved and that the safety device is no longer suitable for this current application.

All work equipment should be inspected if a risk assessment identifies any significant risk of injury occurring to an operator and others in the vicinity during its use. Any equipment which has exceeded its inspection period should not be used, unless there is clear evidence that an inspection has actually taken place during the prescribed time period. This would normally take the form of a colour coded label or tag attached to the equipment, often including the inspection date, which is easily identified, demonstrating that the equipment has been inspected and is valid for use. Inspections should concentrate on the safety-related parts which are necessary for the safe operation of the work equipment. In many cases, this may necessitate testing of these parts at set intervals of time.

Examples of inspections in a cryogenic workplace would include quick checks before use of equipment, such as transfer lines, to ensure that no part of the line is damaged and that the vacuum insulating jacket is intact. More regular inspections, maybe done on a weekly basis, would check the integrity of things like guards, safety devices and the general condition of trolleys, ensuring that the tyre pressures of their wheels are correct. More extensive examinations and inspections, undertaken every six months or longer, would include storage vessel inspections, checking for signs of internal corrosion or other damage, and pressure relief valve inspections, to ensure the continued operation of the safety device in protecting pressurised systems.

### 8.2.4 Health surveillance

Health surveillance is a mechanism by which the effects of exposure to specific hazards over time, which may cause harm to individuals, is routinely monitored. They are often seen as ongoing health checks and in many circumstances may be required by law for individuals who are exposed to hazards such as; frequent noise, vibrations, ionising radiation, solvents, fumes, dusts, biological agents and other substances hazardous to health. Health surveillance is critical for detecting any ill-health effects of the individuals exposed to these hazards at an early stage so that more substantive controls can be implemented to prevent these effects getting worse. In addition, they can highlight any lapses in the workplace control measures, which in turn provides useful feedback in a companies' risk assessment process. This information would then be used to provision more robust hazard controls in those particular working environment. Health surveillance also provides employers with key health data which helps to inform their evaluation of the health risks associated with the processes they undertake. Health surveillance is normally undertaken as a whole group exercise, so that any decline in health experienced by one specific individual, can be compared against the overall decline in health of the team.

Health surveillance serves to protect those employees who, due to other factors, such as existing health impairment, are at an increased risk from what they are being asked to undertake. Someone working with solvents for example with an existing respiratory illness, would be at a greater risk from the process than a healthier member of staff. As such, health surveillance can identify work-related ill health at an early stage so that critical steps can be taken to treat the condition and hopefully prevent further damage occurring. At a safety level, health surveillance can very often give an early warning that any applied protective control measures are no longer effective.

By their very nature of investigating the effects of exposure, health surveillance does not directly protect against detailed hazards in the workplace. Likewise, by having them in place, they do not reduce the need to continually manage health risks. Forming part of the particular legal requirement of any associated regulations and directives, they are completely independent to other forms of health monitoring, such as health checks. Examples of this type of monitoring would be, workplace wellbeing checks, which promote healthy living, along with fitness to work assessments, as adopted to check the suitability of operators to drive forklift trucks.

A typical example of a health surveillance would be where staff are routinely working in noisy environments. In this situation, they would be invited to an annual hearing test, the purpose of which is to quantifiably assess their hearing levels over time, looking for any reduction below that which would be expected due to the normal impairment with age. If detrimental effects are detected, additional strategies, such as the provision of additional hearing protection, or even a change of duties, may need to be considered.

For cryogenic technology users, a typical health surveillance may include a routine check of the lung capacity for staff who are regularly working with liquid cryogens, especially if these cryogens are regularly producing cold vapours as part of the process being undertaken. Clearly, pre-existing medical conditions, such asthma, would need to be factored into the health surveillance. Over time however, any reduction in the lung capacity can be assessed against not only an individual's decline due to age, but also checked against the rest of the team working in the same environment. Any findings from the surveillance, will guide specific control measures required for individuals within the team.

## 8.3 Recording risk assessments

One imperative process which should always be aligned with the application of risk control measures, is the initial preparation of risk assessments for any work activity considered to be hazardous. These assessments not only form the means by which prevalent hazards can be recognised and thereby controlled, but they also should give an indication of which elements of the hierarchy of risk controls need to be applied. By distinguishing hazard severities, high-risk activities can be mitigated by the use of these measures. The overall aim in the procedure being to reduce risks associated with any hazardous processes being undertaken. A formal, written risk assessment should be prepared for all high-risk activities in the workplace.

All formal risk assessments should have a title which clearly defines the activity being assessed. It should also be dated, so that the period corresponds with when the assessment was actually made. Generic descriptions of processes should be avoided, as this can be misleading and vague, making it difficult to apply specific control measures. The assessment needs to then include a list of any significant hazards which have been identified in association with the activity being undertaken, their related hazard severity and any current risk controls which may already be in place. The next step is to then determine the likelihood of these hazard resulting in injury to personnel or damage to equipment. From these factors, the actual risk associated with the activity can be determined.

The method by which the risk is determined can be done by one of two principal ways. The first is where the hazard severity, representing the level of injury that can occur, and the likelihood of the hazard causing harm, are categorised by the use of words and is known as a qualitative method (figure 8.1). For the hazard severity, these words are self-explanatory and would typically be; slight, serious or major, although other variations with a larger range of categories can also be used. For the likelihood, these words would relate to the chances of the hazard causing harm and would typically be; low, medium and high. These can then be used in a risk evaluation table, so that once the hazard severity and the likelihood have been identified, the risk can be plotted on the table and evaluated. The risk itself is then determined as either very low, low, medium, high or very high, depending on where the evaluation lies. For example, a hazard with only a slight severity and a low likelihood would result in a very low risk. In contrast, a hazard with major severity and highly likely to occur would result in a very high risk.

The other method uses numbers to categories the hazard severity and likelihood and is known as a quantitative method (figure 8.2). For the hazard severity these

| | **Likelihood** | | |
|---|---|---|---|
| **Hazard severity** | Low | Medium | High |
| Slight | VERY LOW | LOW | MEDIUM |
| Serious | LOW | MEDIUM | HIGH |
| Major | MEDIUM | HIGH | VERY HIGH |

**Figure 8.1.** Showing the qualitative method of risk evaluation.

| | **Likelihood** | | |
|---|---|---|---|
| **Hazard severity** | 1 | 2 | 3 |
| 1 | VERY LOW (1) | LOW (2) | MEDIUM (3) |
| 2 | LOW (2) | MEDIUM (4) | HIGH (6) |
| 3 | MEDIUM (3) | HIGH (6) | VERY HIGH (9) |

**Figure 8.2.** Showing the quantitative method of risk evaluation.

would range from a 1, which would indicate minor injury or illness, through to 3, which would indicate death or major injury occurring. Likewise, the likelihood would range from a 1, to indicate it is highly unlikely to take place, through to a 3, which would indicate something as being highly likely to occur. These can then be used on a similar risk evaluation table, with the hazard severity and the likelihood being plotted on the table in order to evaluate the risk. The risk itself is then determined by multiplying the hazard severity value with the likelihood value, the result of which relates directly to either; a very low, low, medium, high or very high risk. For example, a hazard with only a hazard severity score of 1 and a likelihood score of 1, would result in a risk value of $1 \times 1 = 1$, which equates to a very low risk. In contrast, a hazard with a severity score of 3 and likelihood score of 3, would result in a risk value of $3 \times 3 = 9$, which equates to a very high risk.

Whichever method is adopted is by individual preference and is not determined by any specific hazard or application. Some assessors for example find that by using numbers, the assessment gives a more direct value in determining the overall risk. Others find that having words classifying the hazard severity and likelihood, is a more obvious way of recognising the prevailing circumstances for the task being assessed.

One central factor to bear in mind when preparing any risk assessment paperwork, is that there can often be a long period of time after it has been initially written before it is subsequently reviewed. In fact, the review may not necessarily be undertaken by the person or persons who originally prepared the document, and therefore, the new assessor may not automatically have the intimate knowledge of the individual task processes involved. For this reason, it is critical that sufficient detail of the activity being considered is included in the assessment for anyone to understand when they read it. It is not uncommon for example where risk assessments are written by one person who subsequently leaves the employment of the company and therefore any following reviews have to be made by a completely different person.

Under many particular national safety regulations, risk assessments must be formally recorded if an organisation employs more than a prescribed number of people. However, informal risk assessments, such as on-the-job assessments for low-risk activities, do not fall into this same requirement. It is important to reflect on the fact that preparing risk assessments should not be about creating huge amounts of paperwork in order to qualify the evaluation. The formal recording of risk assessments should feed into the general process of managing health and safety within organisations. Controlling the risks in the workplace should by definition mean that improvements in safety should be achieved through the identification of sensible measures. By considering what might cause harm to people and recording these findings, organisations can decide whether they are taking reasonable steps to prevent this harm. By writing down these findings, any decisions made are defined and will help organisations decide whether they are suitably protecting their staff from these potentially hazardous situations.

## 8.4 Reviewing the risk assessments

In order to make them effective, risk assessment should be reviewed regularly so that contemporary set ups can be taken into account and also to ensure that the original risk assessment remains relevant to the work processes actually being undertaken. The frequency of these reviews very much depends on the risks associated with the task being assessed and the nature of the task. For example, the fundamental risks associated with general office based work may change very little over a long period of time, meaning that risk assessment reviews can be undertaken over a longer period than other assessments. A manufacturing activity on the other hand, may be changed over shorter timescales due to process revisions. These would typically be things such as the supply and quality variance of raw materials, the product output and quality, and any productivity improvements, such as the installation of new manufacturing equipment. In these cases, risk assessment reviews would need to be undertaken on a more regular basis, especially so where a new process or piece of equipment introduces a significant variance to the activity.

Another significant occasion where a risk assessment review needs to be made is when an accident or incident has occurred in the workplace. In such an event, the premise stands that the original risk assessment should have readily identified the individual hazard which caused the incident and that specific control measures ought to have been put in place to prevent such an event occurring. Clearly, if there has been an incident, this has not happened, and reviewing the assessment with the current knowledge of the cause of the incident will check whether or not the original risk assessment was suitable and sufficient. It may simply be that the identified control measures were not being correctly implemented, and so the risk assessment can remain in its present format, but the application of the control measures need to be reinforced in order to prevent a similar incident occurring again in the future. Where the salient hazard was not originally included in the risk assessment, this will require a more detailed review of the assessment, by which inclusion and evaluation of the hazard needs to be made. Importantly, where this is the case, the likelihood of the hazard causing harm will be based on the accident or incident which prompted the review of the risk assessment in the first place.

## Controlling cryogenic hazards—summary

*Mitigating hazards*— In order to reduce the risks associated with the use of hazardous cryogenic equipment, it is essential that particular control measures are established. By doing so, any unacceptably high risk of harm occurring is moderated by the application of these measures following a tiered, risk management procedure. This procedure is often known as the hierarchy of controls with each tier representing a differing level of mitigation. Typically these would be:

*Eliminate the hazard*— By removing the hazard, all potential effects of that hazard are negated. This method offers the greatest reduction in overall risk.

*Substituting the hazard*—Where elimination is not possible, replacing the hazard with something which is less hazardous in nature is the next alternative. It is important that overall risks are considered when substituting, as alternative solutions may not be as effective and may therefore require a longer exposure to the hazard.

*Combating the risks at source*— This is the process by which risks are reduced through the provision of engineering controls. By combating the risks at source, the hazard severity and/or the likelihood of the hazard causing harm, are effectively reduced. Generally this is implemented through the installation of guards on machinery and providing barriers and warning signs where a hazardous activity is being undertaken.

*Reducing the risk by applying safe systems of work, training and supervision*— Sitting beneath the physical methods of limiting contact with hazards, procedural controls can be implemented in order to provide some level of hazard management. These do not however guarantee that the controls are in place at all time, but offer a degree of procedural assurance if correctly adopted. Typically these procedural controls would include the implementation of safe systems of work, staff training and supervision of high risk tasks.

*Reducing the risk by using personal protective equipment (PPE)*— The final level of risk control is particular to the individual undertaking any task. Here the risks to that individual are reduced through the provision of equipment that specifically protects them from a particular hazard. This is the least effective risk control mechanism as it does not mitigate the hazard, but offers instead direct protection for the person carrying out any hazardous task. The wearing of PPE should therefore be seen as part of an overall risk control strategy employed in a work situation.

*Maintaining and monitoring the risk controls*— Many risk controls have inherent weaknesses and can be accidentally or even deliberately bypassed. It is important therefore that some form of monitoring of risk control systems is employed to ensure that they remain effective. This should be undertaken when they are originally implemented, as well as over the lifetime period when they need to function as a safety control. This monitoring can be undertaken as one of the following:

*Workplace or housekeeping inspections*— Monitoring the general health and safety standards in the workplace through safety tours and specific inspections will evaluate the effectiveness of any established risk controls. Safety tours are typically general inspections of the workplace, whilst specific inspections tend to involve the safety sampling of any particularly dangerous activity.

*Behavioural or activity audits*— This is where the actual process being undertaken is periodically assessed in order to check compliance with the original specification and to ensure that risk controls remain relevant. This type of audit is not a test of the operator or their skills and would not assess factors such as output and performance. They are there to review the hazards and the risk controls which may have an impact on the operator's safety as they undertake a specific task.

*Equipment inspections*— All equipment must be intrinsically safe and suitable for the intended purpose and must remain so over the period of its use. Deterioration due to wear and tear and other forms of damage can however cause equipment to fail. Periodic equipment inspections serve to identify whether work equipment can continue to be used safely. In many countries, certain equipment such as safety devices, require a formal statutory inspection to ensure its safety.

*Health surveillance*— This is the mechanism by which the effects of exposure to specific hazards over time is monitored. They are often seen as ongoing health checks and are often required by law for particular occupational hazards such as noise, vibration, solvents and fumes. Health surveillance is critical for detecting any ill-health effects at an early stage. They are normally undertaken as a whole group exercise, so that any decline in health by an individual can be compared against the overall decline in health of a working team.

*Recording risk assessments*— The implementation of any risk control system for a particular hazard must be supported by a risk assessment which will highlight the level of risk associated with the unmitigated hazard. Any formal risk assessment can be done either as a qualitative method, where words are used to define the hazard severity and likelihood, or a quantitative method, where numbers are used instead.

*Reviewing risk assessments*— Risk assessments must be reviewed regularly to ensure that they remain relevant. The frequency of these reviews will depend on the level of risks and the nature of the task which the assessment relates to. Risk assessments should however be immediately reviewed when an accident or incident has occurred in relation to the task.

## References


[1] The National Institute for Occupational Safety and Health (NIOSH) *Hierarchy of Controls* (www.cdc.gov/niosh/topics/hierarchy/default.html) (Accessed 2020)

[2] hsedocs *Method Statements* (www.hsedocs.com/documents/method-statements/) (Accessed 2020)

[3] Wikipedia *Standard Operating Procedures* (https://en.wikipedia.org/wiki/Standard_operating_procedure) (Accessed 2020)

[4] United Kingdom Health and Safety Executive (HSE) *Permit to Work System* (www.hse.gov.uk/comah/sragtech/techmeaspermit.htm) (Accessed 2020)

[5] BOC *Alarmed Oxygen Level Monitor Systems* (www.boconline.co.uk/en/products-and-supply/cryogenic-equipment/gas-monitors-repeaters/oxygen-monitoring/oxygen-monitoring.html) (Accessed 2020)

# Chapter 9

# Asphyxiation hazards

## 9.1 Asphyxiation hazards general

As a general rule, there is a relatively healthy perception of the workplace risks allied to the hazards associated with the use of cryogenic technology. In fact, hundreds of hours of use of the technology can be undertaken without incidents ever occurring and the systemic awareness of the potential dangers are often manifest with the devout respect they deserve. In particular, facilities where liquid cryogens are stored or used, are commonly provided with such control measure as oxygen monitors in order to protect against low oxygen risks and the appropriate personnel protective equipment (PPE) to mitigate accidental splashes of any cold cryogen liquids. Alongside these measures, safe systems of work are routinely employed, often including details of the necessary training which staff should undertake before attempting cryogenic related work.

All of these procedures are commendable in providing a benign environment to work in, controlling the prevalent risks to acceptable levels where it is deemed safe for personnel to perform their cryogenic duties. As in any workplace though, accidents do happen, and thankfully for the cryogenic industries, the majority are judged as only superficial, causing no serious or permanent injury to the staff working in this particular setting. However, this is not to say that serious incidents never occur in these essentially hazardous locations. A quick scrutiny of related accident records will reveal the facts that injuries are routinely sustained by staff undertaking cryogenic activities, and in certain cases, these incidents are so severe that they result in fatalities. Critically, there is substantial evidence to show that asphyxiation due to the creation of an oxygen depleted atmosphere (ODA) is by far the biggest cause of fatalities in these working environments. Each year, an average of 20 deaths occur in Europe involving people entering an ODA and between 1992 and 2002 in the USA, 85 nitrogen asphyxiation incidents were recorded, in which 80 people were killed [1].

doi:10.1088/978-0-7503-3123-4ch9  

The dictionary definition of asphyxiation is the effect on the body of inadequate oxygen, usually resulting in loss of consciousness or death [2]. The presence of reduced levels of oxygen in the air is an extremely effective phenomena in the cause of work related deaths, literally an invisible hazard, which cannot be perceived by the human senses, but one that is preventable if the correct risk controls are fulfilled. Most cryogenic facilities apply some form of safe systems in order to reduce the likelihood of asphyxiation occurring, as well as providing monitoring system and alarms to give warning when oxygen levels fall below certain concentrations in the workplace. It is surprising therefore to find reports indicating that in spite of suitable training, the provision of safety information, and the correct implementation of risk controls, serious accidents continue to occur, resulting in asphyxiation when using cryogenic liquids [3].

## 9.2 Oxygen

In order to understand the mechanism by which asphyxiation occurs in the human body, it is important to first understand how the oxygen in the air around us is so critical for our very existence. Oxygen is the only gas that supports life on Earth and there is approximately 21% concentration of the gas in the air we breathe, with the majority of the air actually consisting of nitrogen. After hydrogen and helium, oxygen is the third most abundant element in the Universe by mass. In order for it to function in our bodies, as we breathe in the oxygen it is diffused through thin membranes in the lungs, transferring eventually into the red blood cells coursing through our bodies. This blood in turn carries the oxygen to the various cells of our bodies, where it undergoes a chemical reaction with glucose to provide energy to the cells. This overall chemical reaction is called respiration [4], and happens in the body continuously in order for the cells to receive the energy they require. A typical adult human at rest inhales upwards of 2.0 grams of oxygen per minute [5], repeating the process unceasingly, over and over. If there is not enough oxygen in the air however, our bodies will not be able to undertake the respiration process and irrefutably, we will eventually die as a result.

Before this extreme level is reached however, cognitive skills, such as our thinking and decision making, are impaired when the concentration of oxygen in the air falls only slightly below the 21% value. Unfortunately, these effects are not noticeable to the affected individual, as the body's senses, under guidance of the brain, are unable to detect this lowering of the oxygen levels. Through a simple chemical reaction though, the brain automatically starts to reduce its vital functions as it is slowly starved of the important oxygen it requires. This shutting down of the brain's functions has the effect of making the affected individual unaware throughout of what is happening around them. To an observer however, they are likely to display significantly erratic characteristics, clearly appearing to be disorientated and unable to perform even the simplest of tasks when challenged to do so. Outwardly, they would appear to show similar traits to someone intoxicated due to the effects of excessive alcohol.

On this evidence alone, certain national legislation determines that the working environment is considered too hazardous for workers to be in as soon as the oxygen concentration value inhaled is less than 18%. Below this value, there is a significant risk of a sufficient impairment of cognitive skills occurring which will prevent rational thought in order for individuals to make the correct decision, especially where they need to undertake an escape from the rarefied environment. Taken to an absolute extreme, with no oxygen present at all, inhalation of only one or two breaths of the remaining gases in the atmosphere will cause sudden loss of consciousness and can ultimately result in death.

The trigger to this fatal mechanism and the preceding loss of cognitive skills is that the body's drive to breathe is solely in response to the carbon dioxide levels in the blood which would be high if insufficient oxygen is gained with each inhalation. As the human brain does not possess any actual mechanism to detect low oxygen levels, the body will simply respond to the increased carbon dioxide levels in the blood and will attempt to replenish the body with oxygen. This prompts the lungs to function more rapidly in order to gain the sufficient oxygen intake, increasing the rate of breathing as a means of quickly drawing in oxygen to negate the build-up of carbon dioxide. This cardinal necessity of the body to obtain oxygen causes this natural reflex response, compelling the lungs to breath, with the process continuing until sufficient oxygen is gained and the carbon dioxide levels in the blood return to normal. This can be readily seen when any form of exercise is undertaken, thereby using the oxygen, converting it into a power release in the muscles. As such, the body quickly replenishes this energy to the cells by increasing the inhalation rate to gain more oxygen than would normally be required if the body were at rest.

If however, there remains insufficient oxygen gained by each breath, such as that experienced in a low oxygen environment, the body automatically responds further by gradually shutting down brain functions to preserve primary attributes, like breathing and the continuous pumping of blood through the heart. At this stage, disorientation starts to take place, as major characteristics of the body, such as sight, balance and rational thought are impaired by the slowly constricting role of the brain. Ultimately, an unconscious state is experienced as the brain closes more and more of these functions as it gradually becomes starved of oxygen.

This fundamental inability of the human body to detect low oxygen levels is the primary reason why it is imperative to fit oxygen monitors in all areas where cryogenic processes are being undertaken, especially where these processes could create an oxygen depleted atmosphere. Whatever the situation, if there is a risk that the working atmosphere can be affected by the release of inert gases, some form of automatic protection needs to be provided to activate safety controls and to alert all occupants in that location of the potential risks. Low oxygen levels are an invisible facet, appearing no different to a normal environment. In fact, it would only be an individual's sudden difficulty to breath that may give warning to the prevalent dangers of working in an oxygen depleted atmosphere. All other appearances of the workplace would be impossible to distinguish from that of any standard oxygen content in the air.

## 9.3 Causes of oxygen depleted atmospheres

Whilst oxygen depleted atmospheres can be the result of a variety of causes, such as the accidental release of pressurised inert gases into the workplace or the release of gases due to a chemical reaction, there are some very specific reasons why they occur when associated with the use of cryogenic technology. Principally, wherever extremely cold liquid cryogens are being utilised, there is the potential for these liquefied gases to convert back into their natural gaseous state under room temperature conditions. These gases, if released in sufficient quantity, can become a dominant gas in the surrounding air, effectively reducing the oxygen concentration in the atmosphere as they mix with the existing quantities already there. If this release takes place in a relatively small enclosed room with poor ventilation, this effect is further compounded to the point where quite rapidly the resultant oxygen concentration level will become dangerously low. The following key factors which could cause an oxygen depleted atmosphere can be considered as individual risks and assessed accordingly, but should importantly be taken in combination as this is where the overall greater risk is apparent.

### 9.3.1 Evaporation of liquefied gas

Even a small amount of liquid cryogen on release into the working environment will quickly warm up and convert into a significantly large amount of vapour. The addition of this expanding gas mixing into the surrounding air will automatically dilute the localised oxygen concentration value of any given work area. The more commonly used liquid cryogens all have extremely large expansion ratios when they convert from their cold liquid form into their gaseous phase. This fact alone makes it likely that relatively small quantities of cryogen evaporation can dramatically reduce the oxygen levels of the working space around them. Putting this in perspective, a 50 litre container of liquid nitrogen, which is approximately the size of an average large suitcase in volume, would expand on evaporation to upwards of 35 $m^3$. This would be enough to completely fill a small sized bedroom in a typical modern house. In such a situation, the nitrogen gas would be all pervading and any residual oxygen in the room would be diluted down to a dangerously unsafe concentration value.

### 9.3.2 Confined spaces

Confined spaces in the cryogenic workplace occur wherever the working environment is designed in such a way to prevent sufficient dispersal of any evaporated liquefied gas being released from sources such as storage Dewars. Under these circumstances, the resultant lack of oxygen can be profound and more importantly, can often take place rapidly, as there will generally be no process by which the additional gases can be naturally circulated from the workplace.

A confined space is by definition a bounded volume which is substantially enclosed, thereby making the introduction of ventilated air difficult to achieve [6]. Most confined spaces are not generally designed for continuous occupancy and would generally tend to have restrictive means of entry. A typical example would be the inside of a large storage vessel which would have an entry port large enough for a person to enter in order to undertake any necessary inspections or maintenance work. Under normal use, any such storage vessel would not be intended for human occupancy. Generally, this type of confined space would be classified in such a way that a permit system would be required for any entry into the internal working space. By adopting this method, a certain level of assurance is made prior to entry, specifically requiring oxygen levels to be monitored before individuals are allowed access.

However, even normally occupied work areas such as short corridors and small laboratory rooms, especially if they only have a single entry door and limited means of ventilation, would also be considered as potential confined spaces. This is especially so if during their use a detrimental change in the oxygen levels is brought about due to specific activities or processes being undertaken within them. This change could be introduced due to a negative change in the ventilation which provides fresh air into the work area or because of a process, such as decanting liquid cryogens, which could result in the increased release of evaporated gases.

Generally, any work space containing hazardous substances such as liquid cryogens, which under the correct conditions can result in the release of large amounts of gases, should be considered as a confined space. Even if the work space only ever has the potential for these hazards to be present, they must be considered as a worse case situation where the hazardous substances released could engulf an individual working in the area. Likewise, even if the process which can cause the release of gases is only undertaken infrequently within that workplace, it should still be considered to be a confined space during the undertaking of that particular activity.

Any situation where oxygen can be accidentally removed or depleted from a work area also falls under the remit of confined spaces. In most facilities this would only come about due to the accidental release of gases into the workplace atmosphere, which would by default replace the original air present. There are also conditions where a vacuum is deliberately created in large test chambers, pumping out the air from within them. Some of these chambers can be large enough for one, if not more, people to undertake work within them under normal atmospheric conditions. Under vacuum conditions however, oxygen levels would be quickly taken below the 18% concentration value where it is deemed hazardous to be working in that environment.

In addition to the asphyxiation hazards of working in confined spaces, additional hazards such as heat and noise need to be considered as these will have some bearing on the safety controls put in place to make working in a confined space as safe as possible. National regulations controlling work activities undertaken in confined spaces are in place in many countries throughout the world and as such are a legal requirement. As a result, many of these will require a confined spaces entry permit [7] to be in place before entry into the confined space can be made. In addition, there would be a requirement for a written programme of the planned work to be established in advance of any entry permit being issued. This programme of work would have to include details of the designated entrant and any supporting attendants, along with the means of rescue should an injured person become trapped within the confined space.

### 9.3.3 Inadequate ventilation

Inadequate ventilation is where the working environment does not have a continuous replenishment of air to both disperse any released gases and to maintain the necessary oxygen concentration value at safe levels. If the work area has poor ventilation, with no constant supply of fresh air, this will increase the likelihood of an oxygen depleted atmosphere being created. Generally this can be easily remedied with the introduction of local exhaust ventilation, which effectively forces a sufficient quantity of air over a given time, through any particular workplace. This process ensures that the oxygen contained within the ambient air is continuously replenished and any excess gases extracted.

On a cautionary note however, it should not be assumed that by simply adding forced ventilation, that all oxygen depleted atmosphere problems are eliminated. Even with what appears to be good ventilation, localised dead-zones can be created whereby air being drawn into a work area is exhausted from the same work area in such a way that pockets of undisturbed air will remain. The oxygen levels within these pockets over time will gradually reduce, creating an acute oxygen depleted atmosphere in what initially appears to be a well-ventilated space. This is normally caused by having the inlet and outlet of any ventilation system positioned on the same facing wall so that cross movement of air within the working space is effectively prevented (figure 9.1).

### 9.3.4 Spills from open Dewars

Wherever processes are being undertaken with the use of open type Dewars filled with liquid cryogens, there is a greater likelihood of spills occurring within the cryogenic workplace. These spills will rapidly convert to gas, and if sufficient in quantity, can bring about the displacement of the oxygen in the surrounding air and thereby generate an oxygen depleted atmosphere. Typically, this type of Dewar is only used for the transport of relatively small volumes of liquid cryogens and would generally be fitted with some form of retaining lid in order to mitigate the likelihood of these spills during their movement. Even small volumes of cryogens will however

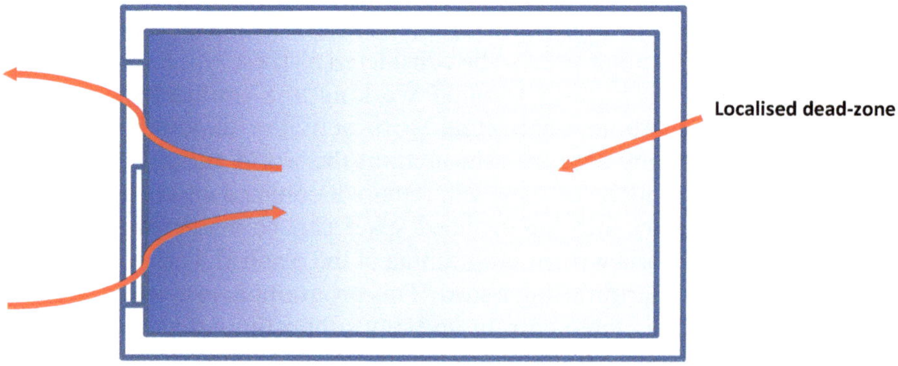


**Figure 9.1.** Showing localised dead-zone where poor ventilation is in place.

convert into a significantly larger amount of gas and procedures to ensure these lids are correctly fitted during transportation should be adopted as normal practice.

Spills from open type Dewars are more likely to occur though during the filling and emptying processes where the lid will for obvious reasons have to be removed. In fact, the overall workplace risks during filling and emptying of any Dewar are higher than normal as a variety of technical aspects could go wrong during the respective processes. Examples would be where the Dewar is accidentally tipped over during filling or emptying, thereby spilling the entire contents, or the Dewar is unintentionally overfilled, wherein the excess quantity will overflow and spill onto the neighbouring floor. Any such releases will subsequently evaporate and contribute this additional gas to the working environment. This will have the effect of producing a rarefied atmosphere immediately in the vicinity of the Dewar.

Filling is typically done by decanting the cryogens from a much larger storage vessel or a supply pipeline, during which losses of the cold liquid are inevitable as they are briefly exposed by the process to the full thermal energy of the surrounding room temperature. This causes the instant evaporation of some of the cryogen, often manifesting itself as a vapour cloud around the Dewar. If the Dewar itself is at room temperature at the start of the decanting process, this will magnify the quantity of boiled-off cryogenic liquid, adding further evaporated gas into the workplace atmosphere.

Likewise, any eventual emptying of the filled Dewar into its intended location is also likely to result in accidental spills and therefore produce further evaporation of any exposed cryogens. Often simply poured into any receiving container, the risks of unintended discharge of the cryogenic liquid from this process are high. Where possible, purpose designed dispensing trolleys and metal funnels should be used to reduce the potential for spillage to occur. It is also worth bearing in mind that where repeated small quantities are being dispensed by this method, each spill and subsequent evaporation causes an accumulation of the vaporised gas into the workplace atmosphere. Even though they may be considered insignificant taken singularly, when combined, they may have an accumulative effect on reducing the oxygen content of the working environment.

### 9.3.5 Leaks from pipework

Where cryogenic liquids are being transferred through purpose-built pipeline systems, leaks from any such assembly should, by association, be reduced to almost zero. Inevitably though, unless the pipeline is a fully-welded structure, flanged joints will be required as part of the construction, especially if the pipeline is of significant length and located at several levels within a facility. As a consequence, each of these joints has the potential to leak over time, in particular where the joint may be subject to a repeated thermal cycle of cryogenic cooling and reheating as the cryogen is routinely passed along the pipeline. The type of sealant system employed for these joints will also determine the integrity of their construction, more so as the effects of this sequence of temperature extremes is experienced by them over a period of time.

The results of any leaks of cryogens from joints in pipework assemblies would in principle be visible as a water vapour plume forming around the leak. This is due to the cold leaking cryogen coming into direct contact with the warm surrounding air and condensing the water out of the atmosphere. The cryogen will also be converting into its gaseous form at this point and in some circumstances, ice will form around the leak. Any gas formed from this evaporating cryogen will disperse into the working environment and in doing so, will displace the oxygen already there. If a leak is sufficiently large, or is undetected for a long period, this gas formed from the leaking cryogen will accumulate and could lower the oxygen concentration value for the work area to a dangerously low level.

Most cryogenic pipelines will have some form of insulation around them in order to reduce the heating effects of the ambient surrounding temperature. Where leaks occur underneath this insulation, they may go unnoticed for long periods, allowing large quantities of cryogenic liquids to leak and convert into gas. One indicator that a leak may be present is the accumulation of ice or water on the insulation in the region where the leak is apparent. This will allow for further investigations to be made in order to locate and repair any seepages. Where there is no obvious signs of anything untoward, leaks of cryogens, which in turn convert to their natural gaseous state, can be a serious hazard if undetected. Even over relatively short time scales, the formed gas could become sufficient in quantity to create an oxygen depleted atmosphere without any evident warning that an asphyxiant is present in the workplace.

Some pipeline systems forming part of a cryogenic assembly will be designed specifically to transfer gases converted from cryogenic liquids. An example of this would be in hospitals where liquid oxygen is stored externally as a means of providing a constant supply of oxygen gas to the various wards. Most gases are invisible and odourless and therefore leaks from systems containing inert gases can quite easily liberate sufficient quantities into the surrounding work area to produce an oxygen depleted atmosphere.

### 9.3.6 Poorly connected hoses

Where the transfer of liquid cryogens is undertaken using transfer lines and hoses, the key feature of any such system is that all associated hose or flange connectors are correctly assembled so that leaks are eliminated. Connection points are the most vulnerable component in any such transferral system and depend on features such as locking threads being correctly fastened in place. The correct assembly of these parts should ensure a sealed system if all the components have been maintained in full working order. Where misalignment of parts takes place however, the incorrect assembly will inevitable create a leak path for cryogens and gases to escape. Assembly of room temperature hose connections can be undertaken with relative ease. When these same connections have to be made under cryogenic conditions and the operator has to wear personal protective equipment (PPE), such as insulated gloves, mistakes can quite easily happen during the assembly process and leaks can result from poorly fitted parts.

Where leaks result from incorrectly connected hoses and transfer lines, the size and eventual quantity of the leak will depend on the misalignment of the parts and the period of time the leak is active. It is conceivable under normal working conditions for a transfer hose to be initially connected and then left unattended until the eventual disconnection needs to be made some time later. During this period, if a leak is present, the surrounding atmosphere will gradually fill with any prevailing gas evaporating from the leaking cryogen. Unless detected by oxygen level sensors, the associated workplace could quite easily become rarefied and represent a serious risk to anyone working in there.

### 9.3.7 Valves left open

Any valve, which is an integral part of a liquid cryogen system, is so designed to provide flow of the cryogen when required. This can be as a means of moving cryogens from one storage vessel to another or simply to flow the cryogens through interconnecting pipework. Where these valves provide an outlet from any such system so that cryogens can be transferred into smaller transport containers for example, these outlets are by their very nature a potential source of leaks of the cryogens. Any significant leaks can result in large quantities of gas being evaporated from the liquid cryogen, which can in turn displace the oxygen in the working environment. As the oxygen concentration value drops to dangerous levels, the air in the workplace becomes an oxygen depleted atmosphere, making it unsafe for personnel to be working there.

Where outlet valves are accidentally left open after use, or incorrectly closed so that a small trace leak through the valve remains apparent, over a period of time large quantities of cryogens can be unintentionally dispensed. As it warms up, this cryogenic liquid will immediately evaporate into its gaseous form, quickly filling any surrounding volume and causing a rarefied air in the adjacent workplace. A valve left fully open would in theory be easy to detect as liquid cryogens would be seen issuing from it. However, where the outlet valve is only part of a sequence of valves needed to dispense the cryogens, operators may not be aware of the valve being in its open state until much later, by which time vast amounts of cryogens may have already poured through the valve. It is wise therefore to have some form of method statement or a checklist when operating cryogenic systems which have a series of valves integral to their assembly. By having these in place, the correct sequence of valve operations can be determined and checks made that outlet valves in particular are in their closed state before the remainder of the system is filled with cryogens.

### 9.3.8 Venting from relief valves or bursting discs

The inclusion of relief valves and bursting discs on cryogenic systems are generally stipulated as a national legal requirement in order to protect any pressurised equipment from failing due to excessive pressures occurring in them during normal use. Where the containing fluid is liquid cryogens, pressure build-up can easily occur due to the natural tendency of the cryogens to warm up as external heat migrates into them over time and gas is evaporated from the liquid. If left unmitigated, this gas can accumulate in sufficient quantities within any containment system to cause an excess of pressure above the liquid surface. By allowing this pressure to be vented through a relief valve or a bursting disc, the equipment is adequately protected from failure occurring as the pressure cannot exceed a defined value set by these devices.

Where a bursting disc operates due to such a build-up of pressure, there is no means of resealing the device. This will effectively release any excess pressure as it is designed to do, but will then continue to vent further gases evolving from the cryogenic liquids until natural evaporation has boiled off all the contained cryogen. Releasing this gas into the working environment will have the result of lowering the oxygen concentration value of the working atmosphere. Where sufficient gas is released, this value can become dangerously low and personnel would need to evacuate the area as soon as possible. Bursting discs tend to operate with a resounding bang, followed by a rapid release of the evaporated gas which will be cold enough to condense water vapour from the air, forming into a cloud or plume around the cryogenic system. Both the audible sound of the bursting disc functioning and the visible water vapour should be sufficient to warn anyone working in the vicinity that they need to be aware of the potential for an oxygen depleted atmosphere being formed adjacent to the cryogenic equipment.

Relief valves are similar to bursting discs in that they vent excess pressure from any contained system. They function as a result of the build-up of pressure pushing against the force of an internal spring which under normal operating conditions holds a sealing plug in place. When the pressure is sufficiently high, the spring is forced backwards, which in turn lifts the sealing plug, thereby allowing a small amount of the excess liquid or gas to be vented from the pressure system. This release will effectively lower the pressure of the system until eventually it is low enough for the spring force to reseal the plug. This process can happen repeatedly as the build-up in pressure periodically lifts the sealing plug and vents sufficient pressure, with the valve then automatically resealing itself as the contained pressure returns to a lower value.

Where this constant venting takes place due to the evaporation of gas from liquid cryogens in storage, these small amounts of gas being released can over time amount to significant quantities. This is made more apparent when multiple cryogen storage containers are housed in one area, with each container potentially venting small amounts of gas over any given period. Added together however, and these seemingly trivial amounts can be sufficient to displace the oxygen in the surrounding air and thereby create an oxygen depleted atmosphere. As the quantity of gas being released from each relief valve will tend to be small, it is conceivable that the actual operating of the valve will go unnoticed, especially if the ambient noise is sufficient to mask the tiny sound of each one venting. More importantly though, if the storage containers are left unattended for any length of time, there is no means of reconciling how much gas has been released into the atmosphere, unless oxygen level monitors are employed.

### 9.3.9 Discharges routed through a confined space (basements, closed laboratories or drainage systems)

Any form of discharged gas from a cryogenic system can over time create an oxygen depleted atmosphere if it is released in sufficient quantities. Discharges can be in the form of venting from the protective systems such as relief valves and bursting disc, or from any specific outlet valve or port which forms part of the overall cryogenic system. Often, relief valves and bursting discs are installed in such a way that they are built as an in-line component, so that any vented gas does not accumulate in the immediate work area, but is diverted through pipework to a more suitable location. Likewise, where localised extract systems are employed in order to reduce the likelihood of a build-up of gases during processes such as decanting, the extracted gas will typically be transferred to the outside or a more suitable place, well away from the workplace.

It is important however, that any such discharge route does not in itself create a hazard, especially where the discharge may not necessarily feed to an external location. If the discharge is into a secondary internal location for example, the atmosphere in there could become rarefied if sufficient quantities of gas are fed into it. This is more so if the discharge setting is also defined as a confined space, especially if ventilation is already inadequate in that locality. However, the actual discharge pipework carrying the gas can also be considered a risk wherever it is routed through an internal setting. Each joint or welded section of the discharge pipe has the potential to leak and any unintended seepage from these sources might go unnoticed.

The types of places which would be prone to suffering an asphyxiation risk if discharged gas is feed directly into them, or if the discharge pipework leaks along its route, would be basement rooms, closed laboratories with sealed enclosures, and drainage wells, pits and trenches. In fact, anywhere that can trap pockets of gas in sufficient concentration to displace the natural oxygen in the air should not have any discharged gas routed to or through it.

## 9.4 Ventilation

For all applications of cryogenic technology, adequate cross-ventilation must be provided in the workplace, especially where cryogenic liquids are being used or stored. The basic requirements of this form of ventilation is that an entrance and an exit for airflow must be present in the working environment, allowing the movement of air between these two locations. In its simplest form, this could be as basic as having two windows open in a room, provided that they are facing in different directions. For many liquid cryogen applications, the air inside the workplace needs to be replenished regularly in order to prevent oxygen levels becoming dangerously low. Without a supply of new air, the gradual build-up of gases released from cryogen storage vessels can become sufficiently high to dilute the oxygen already present in the air. Cross-ventilation provides both the source of new air and the means by which outgoing air can escape. Air will simply be forced in through the entrance route by external winds and convection currents, thereby expelling air already inside the workplace via the exist route (figure 9.2). The position and size of any window, doors and vents provided in the workplace, as well as the path between the different sides of the room where they are located, have significant effect on how well cross-ventilation occurs. It is insufficient for example to have good inlet and outlet vents if the workplace then effectively blocks airflow due to the positioning of equipment, furniture or storage containers within the work area. The rate by which a quantity of air is replaced in any defined workplace volume is normally presented as the number of complete air changes of that volume over a given time period.

As a guidance, for rooms above ground level, with no special air circulation openings, the normal ventilation present will only provide typically one complete air change per hour of the room's gas volume. For an identical room with well-sealed windows, such as double glazing, the ventilation provided here will be notably less and may only provide between 0.7 and 0.9 air changes per hour. Further down the scale, basement rooms with limited window ventilation, would only average around 0.4 air changes per hour [8]. Each of these values can vary from one location to another and are dependent on a variety of factors, such as the layout of the workplace and the external prevailing wind speeds. Therefore, designing ventilation systems which rely on natural movement of air within the working environment need to factor in many aspects of a building's overall dynamics before any specific room can be individually assessed. This would normally necessitate an on-site airflow test to confirm the prevalent conditions and to define the actual air changes per hour for that particular situation and arrangement. Conversely, forced ventilation through the use of devices such as extract fans, can provide significantly larger air changes per hour to the working environment with some level of assurance.

Forced ventilation in the cryogenic workplace needs to be sufficient for controlling any potential exposure to the inert gases being released from cryogen storage vessels. It must also provide the dilution mechanism whereby any build-up of these gases is controlled through the introduction of fresh quantities of air. In this way, dangerously low levels of oxygen can be prevented. Controlling the release of these

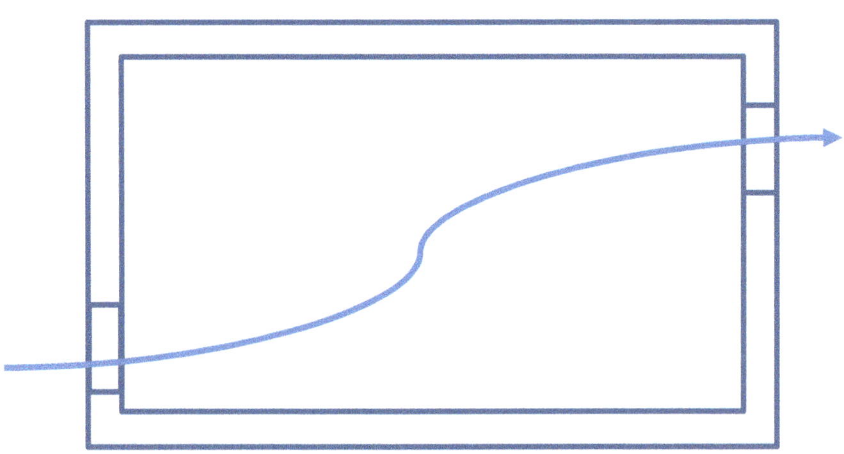

**Figure 9.2.** Showing an example of cross-ventilation through a defined volume.

evaporating gases at source is generally limited as the natural process of cryogens warming up over time will automatically create releases of excess gases. Consideration must therefore be given to the use of oxygen level sensors and alarm systems in areas where the normal ventilation is deemed insufficient for such an event. For general guidance, the following are suggested air change rates per hour for workplace environments (figure 9.3) [9].

For the general handling of cryogens in locations at or above ground level, natural ventilation, typically provided through permanently open vents, is normally deemed sufficient. This is the case, as long as the room is large enough. Many facilities typically construct indoor rooms which are fitted with louvered openings, as well as wall and door vents, where the handling of cryogenic liquids can be managed safely. The recommended guidance is that the total ventilated area should be at least 5% of the floor area of the room in order to provide sufficient dilution of any released gases [10].

Clearly, improving ventilation rates will reduce the likelihood of a build-up of inert gases which could have the effect of reducing the available oxygen in the working environment. However, ventilation systems cannot take into account all possible cryogenic workplace operations, especially where the release of gases may be deemed higher than originally anticipated for a particular process being undertaken. Operations such as decanting from one storage vessel to another, where the likelihood of spills occurring are higher, are an example where momentarily the ventilation system employed may be insufficient to quickly disperse any excess amounts of gas being released into the workplace. It may even be possible to create localised oxygen depleted atmospheres in these situations, even though the ventilation provided to the room has been calculated to be adequate for general applications.

Providing active sensing of air quality to the workplace is one solution to this problem, where any temporary drop in oxygen concentration value would automatically trigger an increase in the ventilation rate. This would resolve the variability of ventilation rate requirements, especially where there are unknowns relating to the hazards being used in the workplace. In the decanting situation for example, where low oxygen levels are sensed, the standard air change rate would be automatically increased proportionally by the ventilation system in order to purge the working environment of the contaminating gases and to bring in new air at the same time. Likewise, when the workplace is running in normal mode, the ventilation rate would self-regulate to a level sufficient to provide the minimum requirement based on the

| **Air change rates per hour** | |
|---|---|
| Basements | 3 to 4 |
| Offices | 6 to 8 |
| Laboratories | 6 to 12 |
| Workshops | 10 |

**Figure 9.3.** Showing suggested air change rates for hour for workplace environments.

oxygen levels being detected by the sensors. Active sensing can be considered as an engineering control which mitigates the risks of oxygen depleted atmospheres being produced.

To emphasis, the human body cannot detect the oxygen levels in the air. It therefore follows that any area which has the potential of becoming an oxygen depleted atmosphere must be correctly fitted with alarmed oxygen monitors. In conjunction with this, the occupants of the work area must be made aware of what their response should be if the alarms are sounded. This could be through regular emergency drills, where the alarms are deliberately tested and any evacuation procedure followed accordingly. The worst situation to be encountered is where all the correct alarms are fitted in a particular work area and yet those working in there are ignorant of what the alarms sounds like and what they have to do in the event of the alarms sounding due to low oxygen levels.

The correct positioning of fixed sensors is equally important in order to capture the highest risk situation. The work area should be reviewed in detail, checking for locations such as trenches or ducts where individuals may occasionally have to work and where heavier than air gases could migrate towards in the event of a release. Awareness of the work area layout needs to also be considered, in particular how workstation dividers are positioned, as these could cause the flow of released gases to collect locally in areas which are not ventilated as well as the rest of the workroom. Ideally, sensors should be located in positions determined by those who have the intimate knowledge of the gases being employed in the work area and how they are likely to disperse, along with the process systems and the equipment being used in that work area. In general, for gases lighter than air, the sensors should be above the area where leaks are most likely. For gases heavier than air, the sensors must be at floor level or in inspection pits or ducts into which the heavy gas may naturally flow. It is worth noting here, that even lighter than air gases will be extremely cold on their initial release from their liquid form. This will make them temporarily heavier that the air until they have gained sufficient heat to attain their natural buoyancy.

Ventilation should even be considered for outdoor areas, even though at first sight the natural flow of air through the work area would appear adequate. Any such location must not be enclosed by walls as these could effectively function as bunds, trapping any collected vapour released during operations such as cryogen transfers. Even with natural ventilation helping to disperse any released gases from any outdoor areas surrounded by walls, there would remain the risk of localised oxygen depleted atmospheres being formed.

It is also worth considering the consequences of spillages of cryogens in any such outdoor areas, as the release of extremely cold liquids onto the infrastructure work surfaces or even into normal domestic drains, could have severe effects. Most modern drainage systems are built from plastic, a material which is adequate for most general applications, but specifically not suitable for cryogenic liquids to be running through. It may therefore be prudent to have any external drainage systems repositioned so that they are well away from any external work area where there is likely to be a spillage of these extremely cold liquids.

## 9.5 Risk assessment

As a general requirement of any risk management strategy, the purpose of preparing a risk assessment is to instigate a proactive method of improving the health and safety in the workplace. By doing so, this will allow actions to be taken in order to prevent anyone getting hurt when they are undertaking their particular work activities. Risk management can therefore be considered as a method by which hazards are systematically identified and likely injuries from those hazards are suitably prevented. By definition, a hazard is something that can inherently cause harm. The harm in this context includes not only possible injuries but also other types of harm, such as diseases or even damage to equipment. To be most effective, any risk assessment undertaken should be task specific, taking into account not only the process being undertaken but also who is involved in doing this work. The aim of the assessment is to reduce the overall risk of harm occurring by the implementation of apposite control measures. The aim is not to prevent certain activities from being enacted, but to provide protection to those involved in the activity so that they are not at risk of harm.

Often perceived as bureaucratic in nature, risk assessments should be seen as a trenchant method of managing the health and safety within an organisation, rather than a paper-exercise which simply ticks the boxes against a formal checklist of requirements. The process should enable managers to look critically at what takes place within their sphere of influence and to modify processes in order to make the workplace a safer place to be in.

In practical terms, any cryogenic related work area which has the potential to create an oxygen depleted atmosphere should have a risk assessment prepared for it so that the latent hazard severity can be sufficiently determined and risk control measures applied if the overall risk is considered too high. This would specifically include workplace environments defined as confined spaces, areas with insufficient ventilation, places where cryogenic liquids are stored, and all areas where the processing of cryogenic liquids, such as decanting, are periodically undertaken. A risk assessment needs to be made for all the types of cryogenic work carried out in a work area involving the use of liquid cryogens, no matter how infrequent. Even if a particular activity is only undertaken once a year, the overall risks associated with that task needs to be assessed. The risk is determined by considering the likelihood of an oxygen depleted atmosphere being created by the release of inert gases into the workplace, and the consequence if it is.

Any such risk assessments should include oxygen concentration calculations for two principal situations:

- *The normal evaporation of cryogens*—Where the consequential gas released from all the cryogens stored in a particular work area is assessed in order to determine if an oxygen depleted atmosphere is created by the gas.
- *The sudden release of cryogens from a storage vessel*—Where the sudden release of the entire cryogen content of the largest storage vessel in that same work area is assessed in order to determine if an oxygen depleted atmosphere is created by the gas. Typically, this latter release represents a worst case

situation where there has been the complete failure of the largest storage vessel due to it being ruptured or having been accidentally toppled over.

### 9.5.1 Normal evaporation of cryogens

In this situation, the overall quantity of gases being constantly released from any stored cryogens is considered. This gas build-up will come about through the normal evaporation of the stored liquid cryogens over time as they progressively warm up and will become manifest through the constant opening of pressure relief valves on the storage vessels as they vent any excess pressure into the surrounding workplace.

To calculate the remaining oxygen concentration relating to this normal evaporation of the stored cryogens, the first step is to calculate the actual concentration of gas released from the storage vessels over a given period of time. In equation (9.1), this is represented as $C$ and is derived from the volume of cryogen gas released $L$ (Volume/Time), divided by the multiplication of the workplace volume $V$ (Volume) and the air changes per hour, which is represented as $n$ (No units). This air change value will clearly depend on the presence of natural or forced ventilation into the work area and how well a room is sealed or provided with cross-ventilation.

$$\text{Gas concentration} = C = \frac{L}{V \times n} \tag{9.1}$$

where

$L$ = Gas release ($m^3$/h);
$V$ = Room volume ($m^3$); and
$n$ = Air changes per hour.

In order to derive the gas concentration value $C$, equation (9.2) is used to first calculate the cryogen gas released $L$. This is done by multiplying the number of storage vessels in the work area $N$, the *expansion ratio* of the particular cryogen being subsequently released as gas and the normal evaporation rate of the cryogen for any particular storage vessel $V_r$. This total is then divided by the *time period* over which the gas is being released. $V_r$ is often known as the Dewar manufacturer quoted evaporation rate and is typically around 2.5% of the liquid capacity per day for new storage vessels and would normally be marked on the Dewar or supplied as part of the technical specification documentation [11]. It is prudent however to assume deterioration of the vessel insulation over time by doubling this quoted value to 5% of the liquid capacity per day for older, well-used storage vessels. Obviously, this will result in a higher overall gas release value being calculated, but should credibly reflect the amount of gas actually being released. The factor of **1000** in the equation is used when the volume of release is expressed in litres.

$$\text{Gas released} = L = \frac{N \times V_r \times \text{Expansion ratio}}{\text{Time period} \times 1000} \tag{9.2}$$

where

$N$ = Number of Dewars;

$V_r$ = Volume released per Dewar—Dewar quoted evaporation rate. It is prudent to double this value to allow for the deterioration of the Dewar insulation over time.

Calculating the volume of the work area room is done by using equation (9.3). Here, the primary work area dimensions $V_w$ are derived by multiplying the height, width and length. Subtracted from this value is any large volume of equipment housed in the room, $V_i$, as this represents lost air content in the room for the purposes of the calculation. This could include storage cupboards, pump assemblies or even the cryogen storage Dewars themselves if they are deemed large enough to be considered.

If the stored cryogens being assessed are dominantly liquid nitrogen, a further correction to the volume of the work area equation needs to be applied. As any released nitrogen gas will be heavier than air, it will progressively fill the room from the ground upwards, especially in situations where the movement of the air content in the room is sufficiently low. This will initially create a band of nitrogen rich air at ground level, which will as a consequence have a very low oxygen concentration value. As more nitrogen gas is progressively released, this layer will eventually become deeper, until it has created an oxygen depleted atmosphere all the way up to head height, and more specifically, up to the height of the mouth and nose of any individuals standing upright in the storage area. If the nitrogen gas extends above this point, these individuals will be at an immediate risk of suffering asphyxiation due to breathing in the low levels of oxygen remaining in the resultant nitrogen-air mixture.

On average, the nose and mouth of a typical person will only be approximately 1.8 metres from the ground. Therefore, as soon as the nitrogen gas layer has reached this height, the risks of asphyxiation becomes magnified and breathing in the nitrogen rich air is likely to result in individuals fainting due to the lack of oxygen. For this reason, any air above 2.0 metres for the purposes of the room volume calculation is considered unavailable to the majority of people as they would not normally be able to make use of it in the event of a liquid nitrogen gas build-up.

As helium gas is lighter than air, the contrary applies for a room containing predominantly liquid helium storage vessels, in that any released gas will float upwards and begin to fill the room from its highest point downwards. The full height of the room should therefore be included in the volume calculation for the work area as the air above 2.0 metres remains available until the oxygen depleted atmosphere created by the released gas eventually descends to the height at which the nose and mouth of the people inside the room are positioned. In fact, in the event of an individual initially becoming unconscious due to the helium gas causing low levels of oxygen above them, they are likely to collapse back into a layer of air which should in theory have sufficient oxygen content remaining for them to make an immediate recovery and crawl away to safety.

$$V = V_w - V_i \tag{9.3}$$

where

$V_w$ = Volume of workplace (maximum height of 2 m for liquid nitrogen which is a little greater than a typical person's height);

$V_i$ = Volume of objects/equipment within the workplace.

Once the gas concentration value $C$ has been calculated, the remaining oxygen concentration value is derived by using equation (9.4). This will be determined as a percentage of the air in the work area room and needs to be at least 18% in order to satisfy the safe use of the room over the period of time which the calculations have been based upon. If the resultant percentage is below this limiting value, figure 9.4 [12] is used to reconcile the severity of the hazard in what will be an oxygen reduced environment.

$$\text{Oxygen concentration} = \frac{21}{(100 + C)} \times 100 \tag{9.4}$$

where $C$ = Gas concentration.

The way in which the gases produced from cryogenic liquids are diluted into the surrounding air can be likened to the way in which an artist would add certain

| | |
|---|---|
| 18% - 19.5% | May affect physical and intellectual performance without person's knowledge. |
| 15% - 18% | Decreased ability to work strenuously. May impair co-ordination and may induce symptoms in persons with coronary, pulmonary, or circulatory problems. |
| 12% - 15% | Respiration deeper, increased pulse rate, and impaired coordination, perception and judgement. |
| 10% - 12% | Further increase in rate and depth of respiration, further increase in pulse rate, performance failure, giddiness, poor judgement, blue lips. |
| 8% - 10% | Mental failure, nausea, vomiting, fainting, ashen face, blue lips. |
| 6% - 8% | Loss of consciousness within a few minutes, resuscitation possible if carried out immediately. |
| 0% - 6% | Loss of consciousness almost immediately, death ensues, brain damage, even if rescued. |

**Figure 9.4.** Showing the effects on the human body in reduced oxygen concentration atmospheres.

pigments to their palette in order to change the base colour they are working with. For example, if the artist were to add a very small amount of white to red paint, this would change the tone very little and in principle the main intensity of the red paint would remain intact. Adding a larger amount of white paint to the red however, will change the hue completely, transforming it into a shade of pink, a tone completely different to the original colour. Likewise, if a small amount of cryogen evaporated gas is released into the surrounding air in any particular working environment, it will change the oxygen content very little. It would be deemed safe to continue working in that environment even though the small amount of additional gases have been released. If however, a larger amount of cryogen evaporate gas is released into the same working area, the makeup of the air is dramatically changed and the oxygen concentration will be significantly altered. Under these conditions it would be considered unsafe to continue working in this environment.

### 9.5.2 Sudden release of cryogens from a storage vessel

For the situation where a sudden release of a large quantity of cryogen evaporated gas has occurred, equation (9.5) is used directly to calculate the volume of oxygen $V_o$ remaining after the rapid discharge. $V$ in the equation is the room volume already used in the normal evaporation calculations, corrected depending on whether the cryogen evaporated gas is lighter or heavier than air. Subtracted from this value is the amount of cryogen evaporate gas released by any spillage or rupture of the entire contents of a Dewar. This is calculated using $V_D$, which is the capacity of the largest storage vessel in the working area, making the assumption that in the event of a sudden release, the worst case would be when the largest vessel provides that quantity of liquid cryogen which in turn converts into gas. This volume is multiplied by $f_g$ which is the gas factor (expansion ratio) pertaining to the liquid cryogen being used. The resultant value gives the amount of gas released as the liquid cryogen immediately warms up and expands into its gaseous state. This value is divided by **1000** which expresses the volume of release in litres.

$$\text{Volume of oxygen} = V_o = 0.21\left[V - \left(\frac{V_D \times f_g}{1000}\right)\right] \tag{9.5}$$

where

$V_D$ = Capacity of largest Dewar (litres);
$V$ = Room volume ($m^3$);
$f_g$ = Gas factor, i.e. 710 for nitrogen and 780 for helium.

To calculate the resultant oxygen concentration $C_o$ in any particular work area following a sudden release of cryogenic liquid, equation (9.6) is used. This is basically the ratio of the remaining oxygen calculated in equation (9.5) and the original volume of the work area. Expressed as a percentage, the oxygen concentration chart (figure 9.4) can then be used to determine the hazard severity and hence the risk of working in an environment with the resultant oxygen concentration value.

$$\text{Oxygen concentration} = C_o = \frac{100 \times V_o}{V} \tag{9.6}$$

As a cautionary warning when calculating the oxygen concentration values for both the normal evaporation and sudden release situations, it becomes clear that they provide only an indication of the absolute release of gases into the environment and therefore do not represent the reality of the circumstances. Both sets of equations make the assumption that there is a certain level of airflow in the room being assessed and that the releases have distributed themselves uniformly around that room to the point where the assessment is being made. In the case of the sudden release in particular, there is a further supposition that the total quantity of evaporated gas has instantly converted from the liquid state and has equally rapidly dispersed and mixed with the air throughout the enclosed volume of the room. In fact, neither of the equations take into account the time-dependent flow, mixing and dispersion of the released gas with the existing surrounding air. In reality, there is likely to be associated factors in operation over a given timeframe where only partial evaporation has taken place of any released cryogen. Under these conditions, oxygen levels may well remain sufficiently high for a reasonable amount of time in the majority of the room and only very localised zones would be subject to a true oxygen depleted atmosphere. During this interval, it is conceivable for an evacuation of the affected work area to be made, for extra ventilation to be introduced or other safety controls to be ostensibly provided.

The easiest way to imagine this is to consider a long tunnel, sealed at one end, at which point there is located a cryogen storage vessel. In the event of a sudden release of the entire contents of this storage vessel, directly following this occurrence, an oxygen depleted atmosphere is highly likely to be created immediately in the vicinity of the storage vessel. Initially, the remainder of the tunnel will be unaffected as the gas has not as yet permeated along its length. This region of the tunnel will have sufficient residual oxygen in the air to provide personnel ample time to evacuate safely from the tunnel on seeing the sudden release occurring. Obviously, as time progresses, more of the cryogen from the storage vessel will be converted into gas and this, along with the gas already evaporated, will begin to mix with the surrounding air further along the tunnel. If it were possible to see the cryogen gas as a different colour to the surrounding air, over time, a coloured band would be seen steadily progressing the length of the tunnel from the storage vessel location, all the way to the opposite end (figure 9.5).

In reality, as both the oxygen within the air and any cryogen gas released are invisible to the human eye, a series of oxygen monitors positioned at equal distances along the length of the tunnel would effectively give a representative image of this occurring. As the released gas mixes with the ambient air and thereby locally reduces the oxygen concentration, the nearest oxygen monitor would register this drop in value. As this oxygen depleted cloud progresses, more of the oxygen monitors further down the tunnel would indicate a dangerously low oxygen concentration value. Eventually, all the monitors would display a reading below the 18% value as the entire tunnel gradually becomes full of the released cryogen gas.

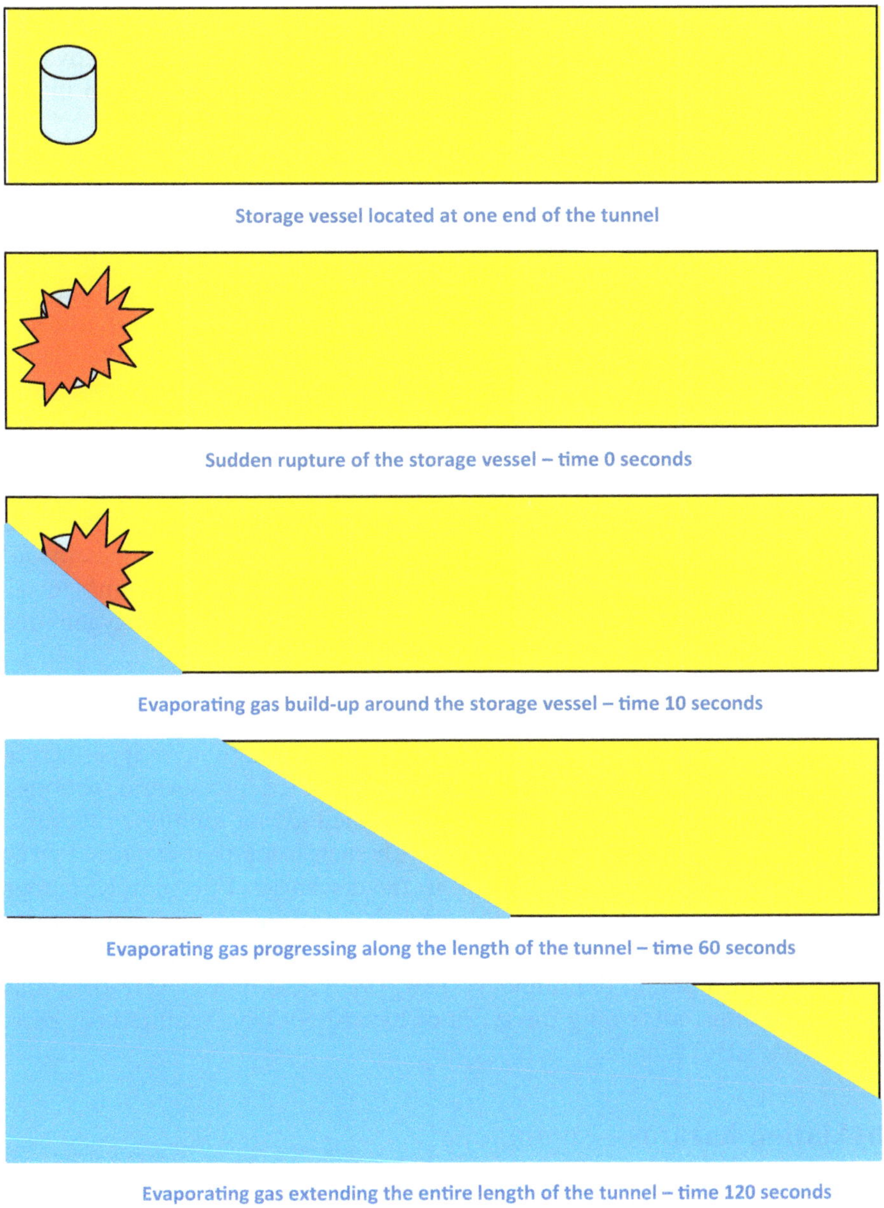


**Figure 9.5.** Showing the different phases of released gas permeating along a tunnel.

This entire process will take time to complete and during the initial period of gas release, there will still exist regions of the tunnel which have sufficient oxygen content in the air. This will allow two things to happen. The first, is the immediate evacuation of personnel working in the unaffected region of the tunnel, who will be able to safety breathe throughout as the oxygen concentration value of the air they inhale will still be above the critical value of 18%. The second, is that control measures, such as the operating of emergency ventilation or gas extraction from the tunnel can be implemented. By triggering these measures, the removal of the released inert gas from the tunnel can be achieved in a timely fashion in order to reinstate the necessary oxygen concentration level to a safe value.

Whilst rare in occurrence within most cryogenics facilities, there are recorded cases where cryogen spills within significantly long tunnels have taken place and as a result additional risk management control measures have been implemented at these locations. At the Fermi National Accelerator Laboratory in the USA [13], a defined moderate helium spill took place in 2003 along the unmanned 6.3 Km tunnel system of the Tevatron circular particle accelerator [14]. Data from oxygen monitors along the length of the tunnel provided evidence of the progression of the evaporated helium gas as it mixed with the ambient air. From this information, it was establish that the lowest recorded oxygen concentration level throughout the duration of the spill was 19.5%. It was also recognised that at this value automatic exhaust fans did not activate as they were set to initiate at 18.0%. In all, it took the helium gas nearly 5 hours to travel around the complete tunnel system. As a result of the spillage, a series of gas diffusion tests were undertaken on long, cylindrical pipes in order to better understand how released cryogen gases would mix with the ambient air [15, 16]. The results from the findings were then used to inform oxygen depleted atmosphere mitigations for personnel working in the tunnels.

However, probably the most noteworthy cryogen spill within a tunnel took place at the European Organization for Nuclear Research (CERN) [17] during 2008, in their Large Hadron Collider (LHC) facility [18]. Due to a magnet quenching incident, upwards of 6 tonnes of liquid helium was vented into several sections of the tunnel which house the LHC. Apart from producing vast amounts of evaporated helium, the escaping vapour was released with such force that it caused structural damage to certain component parts local to the discharge. Fortunately, recognising the fact that the LHC contains upwards of 100 tonnes of liquid helium, a series of scale model tests had been previously undertaken in order to understand the likely helium propagation within the tunnel system [19]. These results were then scaled to the LHC conditions, informing the types of hazard controls required for personnel working in the LHC tunnels.

## Asphyxiation hazards—summary

- *General*—Asphyxiation is the effect on the body of inadequate oxygen, usually resulting in loss of consciousness or death. Asphyxiation due to the creation of an oxygen depleted atmosphere (ODA) is by far the biggest cause of fatalities in the cryogenic working environment.
- *Oxygen*—Oxygen is the only gas that supports life and there is approximately 21% concentration of the gas in the air we breathe. The working environment is considered too hazardous to be working in as soon as the oxygen concentration value is less than 18%. Below this value, there is a significant risk of the impairment of cognitive skills occurring which will prevent rational thought.

With no oxygen present, inhalation of one or two breaths will cause sudden loss of consciousness and ultimately can result in death.

- *Causes of oxygen depleted atmospheres*—There are generally specific reasons why oxygen depleted atmospheres are created when associated with the use of cryogenic technology. Wherever liquid cryogens are being utilised, there is the potential for these liquefied gases to convert back to their gaseous state. If released in sufficient quantity, these gases can become the dominant gas in the surrounding air, displacing the oxygen to dangerously low levels.
- *Evaporation of liquefied gas*—The commonly used liquid cryogens all have significantly large expansion ratios when they convert from their liquid into their gaseous phase. Even small amounts of liquid cryogens on release will quickly warm up and convert into a large amount of vapour. The addition of this expanding gas mixing into the surrounding air, will dilute the oxygen concentration value of any given work area.
- *Confined spaces*—A confined space is defined as a bounded volume which is substantially enclosed, thereby making the introduction of ventilated air difficult to achieve. Confined spaces can be created in the cryogenic workplace wherever the working environment is designed in such a way that it prevents sufficient dispersal of any liquefied gas being released.
- *Inadequate ventilation*—Inadequate ventilation in the cryogenic workplace is defined as an environment that does not have a continuous replenishment of air to both disperse any released gases and to maintain the necessary oxygen concentration value at safe levels. Ventilation can be improved by adding forced ventilation, although if not correctly installed, localised dead-zones can be created whereby air being drawn into the work area is exhausted in such a way that pockets of undisturbed air will remain.
- *Spills from open Dewars*—There is a greater likelihood of cryogen spills occurring where open type Dewars are used, especially during the filling and emptying processes. These spills will rapidly convert to gas, and if sufficient in quantity, can bring about the displacement of the oxygen in the surrounding air.
- *Leaks from pipework*—Where cryogenic liquids are being transferred through pipeline systems, each joint, valve or outlet has the potential to leak, especially where they are subject to repeated thermal cycles of cryogenic cooling and reheating. Where pipelines are insulated, leaks can go undetected, allowing large quantities of cryogenic liquids to be discharged and convert into gas.
- *Poorly connected hoses*—Connection points in any cryogen transfer systems are the most vulnerable in terms of eliminating potential leaks. The correct assembly of parts should ensure a sealed system, whereas misaligned parts can create leak paths for cryogens and gases to escape. Where leaks occur, the surrounding atmosphere will gradually fill with any discharging gas evaporating from the system, creating a rarefied air and representing a serious risk to anyone working near to the leak.
- *Valves left open*—Valves are designed to provide flow of liquid cryogens when required. Where they provide an outlet from a cryogen system, they are by their very nature a potential source of leaks if not correctly closed. Any significant discharges from a valve left open can result in large quantities of gas being evaporated into the working environment.
- *Venting from relief valves or bursting discs*—Relief valves and bursting discs are provided in order to protect cryogen systems from potential over pressurisation. When a bursting disc operates, there is no means of resealing the device and it

will continue to vent evaporated gas. In contrast, a relief valve functions by any excess pressure pushing against the force of an internal spring and seal in order to vent. The relief valve will therefore reseal as soon as the excess pressure has been vented. Evaporated gas venting through both types of device will displace the oxygen in the surrounding air.

- *Discharges routed through a confined space*—Any form of discharged gas from cryogen systems can over time create an oxygen depleted atmosphere. Often these discharges are therefore extracted by forced ventilation away from the working area. These discharge systems can however leak and if routed through confined spaces can represent a more significant hazard. Typical locations to avoid routing these discharges through would be basement rooms, closed laboratories, drainage wells, pits and trenches.
- *Ventilation*—For all applications of cryogenic technology, adequate cross-ventilation must be provided in the workplace. Entrance and exit airflow must be present in order to replenish the air to prevent oxygen levels becoming dangerously low. Typically, rooms above ground level with no special air circulation openings, will provide one complete air change per hour. Forced ventilation will provide higher values in order to dilute any released gases evaporated from the use of cryogens.
- *Risk assessment*—Any cryogenic work area which has the potential to create an oxygen depleted atmosphere should have a risk assessment prepared for it so that suitable risk control measures can be applied. Any such risk assessment should include oxygen concentration calculations for two principal situations:
  - The normal evaporation and consequential gas released from all the cryogens stored in a particular work area.
  - The sudden release of the entire cryogen content of the largest storage vessel in that same work area.

## References

[1] Chemical Safety and hazard investigation Board (CSB) *Safety Bulletin on Nitrogen Asphyxiation Hazards* (www.csb.gov/board-releases-safety-bulletin-on-nitrogen-asphyxiation-hazards-underscores-need-for-good-safety-practices/) (Accessed 2020)

[2] Merriam-Webster.com *Asphyxiation* (www.merriam-webster.com/dictionary/asphyxiation) (Accessed 2020)

[3] Asia Industrial Gases Association *Hazards of Inert Gases and Oxygen Depletion—AIGA* 008/11 (www.asiaiga.org/uploaded_docs/aiga%20008_11_hazards%20of%20inert%20gases%20and%20oxygen%20depletion.pdf) (Accessed 2020)

[4] Martini F H 2004Fundamentals of Anatomy and Physiology 6th edn (New York: Benjamin Cummings)

[5] LumenLearning.com *Uses of Oxygen* (https://courses.lumenlearning.com/introchem/chapter/uses-of-oxygen/) (Accessed 2020)

[6] The United Kingdom Health and Safety Executive (HSE) *Confined Spaces* (www.hse.gov.uk/confinedspace/) (Accessed 2020)

[7] Occupational Safety and Health Administration (OSHA) *Confined Spaces Entry Permit* (www.osha.gov/sites/default/files/2018-12/fy11_sh-22318-11_Mod_6_CSPermit.pdf) (Accessed 2020)

[8] J Few J, Allinson D and Elwell C *Airtightness and Non-uniformity of Ventilation Rates in a Naturally Ventilated Building with Trickle Vents* (www.aivc.org/sites/default/files/D1_S4C-03_0.pdf) (Accessed 2020)
[9] Wikipedia *Air Changes per Hour* (https://en.wikipedia.org/wiki/Air_changes_per_hour) (Accessed 2020)
[10] Planningportal *Building Regulations: Ventilation* (www.planningportal.co.uk/info/200130/common_projects/17/extensions/7) (Accessed 2020)
[11] Technifab.com *Normal Evaporation Rate (NER) for Dewars* (https://technifab.com/normal-evaporation-rate-ner-for-dewars/) (Accessed 2020)
[12] NorthWest Occupational Health and Safety North Vancouver, Canada *Physiological and Toxicological Effects of Oxygen Deficiency and Enrichment—Neil McManus—(2009)* (www.nwohs.com/Oxygen%20Regulatory%20Limits%20I.pdf) (Accessed 2020)
[13] *The Fermi National Accelerator Laboratory (Fermilab)* (www.fnal.gov/) (Accessed 2020)
[14] Fermi National Accelerator Laboratory (Fermilab) *The Tevatron Circular Particle Accelerator* (https://fnal.gov/pub/tevatron/tevatron-accelerator.html) (Accessed 2020)
[15] Theilacker J C and White M J 2005 Diffusion of gases in air and its affect (sic) on oxygen deficiency hazard abatement *Fermilab Cryogenics Conference* (https://lss.fnal.gov/archive/2005/conf/fermilab-conf-05-635-ad.pdf) (Accessed 2021)
[16] Soyars W M and Schiller J L 2005 Fermilab open channel helium flow during rupture event *Cryogenics Conference* (https://lss.fnal.gov/archive/2002/pub/fermilab-pub-02-424-ad.pdf) (Accessed 2021)
[17] The European Organization for Nuclear Research (CERN) (https://home.cern/) (Accessed 2020)
[18] The European Organization for Nuclear Research (CERN) *The Large Hadron Collider (LHC) Facility* (https://home.cern/science/accelerators/large-hadron-collider (Accessed 2020)
[19] Chorowski M, Fydrych J, Konopka-Cupial G and Riddone G 2003 Summary of the experimental studies of cold helium propagation along a scale model of the LHC tunnel *CEC/ICMC* (www.researchgate.net/publication/42854103_Summary_of_the_Experimental_Studies_of_Cold_Helium_Propagation_along_a_Scale_Model_of_the_LHC_Tunnel) (Accessed 2021)

**IOP** Publishing

The Safe Use of Cryogenic Technologies

A handbook for best practice and training

**Robert Done**

# Chapter 10

# Case studies

## 10.1 Worked examples

As a means of putting into perspective the method of calculating the oxygen concentration values for particular work areas, the following case studies are provided as a constructive insight into the detailed process. It is important to recognise however that these calculations should form part of an integrated risk assessment which needs to be prepared for the work area if cryogenic technology is being used within it. This would include identifying all appropriate hazards and implementing control measures and strategies in order to reduce the overall risks associated with them. Clearly, where liquid cryogens are being stored or used in the workplace environment, the associated asphyxiation hazard will be the dominant factor in this assessment. This is not to say that other prevailing hazards, such as the extremely low temperatures involved and the general hazards pertinent to the workplace should be given any less credence. A risk assessment of the working environment must consider all germane hazards in association with their likelihood of causing an incident, no matter how trivial or insignificant they initially may seem.

### 10.1.1 Case study 1

This case study looks at a hypothetical situation where a temporary liquid cryogen storage facility has to be established in a university lecture theatre. As this represents a change of use of the room, a thorough assessment of the risks which this will generate will need to be undertaken. Part of this assessment will be the evaluation of the oxygen concentration levels in the lecture theatre during the period when the liquid cryogens are being temporarily stored there. The aim of the risk assessment is to identify the likelihood that the oxygen concentration levels can become dangerously low for given circumstances and what control measures need to be implemented to make this workplace safe. Figure 10.1 shows the proposed storage vessel layout in the university lecture theatre during the planned period of use.

doi:10.1088/978-0-7503-3123-4ch10  

**Storage vessel information**

| Cryogen | Quantity | Size | Age | Evaporation rate |
|---|---|---|---|---|
| Helium | 1 | 1000 L | New | 25.0 L/day |
| Helium | 5 | 100 L | New | 2.5 L/day |
| Nitrogen | 2 | 1000 L | 10 years | 25.0 L/day |
| Nitrogen | 10 | 100 L | 10 years | 2.5 L/day |

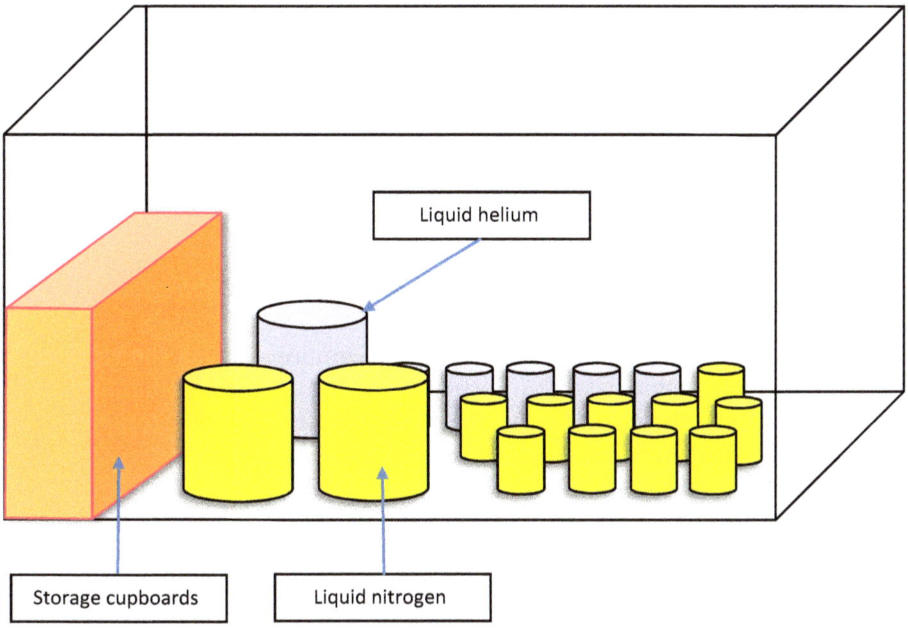


**Figure 10.1.** Showing the proposed storage vessel layout within the lecture theatre.

*Specification information*

- Due to major construction work being undertaken on the campus, a United Kingdom university lecture theatre has to be temporarily used as a liquid cryogen storage area in order to provide continued supplies of cryogens to the research laboratories on the campus.
- The lecture theatre will be commandeered for this purpose over the period 1st March to the 31st May.
- Filling of the storage vessels will be undertaken immediately outside the lecture theatre's rear door where there is a small paved area adjacent to a service road where the supply tanker can be parked for the duration of the transfers.

- There will be routine decanting of the cryogens within the lecture theatre from the storage vessels into smaller transportable Dewars.
- The lecture theatre has the overall floor dimensions of 20 m × 25 m.
- The lecture theatre is 3.5 m high.
- There are a set of large storage cupboards at the far end of the lecture theatre taking up approximately 5% of the room's volume.
- The lecture theatre has been selected as it is located on the ground floor, thereby negating the need of transporting the large storage Dewars in personnel lifts after they have been filled.
- The lecture theatre has recently been refurbished with double glazing windows installed throughout.
- The lecture theatre has air-conditioning, although this is currently only working on low flow rate due to a fault on the main fan motor.
- The proposal is to have up to 17 cryogen storage vessels in the lecture theatre at any one time.
- These storage vessels will be filled with either liquid helium or liquid nitrogen.
- The liquid helium storage vessel have only recently been purchased.
- The liquid nitrogen storage vessels were bought at least 10 years ago.

**Storage vessel information**

*Calculation 1—Considering the oxygen levels in the lecture theatre following normal evaporation of the cryogens*

The first calculation undertaken makes the assumption that all the storage vessels are fully charged and that standard atmospheric conditions apply within the University lecture theatre. Under these conditions, the contents of each of the storage vessels will warm up over time and in doing so, progressively release small amounts of gas from the liquid phase. This will build-up the pressure inside the storage vessels which in turn is automatically controlled by spring operated relief valves which will periodically release a small amount of this generated gas into the surrounding air in the lecture theatre. Over a period of time, this gas will accumulate, and if the air is not replenished, the accumulation could become a hazard for anyone working in the temporary cryogen storage area. It is noted that the air-conditioning in the lecture theatre is currently only operating on a low flow setting and this may result in some higher than normal temperatures existing in the lecture theatre. This would marginally accelerate the normal evaporation rates for each of the storage vessels. However, as the temporary use of the lecture theatre is during the early part of the year when the weather conditions are generally mild, it is not envisaged that temperature extremes will be experienced and these can be effectively ignored from the calculations.

If a situation ever existed where the storage vessels are likely to experience higher than ambient temperatures, the quoted evaporation rate per vessel should be increased in order to take into account the higher quantities of gas which would be evaporated. Doubling these values would adequately compensate for any temperature effect experienced by the storage vessels. An example where this may occur is where the storage room has other functioning equipment, such as vacuum pumps and water-baths, each producing a quantity of heat from their operation.

Collectively, this generated heat may be sufficient to raise the ambient temperature, which in turn will accelerate the evaporation rate of the stored cryogens.

Equation (10.1): The first equation to be considered is that for determining the total gas released from the entire storage vessels over a given time period:

$$\text{Gas released} = L = \frac{N \times V_r \times \text{Expansion ratio}}{\text{Time period} \times 1000}$$

Where:

The number of liquid helium storage vessels of a particular size is multiplied by the relevant evaporation rate and the expansion ratio of 780 for helium.

The number of liquid nitrogen storage vessels of a particular size is multiplied by the relevant evaporation rate and the expansion ratio of 710 for nitrogen. As these vessels are all 10 years old, the evaporation rate has been doubled to take into account any likely degradation of the vessel insulation over that duration.

The manufacturer's evaporation rates are quoted in litres per day, so the equation is divided by 24 to obtain an answer based on an hourly gas release rate.

Therefore:

$$L = \frac{(1 \times 25 \times 780) + (5 \times 2.5 \times 780) + (2 \times (2 \times 25) \times 750) + (10 \times (2 \times 2.5) \times 710)}{24 \times 1000}$$

$$L = 5.66 {}^{m^3}\!/_{h}$$

Equation (10.2): The next calculation considers the volume of the room, assuming initially that only **helium** gas would be released over the defined time period:

Where:

$$V = V_w - V_i$$

The principal dimensions of the lecture theatre are included in the calculation, taking into account the full height of the room in this version as the evaporated helium gas would be lighter than air and would preferentially fill the available space from the top downwards. In the lecture theatre specification, it was defined that 5% of the available space is occupied by a set of storage cupboards. The lecture theatre volume can therefore simply be multiplied by 95% in order to determine the actual volume of air existing prior to any evaporated gas being released.

Therefore:

$$V = (3.5 \times 20.0 \times 25.0) \times 0.95$$

$$V = 1662.5\ \text{m}^3$$

The lecture theatre specification also defines several indicators for determining the total air changes per hour moving through the room. The first is that it is on the ground floor, which normally provides at least 1.0 complete air change per hour due to natural ventilation. Sealed, double glazing has however recently been added, which would lower this value to something in the region of 0.7 air changes per hour. Unless there are safeguards where windows and doors in the lecture theatre are fixed in their open position this value is a realistic rate. There is however air-conditioning in the lecture theatre which would add to the quantity of gas being exchanged every hour, albeit this is currently set at a low flow rate. The assumption made for the purpose of this case study therefore is, that at its best there would be no greater than 0.8 complete air changes per hour for the lecture theatre.

Therefore:

$$n = 0.8$$

Equation (10.3): Using the same volume equation, the next calculation considers the volume of the room, assuming this time that only *nitrogen* gas would be released over the defined time period:

$$V = V_w - V_i$$

The same principal dimensions of the lecture room are included in the calculation, only this time the height of the room is limited to 2.0 metres, as the evaporated nitrogen gas would be heavier than air and would preferentially fill the available space from the bottom upwards, which would make the air above the 2.0 metres height unavailable for breathing. The same 5% of the available space occupied by the set of storage cupboards needs to be subtracted from this overall volume to determine the actual volume of air existing prior to any evaporated gas being released.

$$V = (2.0 \times 20.0 \times 25.0) \times 0.95$$

$$V = 950.0 \text{ m}^3$$

The same air changes rate as previously determined of no greater than 0.8 complete air changes per hour for the lecture theatre can again be considered for this room.

$$n = 0.8$$

Equation (10.4): The next calculation, which is for the gas concentration released into the lecture theatre, needs to be based on the smaller of the two room volumes calculated. This is because liquid nitrogen is one of the cryogens stored in the lecture theatre and therefore this poses a greater risk of causing an oxygen depleted atmosphere from the ground upwards to the limiting height of 2.0 metres. This value is multiplied by 100 to represent it as a percentage of the total room volume.

$$C = \frac{L}{V \times n}$$

$$C = \frac{5.66}{950.0 \times 0.8} \times 100$$

$$C = 0.74\ \%$$

The helium/nitrogen concentration of the room has therefore been increased by 0.64%. Since the normal oxygen content is approximately 21%, the resultant oxygen in the lecture theatre after this gradual release over time can now be calculated using the next equation.

$$\text{Oxygen concentration} = \left(\frac{21.0}{100 + 0.74}\right) \times 100 = 20.8\%$$

This value is then referenced to the hazard severity chart (figure 9.4) which lists the various levels at which the lack of oxygen in an atmosphere becomes an issue. As the reduction in oxygen content in the lecture theatre due to the gradual release of gas from the storage vessels is small, there is a negligible consequence to anyone working in the environment. At 20.8%, the oxygen concentration remains above the 18.0% to 19.5% range where some impairment of physical and mental performance begins to take place.

| | |
|---|---|
| 18% – 19.5% | May affect physical and intellectual performance without person's knowledge. |

It would therefore be deemed safe for University staff to work in the lecture theatre for this consideration, even when all the storage vessels are full. Under these conditions, the provision of additional forced ventilation is not required. As the room has sealed double glazing installed, it may however be prudent to have a certain number of windows permanently locked in their open position by way of providing additional natural ventilation into the room.

*Calculation 2—considering the oxygen levels in the lecture theatre due to the sudden release of helium*

The next calculation makes the assumption that there has been a sudden, catastrophic failure of the largest liquid helium storage vessel and that its entire contents are evaporated into the lecture theatre. This works on the basis that the storage vessel has been damaged sufficiently to release all of the cryogen into the available volume of the room. Given the temporary nature of the lecture theatre being used as a storage facility, it is conceivable that immediately after the filling of the storage vessel outside of the lecture room, on wheeling it back inside, some event could occur which causes the damage. The likeliest scenario, is that the storage vessel is accidentally tipped over and on contact with the ground, ruptures both the insulating jacket around it and the inner containment vessel wall. This would not only liberate a large amount of liquid cryogen from the vessel, but would also introduce an enormous amount of heat into the remaining cryogen still contained within the vessel. Both the leaked cryogen and the quantity still inside the vessel would rapidly boil-off, filling the lecture theatre with a cloud of helium gas. As this cloud is lighter than air, initially it will begin to fill the room from the ceiling downwards, gradually descending to a critical height where individuals still within the room would be at risk of breathing in the concentrated gas.

Equation (10.5): The initial calculation determines the volume of oxygen within the room when this sudden release has taken place. It uses the following equation:

$$\text{Volume of oxygen} = V_o = 0.21\left[V - \left(\frac{V_D \times f_g}{1000}\right)\right]$$

Where:

$V$ is the effective volume of the room calculated in equation (10.2), which is specific to the release of liquid helium. In this case 1662.5 m$^3$.

$V_D$ is the largest liquid helium storage vessel size, which in this case is 1000 litres.

$f_g$ is the expansion ratio when the liquid helium converts into gas which is 780.

$$V_o = 0.21\left(1662.5 - \left(\frac{1000 \times 780}{1000}\right)\right)$$

$$V_o = 185.3 \text{ m}^2$$

Equation (10.6): This volume of oxygen is now converted into the oxygen concentration value, expressed as a percentage of the total gas within the room. This oxygen concentration value is then referenced to the hazard severity chart (figure 9.4) and the appropriate level determined:

$$\text{Oxygen concentration} = C_o = \frac{100 \times V_O}{V}$$

$$C_o = \frac{100 \times 185.3}{1662.5} = 11.1\,\%$$

| | |
|---|---|
| 10% – 12% | Further increase in rate and depth of respiration, further increase in pulse rate, performance failure, giddiness, poor judgement, blue lips. |

As can be seen, a sudden release of this amount of liquid helium into the lecture theatre, with the rate of ventilation already determined as 0.8 air changes per hour, will result in a significant drop in the oxygen concentration within the room. The calculated 11.1% value is well below the recognised 18% when action should be taken to evacuate the enclosed area. Anyone remaining in this oxygen depleted atmosphere would be at an extreme risk of suffering asphyxiation. Being within the 10% to 12% range indicates that individuals would already be suffering the severe effects of the reduced oxygen levels and as a result would be unlikely be able to make the correct judgements in order to escape the room. They may even at this stage have lost the use of their limbs as the process of the brain being starved of vital oxygen would trigger a shutdown of nonessential body functions. In this situation, even if the individual were conscious of the urgent need to evacuate the dangerous environment, they would be unlikely to have the capability of doing so.

It would therefore be deemed extremely unsafe for University staff to be in the lecture theatre if the largest liquid helium storage vessel catastrophically failed. Under these conditions, the provision of additional forced ventilation is essential and, ideally, should be provided by having an automatic system installed which immediately increases the air changes per hour if there is a reduction in the oxygen concentration below the 18% limit.

*Considering the oxygen levels in the lecture room due to the sudden release of nitrogen*

This calculation makes the assumption that there has been a sudden, catastrophic failure of the largest liquid nitrogen storage vessel and that its entire contents are evaporated into the lecture theatre. As considered with the failure of the liquid helium storage vessel, this works on the basis that the liquid nitrogen storage vessel has been damaged sufficiently to release all of the cryogen into the lecture theatre over a given time frame. Once again the scenario considers what would happen if immediately after the filling of the storage vessel outside of the lecture theatre, an event occurs which causes damage to both the insulating jacket around it and the inner containment vessel wall. This would not only liberate a large amount of liquid nitrogen from the vessel, but would also introduce an enormous amount of heat into the remaining cryogen still contained within the vessel. Both the leaked cryogen and the quantity still inside the vessel would rapidly boil-off, filling the lecture theatre with a cloud of nitrogen gas. As this cloud is heavier than air, it will immediately begin to fill the room from the floor upwards, gradually ascending to a critical height where individuals still within the room would be at risk of breathing in the concentrated gas.

Equation (10.7): The initial calculation determines the volume of oxygen within the room when this sudden release has taken place. It uses the following equation:

$$\text{Volume of oxygen} = V_o = \left[V - \left(\frac{V_D \times f_g}{1000}\right)\right]$$

Where:

$V$ is the effective volume of the room calculated in equation (10.3), which is specific to the release of liquid nitrogen, limiting the height of the lecture room to 2.0 metres. In this case 950.0 $m^3$.

$V_D$ is the largest liquid nitrogen storage vessel size, which in this case is 1000 litres.

$f_g$ is the expansion ratio when the liquid nitrogen converts into gas which is 710.

$$V_o = 0.21\left[950.0 - \left(\frac{1000 \times 710}{1000}\right)\right]$$

$$V_o = 50.4\ \text{m}^2$$

Equation (10.8): This volume of oxygen is now converted into the oxygen concentration, expressed as a percentage of the total gas within the room. This oxygen concentration value is then referenced to the hazard severity chart (figure 9.4) and the appropriate level determined:

$$\text{Oxygen concentration} = C_o = \frac{100 \times V_o}{V}$$

$$C_o = \frac{100 \times 50.4}{950.0} = 5.3\,\%$$

0% – 6% Loss of consciousness almost immediately, death ensues, brain damage, even if rescused.

The result shows that with a sudden release of this amount of liquid nitrogen into the lecture theatre, with the rate of ventilation already determined as 0.8 air changes per hour, will result in a substantial drop in the oxygen concentration within the room. The calculated 5.3% value is significantly below the recognised 18% when action should be taken to evacuate the area. Anyone remaining in this oxygen depleted atmosphere would be at an extreme risk of suffering immediate asphyxiation. Being within the range 0% to 6% means that individuals would instantly fall unconscious, with a high risk of death ensuing. Even if individuals were rescued from this situation and successfully resuscitated, permanent and irreversible damage to the brain may already have taken place.

It would therefore be deemed extremely unsafe for University staff to be in the lecture theatre if the largest liquid nitrogen storage vessel catastrophically failed. Once again, under these conditions, the provision of additional forced ventilation is essential and, ideally, should be provided by having an automatic system installed which immediately increases the air changes per hour if there is a reduction in the oxygen concentration below the 18% limit.

Compared to the sudden release of the liquid helium, the sudden release of liquid nitrogen scenario carries a greater risk of fatalities resulting from such an event. Even though the liquid nitrogen cryogen has a smaller expansion ratio than the liquid helium, and therefore produces less gas on conversion, it is the fact that the breathable air in the lecture theatre for the liquid nitrogen case is initially smaller because of the 2.0 metre limit on height. This is an important fact to recognise, as many facilities would intuitively believe the liquid helium to be a more significant hazard, which is not necessarily the case.

*Conclusions and recommendations*

Having undertaken the comprehensive oxygen concentration calculations, the following recommendation would be made:

- Alarmed oxygen monitors should be fitted to the lecture theatre during its temporary use as a cryogen storage area. One set of monitors should be fitted at ceiling height and the second set at ground level. Both sets should be set with two levels of alarm;
  - 19.5%—which should lead to the urgent investigation of the cause of the alarm and corrective action undertaken
  - 18.0%—which should cause the immediate evacuation of the lecture theatre and should also automatically start a system of additional forced ventilation into the room in order to introduce additional air changes per hour
- These alarms should be tested every month to ensure they remain in a fully functioning state.
- A procedure should be established by which the oxygen concentration levels within the lecture theatre can be evaluated following the event of the evacuation alarm sounding. This could be as simple as attaching a personal oxygen monitor with its own audible alarm to the end of a long pole and inserting this through an open door into the lecture theatre. This could also be employed as a means of checking any false alarms.

- Additional forced ventilation should also be provided locally during any decanting processes from the storage vessels and when the storage vessels are moved.
- Periodically, all relevant staff should undertake a practice spillage drill so that they become familiar with the sounds of the two levels of alarms and the process of evacuating the working area.
- Establish an all-clear procedure so that personnel are not able to immediately return into a hazardous environment if the alarms have been triggered.
- Check that any local drains are protected in the event of accidental spillage of the cryogens.
- Check that any flooring in the lecture theatre is free from damage which could cause movement of the cryogen storage vessels to be impaired.

### 10.1.2 Case study 2

This case study looks at a hypothetical situation where a small basement room at a research facility is being used as a cryogenic liquid decanting station. A set of storage Dewars are filled elsewhere at the facility and moved to the basement room, where users can decant small quantities of the cryogens into portable 5 litre open-topped cryogenic storage containers. During any decanting process, there is a current policy which stipulates that the door to the decanting room is held open in order to provide some level of ventilation to the room. There is no additional ventilation in the room and therefore as a precaution, staff entering the room must wear a personal oxygen monitor which emits an audible alarm when the oxygen concentration falls below 19%. There have been several occasions recently when the personal oxygen monitor alarms have sounded during the decanting process, indicating that the oxygen levels in the immediate vicinity have dropped to a dangerously low value. Notably, these alarms do however reset themselves as soon as the decanting process has been halted, indicating that the oxygen levels have sufficiently recovered. As the alarms have been frequently triggered it has been decided that an evaluation of the oxygen concentration levels in the decanting room will be undertaken with some urgency and any additional control measures implemented to make the workspace safer. Figure 10.2 shows the current storage vessel layout in the decanting room during routine use.

*Specification information*

- A basement room containing six 25 litre Dewars of liquid nitrogen.
- The Dewars are all 15 years old.
- The basement room floor dimensions are 3.0 m × 6.0 m.
- The room is 2.5 m high.
- There are a set of equipment shelves, 3.0 m long × 0.5 m deep × 2.0 m high, at one end of the room. These are general full with stored equipment.
- The manufacturers quoted evaporation rate for the Dewars is 0.625 litre per day.

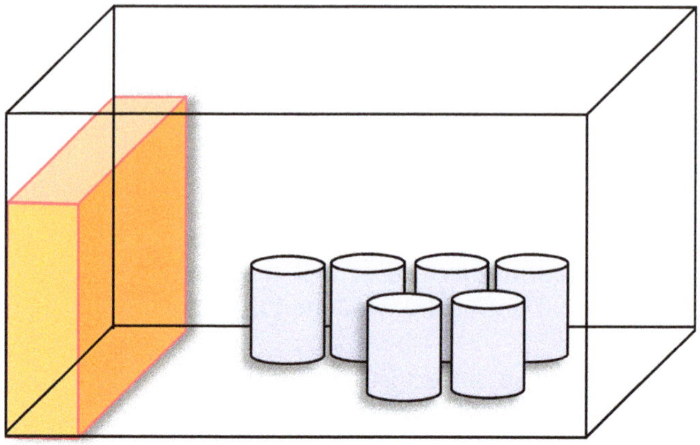

**Figure 10.2.** Showing the storage vessel layout within the decanting room.

*Calculation 1—Considering the oxygen levels in the decanting room following normal evaporation of the cryogens*

The first calculation undertaken makes the assumption that all the storage vessels are fully charged and that standard atmospheric conditions apply. Under these conditions, the contents of each of the storage vessels will warm up over time and in doing so, progressively release small amounts of gas from the liquid phase. This will produce a build-up the pressure inside the storage vessels which in turn will be automatically controlled by spring operated relief valves periodically releasing a small amount of this generated gas into the surrounding air in the decanting room. Over a period of time, this gas will accumulate, and if the air is not sufficiently replenished, could become a hazard for anyone working in the decanting room.

Equation (10.1): The first equation to be considered is that for determining the total gas release from the entire storage vessels over a given time period:

$$\text{Gas released} = L = \frac{N \times V_r \times \text{Expansion ratio}}{\text{Time period} \times 1000}$$

Where:

The number of liquid nitrogen storage vessels of a particular size is multiplied by the relevant evaporation rate and the expansion ratio of 710 for nitrogen. As these vessels are all 15 years old, the evaporation rate has been doubled to take into account any likely degradation of the vessel insulation over that duration.

The manufacturer's evaporation rates are quoted in Litres per day, so the equation is divided by 24×1000 to obtain an answer based on an hourly gas release rate.

$$L = \frac{6 \times 0.625 \times 2 \times 710}{24 \times 1000} = 0.2219 \text{ m}^3 \text{ h}^{-1}$$

Equation (10.2): The next calculation considers the volume of the decanting room, when *nitrogen* gas is released over the defined time period:

Where:

$$V = V_w - V_i$$

The principal dimensions of the lecture room are included in the calculation, with the height of the room limited to 2.0 metres as the evaporated nitrogen gas would be heavier than air and would preferentially fill the available space from the bottom upwards making the air above the 2.0 metres height unavailable for breathing. Subtracted from this principal volume is the space occupied by the equipment shelves, making the assumption that they are full at the time of the assessment.

Therefore:

$$V = (3.0 \times 6.0 \times 2.0) - (0.5 \times 3.0 \times 2.0) = 33.0\ \text{m}^3$$

The decanting room specification defines it as a basement facility, with very limited ventilation. It is reasonable therefore to make the assumption for the purpose of this case study that at its best, there would be no greater than 0.4 complete air changes per hour for the decanting room.

Therefore:

$$n = 0.4$$

Equation (10.4): The next calculation, which is for the gas concentration released into the room, needs to be based on the calculated room volume. This value is multiplied by 100 to represent it as a percentage of the total room volume.

$$C = \frac{L}{V \times n}$$

$$C = \frac{0.2219}{33.0 \times 0.4} \times 100 = 1.68\ \%$$

The nitrogen concentration of the decanting room has therefore been increased by 1.68%. Since the normal oxygen content is approximately 21%, the resultant oxygen in the lecture room after this gradual release over time can now be calculated using the next equation.

$$\text{Oxygen concentration} = \left(\frac{21.0}{100 + 1.68}\right) \times 100 = 20.65\%$$

This value is then referenced to the hazard severity chart (figure 9.4) which lists the various levels at which the lack of oxygen in an atmosphere becomes an issue. As the reduction in oxygen content in the decanting room due to the gradual release of nitrogen gas from the storage vessels is small, there is a negligible consequence to anyone working in the environment. At 20.65%, the oxygen concentration remains above the 18.0% to 19.5% range where some impairment of physical and mental performance begins to take place.

| | |
|---|---|
| 18% – 19.5% | May affect physical and intellectual performance without person's knowledge. |

It would therefore be deemed safe for the research facility staff working in the decanting room for this consideration, even when all the storage vessels are full. Under these conditions, the provision of additional forced ventilation is not required. As the room is in a basement however, it may be prudent to have permanently open louvre ventilation provided in the door in order to ensure some level of ventilation is acquired during the decanting process should the door ever be accidentally closed. In the longer term, it may also be prudent to install some form of additional ventilation to the room in order to increase the air changes per hour coming into the decanting room.

*Calculation 2—considering the oxygen levels in the decanting room due to the sudden release of nitrogen*

The next calculation makes the assumption that there has been a sudden, catastrophic failure of the largest liquid nitrogen storage vessel and that its entire contents are evaporated into the decanting room. This works on the basis that the storage vessel has been damaged sufficiently to release all of the cryogen into the available volume of the room. As the storage vessels are filled elsewhere and moved to the decanting room, the likeliest scenario, is that one of these vessels is accidentally tipped over as it is being delivered, and on contact with the ground ruptures both the insulating jacket around it and the inner containment vessel wall. This would not only liberate a large amount of liquid cryogen from the vessel, but would also introduce an enormous amount of heat into the remaining cryogen still contained within the vessel. Both the leaked cryogen and the quantity still inside the vessel would rapidly boil-off, filling the decanting room with a cloud of nitrogen gas. As this cloud is heavier than air, it will immediately begin to fill the room from the floor upwards, gradually ascending to a critical height where individuals still within the room would be at risk of breathing in the concentrated gas.

Equation (10.5): The initial calculation determines the volume of oxygen within the room when this sudden release has taken place. It uses the following equation:

$$\text{Volume of oxygen} = V_o = \left[V - \left(\frac{V_D \times f_g}{1000}\right)\right]$$

Where:

$V$ is the effective volume of the room calculated in equation (10.3), limiting the height of the lecture room to 2.0 metres. In this case 33.0 m$^3$.

$V_D$ is the largest liquid nitrogen storage vessel size, which in this case is 25 litres.

$f_g$ is the expansion ratio when the liquid nitrogen converts into gas which is 710.

$$V_o = 0.21\left(33.0 - \left(\frac{25 \times 710}{1000}\right)\right) = 3.203 \text{ m}^3$$

Equation (10.6): This volume of oxygen is now converted into the oxygen concentration, expressed as a percentage of the total gas within the room. This oxygen concentration value is then referenced to the hazard severity chart (figure 9.4) and the appropriate level determined:

$$\text{Oxygen concentration} = C_o = \frac{100 \times V_O}{V}$$

$$C_o = \frac{100 \times 3.203}{33.0} = 9.70\,\%$$

| | |
|---|---|
| 8% – 10% | Mental failure, nausea, vomiting, fainting, ashen face, blue lips. |

The result shows that with a sudden release of this amount of liquid nitrogen into the decanting room, with the rate of ventilation already determined as 0.4 air changes per hour, will result in a substantial drop in the oxygen concentration within the room. The calculated 9.70% value is significantly below the recognised 18% when action should be taken to evacuate the area. Anyone remaining in this oxygen depleted atmosphere would be at an extreme risk of suffering the early effects of asphyxiation. Being within the range 8% to 10% means that individuals would begin to feel drowsy, and may also feel nauseous. Visible indicators of their condition would be a pale, ashen face and possible blue lips.

It would therefore be deemed extremely unsafe for research facility staff to be in the decanting room if the largest liquid nitrogen storage vessel catastrophically failed. Under these conditions, the provision of additional forced ventilation is essential and, ideally, should be provided by having an automatic system installed which immediately increases the air changes per hour if there is a reduction in the oxygen concentration below the 18% limit.

*Calculation 3—considering the oxygen levels in the decanting room due to the release of nitrogen during the decanting process*

The next calculation makes the assumption that a certain amount of nitrogen gas is released into the decanting room during the actual decanting process. As there is recorded evidence of personal oxygen monitor alarms sounding during this procedure, it is a realistic assumption that a sufficient amount of gas is being released to lower the oxygen concentration value in the air immediately around the person undertaking the decanting. As the transfer of liquid nitrogen from the storage vessels is generally into small 5 litre, portable open-topped storage containers, it is feasible that at least 10% of the liquid nitrogen being decanted is lost into the atmosphere due to evaporation. This vaporised gas will form a cloud directly around the storage Dewars and the small portable storage containers into which the liquid nitrogen is being filled. For a few seconds at least, this immediate region is likely to experience a drop in the oxygen concentration value as the nitrogen gas dominates the atmosphere where the operator is standing. For this purpose, the calculation will therefore consider only a localised volume of air which will be momentarily affected by this drop in oxygen content.

Equation (10.7): The initial calculation considers the volume of air immediately around the person undertaking the decanting when nitrogen gas is released over a defined time period. It uses the following factors:

For the purposes of estimating the oxygen concentration value of the air close to the decanting process, it is assumed that a particular defined volume of the decanting room will only be affected by the evaporating gas. This volume can therefore be determined as the space in which the person undertaking the decanting is occupying during the process. For this reason, it is determined that the whole decanting process will occupy only a small, 1.25 m by 1.25 m area. In addition, the person decanting the liquid nitrogen will be wearing a personal oxygen monitor. It is normal to wear this at about chest height so that any evaporating gases heavier than air would trigger the alarm as they accumulate from the ground upwards, but essentially, before they reach the height of the nose and mouth. It is assumed that the oxygen monitor would be worn at no greater than 1.75 m from the ground, which allows this isolated volume to be determined.

Therefore:

$$V = 1.25 \times 1.25 \times 1.75 = 2.734 \text{ m}^3$$

Equation (10.8): The initial calculation determines the volume of oxygen within this decanting volume when any escaped nitrogen release has taken place. It uses the following equation:

$$\text{Volume of oxygen} = V_o = \left[V - \left(\frac{V_D \times f_g}{1000}\right)\right]$$

Where:

$V$ is the effective volume of the decanting volume calculated in equation (10.7). In this case 2.734 m$^3$.

$V_D$ is the size of the small portable open-topped storage containers into which the liquid nitrogen is being decanted, which in this case is 5 litres. This value is multiplied by 10% which represents the estimated spillage of liquid nitrogen lost during the decanting process.

$f_g$ is the expansion ratio when the liquid nitrogen converts into gas which is 710.

Therefore:

$$V_o = 0.21\left(2.734 - \left(\frac{0.1 \times 5 \times 710}{1000}\right)\right)$$

$$V_o = 0.50 \text{ m}^2$$

Equation (10.9): This volume of oxygen is now converted into the oxygen concentration, expressed as a percentage of the total gas within the decanting volume. This oxygen concentration value is then referenced to the hazard severity chart (figure 9.4) and the appropriate level determined:

$$\text{Oxygen concentration} = C_o = \frac{100 \times V_O}{V}$$

$$C_o = \frac{100 \times 0.50}{2.734} = 18.3\ \%$$

| | |
|---|---|
| 18% – 19.5% | May affect physical and intellectual performance without person's knowledge. |

The result shows that with a standard decanting process losing upwards of 10% of the liquid nitrogen being transferred into the immediate space around the small portable open-topped storage containers, will result in a marked drop in the oxygen concentration within that volume. The calculated 18.3% value is very close to the recognised 18% when action should be taken to evacuate the area. Anyone remaining in this oxygen depleted atmosphere would be at a risk of suffering the early effects of asphyxiation where they will not be aware of the impairment of their judgement in what they are doing. Being within the range 18.0% to 19.5% also means that individuals would begin to feel drowsy, and may demonstrate signs of being forgetful and disorientated.

This value also confirms the fact that there have been circumstances where the personal oxygen monitors being worn during the decanting process have had their alarm triggered, as locally, the oxygen concentration value has become low. It would not however be deemed unsafe for research facility staff to remain in the decanting room, but a policy should be adopted which recommends halting of the decanting and withdrawing from the room until the oxygen levels have sufficiently recovered. A more robust solution would be to provide additional forced ventilation to the decanting room, especially during the actual decanting process, and more specifically, local to the containers being filled where the greatest loss of evaporate nitrogen gas will be experienced

*Conclusions and recommendations—general storage of liquid nitrogen*

Having undertaken the comprehensive oxygen concentration calculations, the following recommendation would be made:

- Alarmed oxygen monitors should be fitted to the decanting room with an oxygen concentration readout located by the entrance door, external to the room. At least two oxygen monitors should be set at ground level, with two levels of alarm;
    - 19.5%—which should lead to the urgent investigation of the cause of the alarm and corrective action undertaken
    - 18.0%—which should cause the immediate evacuation of the decanting room and any adjacent corridor and should also automatically start a system of additional forced ventilation into the room in order to introduce additional air changes per hour
- These alarms should be tested every month to ensure they remain in a fully functioning state.
- A procedure should be established by which the oxygen concentration levels within the decanting room can be verified following the event of the evacuation alarm sounding. This could be as simple as attaching a personal oxygen monitor with its own audible alarm to the end of a long pole and inserting this through an open door into the decanting room. This could also be employed as a means of checking any false alarms.

- Additional forced ventilation should also be provided locally during all decanting processes from the storage vessels and when the storage vessels are moved.
- Periodically, all relevant staff should undertake a practice spillage drill so that they become familiar with the sounds of the two levels of alarms and the process of evacuating the working area.
- Establish an all-clear procedure so that personnel are not able to immediately return into a hazardous environment if the alarms have been triggered.
- Check that any local drains are protected in the event of accidental spillage of the cryogens.
- Check that any flooring in the decanting room is free from damage which could cause movement of the cryogen storage vessels to be impaired.

*Conclusions and recommendations—decanting process*

Having undertaken the comprehensive oxygen concentration calculations, the following recommendation would be made:

- Personal, alarmed oxygen monitors should be provided to all research facility staff undertaking decanting in the decanting room. These monitors should have two levels of alarm;
    - 19.5%—which should lead to the immediate halting of the decanting process and associated corrective action undertaken, such as moving away from the container being filled
    - 18.0%—which should cause the immediate evacuation of the decanting room
- These monitors should be tested every month to ensure they remain in a fully functioning state.
- Adopt a lone working policy so that supporting staff are always available should any personal oxygen monitor alarms be triggered.
- Additional forced ventilation should be provided locally during all decanting processes from the storage vessels. This can be a standalone system, or be automated so that as the door to the decanting room is opened, the ventilation rate is increased.

# Chapter 11

## Conclusion

The patois of cryogenic technology is rich with technical acronyms, opaque product-specific references and strange, unfamiliar names attributed to the materials utilised for the sole aim of achieving near absolute temperatures. Yet the associated lexicon of safety terms accredited to the technology differs only marginally from that adopted by many of the generic industrial activities throughout the world. Terms such as 'risk assessments' and 'risk control systems' are a common denominator in achieving safety in the general workplace, and these too can likewise be adopted in any cryogenic facility. Injuries to personnel and damage to property are just as possible in the cryogenic environment as say that encountered in a foundry, even though their respective temperature regimes are at the opposite ends of the thermal spectrum. Similarly, risks are determined in exactly the same manner, whatever the activity, albeit the management of these risks in the cryogenic context will be more specifically determined by the thermodynamics taking place to achieve the extremely cold temperatures of its making.

Whether we accept that accidents have a certain inane inevitability or not, our view of the functioning cryogenic technology workplace must be guided by the knowledge of the associated risks and the wealth of legislation set in place to protect both the empiricists and the technicians working there. We must learn to regard the cryogenic working environment not as a secure, abiding facility, but one full of anticipated dangers and potential hazards. It may be, that in the larger scope of the technology, that some level of risk must be inevitable in order to achieve discovery in its purist sense. But this must not come at the expense of the wellbeing of those who work in that same enterprising field. Pragmatically, risks must be minimised, so that cryogenic technology can achieve the distinction and merits it deserves without causing life changing injuries in the process. Simply put, there is no quick and easy panacea to making the cryogenic workplace safe. The very nature and low temperature range of the equipment used throughout cryogenics manifests these risks, and as such, each individual article within the workplace needs to be addressed on its

doi:10.1088/978-0-7503-3123-4ch11

own value and the hazards it represents recognised. On occasions, these hazards can be accumulative, adding to their associated neighbour in a way that magnifies the overall significance. Many are profoundly difficult to comprehend, metaphorically waiting in the wings to become dangerous when least expected. Whatever their character though, the risks associated with cryogenic technology are very real and must always be reconciled as such when any work is undertaken which involves performing tasks at the incredibly low temperatures of its function.

Whatever safety devices and procedures put in place to protect people and equipment, can only ever provide that function if they themselves form part of the overall safety culture of any cryogenics facility. It is seemingly futile to employ expensive protective systems if they are not maintained in good working conditions, or are intentionally disregarded for the sake of economics. Frequently, the implementation of many of these safety devices forms part of the obligations within national legal requirements. Fundamentally, the law states that they must be applied, and as such, they serve the direct purpose of protecting the health and wellbeing of staff working in the cryogenics industry. These are not mere occlusions stipulated by the whim of legislators, but are inevitably based on meaningful evidence of the likelihood and type of accident that would happened if there were not in place.

Appreciatively, in many industries, both the employer and the employee have long ago come to a tacit understanding that workplace health and safety is implicit and that all necessary safeguards are in place to protect them from harm. Yet all too routinely we are informed of grave situations where this is evidently not so, with often tragic consequences resulting from the divergence from this philosophy. This realism is often brought about by an otherwise apathetic and indolent approach towards health and safety from all involved. Under these circumstances, complacency in the workplace can materialise in the guise of misguided principles and the belief that the ownership of health and safety belongs to someone else and that everything being used or undertaken within the working environment is readily safe. Likewise, the anthropology of human risk taking and the accompanying instructive training which teaches how best to reduce those risks, can often be lost amidst the matrix of other more powerful influencers. A companies' pursuit of better profit and output for example, can both have a greater impetus within an organisation than the need to continually review all that encompasses the protection of life, property and the environment.

It is therefore vitally important that organisations working under the extensive title of cryogenics have a combination of fail-safe systems in place to protect their staff and their property. These systems must operate independent of human interaction in order to protect against cryogenic related accidents, and where possible, should be supported by strong safety procedures which in turn adds value to these systems. Over time though, cryogenic technology organisations should also invest in a consistent approach to the health and safety training of their staff so that a desire to work safely becomes endemic throughout their facilities. By providing their staff with the necessary health and safety knowledge associated with the technology, both employers and employees will be able to develop a conviction and a passion for preventing cryogenic related accidents from ever occurring. Without doubt, cryogenics is a hazardous technology, but with careful scrutiny and diligent consideration, many of these hazards can be mitigated to the point where the associated risks are deemed to be minimal. This is the ideal place to be at, with the working environment providing the necessary cryogenic temperature regime, as close to absolute zero as possible, yet doing this with an accident rate tending to its own zero, where safety is paramount and by association, nobody gets hurt.

www.ingramcontent.com/pod-product-compliance
Ingram Content Group UK Ltd.
Pitfield, Milton Keynes, MK11 3LW, UK
UKHW051903100926
8556IPUK00005B/745

9 780750 331210